MW01633795

QUANTUM TURBULENCE

Turbulence plays a crucial role in contexts ranging from galaxy formation to heavy atomic nuclei, from jet engines to arterial blood flow, challenging engineers, physicists, and mathematicians. Recently, turbulence of quantum fluids displaying superfluidity has emerged as an exciting area of interdisciplinary research that spans fluid dynamics, low-temperature physics, and Bose–Einstein condensation.

The first book on quantum turbulence, this work describes state-of-the-art results and techniques, stressing analogies and differences with classical turbulence. The authors focus in particular on low temperature phases of liquid helium, drawing on evidence from recent experiments, theory, and numerical simulations. Written by leading figures in the field, this is a go-to reference for students and researchers at all levels.

CARLO F. BARENGHI is Professor of Fluid Dynamics at Newcastle University. He has authored over 200 articles and three previous books, including *A Primer of Quantum Fluids* (2016).

LADISLAV SKRBEK is Professor of Physics at Charles University in Prague. He has four decades' experience of research and teaching on low-temperature physics, superfluidity, and quantum fluid dynamics.

KATEPALLI R. SREENIVASAN holds professorships in the Department of Physics, Department of Mechanical and Aerospace Engineering, and the Courant Institute of Mathematical Sciences at New York University. He has published extensively on turbulence and edited books including *A Voyage Through Turbulence* (2011) and *Ten Chapters in Turbulence* (2012).

QUANTUM TURBULENCE

CARLO F. BARENGHI
Newcastle University

LADISLAV SKRBEK
Charles University

KATEPALLI R. SREENIVASAN
New York University

CAMBRIDGE
UNIVERSITY PRESS

Shaftesbury Road, Cambridge CB2 8EA, United Kingdom

One Liberty Plaza, 20th Floor, New York, NY 10006, USA

477 Williamstown Road, Port Melbourne, VIC 3207, Australia

314–321, 3rd Floor, Plot 3, Splendor Forum, Jasola District Centre, New Delhi – 110025, India

103 Penang Road, #05–06/07, Visioncrest Commercial, Singapore 238467

Cambridge University Press is part of Cambridge University Press & Assessment, a department of the University of Cambridge

We share the University's mission to contribute to society through the pursuit of education, learning and research at the highest international levels of excellence.

www.cambridge.org
Information on this title: www.cambridge.org/9781009345668

DOI: 10.1017/9781009345651

© Carlo F. Barenghi, Ladislav Skrbek, and Katepalli R. Sreenivasan 2024

This publication is in copyright. Subject to statutory exception and to the provisions of relevant collective licensing agreements, no reproduction of any part may take place without the written permission of Cambridge University Press & Assessment.

First published 2024

Printed in the United Kingdom by TJ Books Limited, Padstow Cornwall

A catalogue record for this publication is available from the British Library

A Cataloging-in-Publication data record for this book is available from the Library of Congress

ISBN 978-1-009-34566-8 Hardback

Cambridge University Press & Assessment has no responsibility for the persistence or accuracy of URLs for external or third-party internet websites referred to in this publication and does not guarantee that any content on such websites is, or will remain, accurate or appropriate.

Contents

Preface

There was a time when turbulence mostly meant the erratic motion of air and water. This was natural because it was the sum of our common experience for a long time. Over time, this specific form of turbulence, called classical turbulence in this book, has come to be known to share some of its characteristics with other forms of turbulence: magnetohydrodynamic turbulence, plasma turbulence, scalar turbulence, two-dimensional turbulence, weak turbulence, etc. There are similarities and differences but there is little doubt that the study of one form of turbulence enriches the understanding of all others. Classical turbulence, also called hydrodynamic turbulence or fluid turbulence, particularly to distinguish it from the stochastic state of nonlinear dynamical systems with spatial complexity, is thought to be connected to the behavior of the Navier–Stokes equation in three dimensions. The other forms of turbulence just mentioned include new phenomena that contain information additional to the Navier–Stokes (e.g., magnetic fields in magnetohydrodynamic turbulence) or result from simplifications (e.g., the reduction of dimensionality in the case of two-dimensional turbulence). The latest addition to the family of turbulence is quantum turbulence, which is the subject of this book. What quantum turbulence means, especially the quantum part, and why anyone seriously interested in the broad field of turbulence should pay attention to it, we shall explain here.

Quantum turbulence is the turbulent state of quantum fluids, some of whose properties depend on quantum mechanics. However, just as the properties of air or water appear in the Navier–Stokes equation essentially through viscosity and density (for the most part with no cognizance of the detailed molecular structure), only some macroscopic properties of quantum fluids are necessary for a description, up to a certain depth of understanding of the phenomena; the amount of quantum mechanical knowledge required to understand quantum turbulence is rather basic and not extensive. If one pushes classical turbulence to various limiting states, one does better if equipped with the underlying molecular structure of classical fluids;

just the same way, one needs to know quantum mechanics more extensively when one desires a more in-depth understanding of quantum turbulence. The similarities with classical turbulence are quite impressive: While the properties of individual quantized vortices are, in fact, different in some crucial respects from hydrodynamic vortex lines, a bundle of quantum vortex lines behaves in much the same way as a vortex tube; under certain circumstances, the collective interaction of vortex lines produces turbulence energetics that are very similar to Kolmogorov turbulence in classical turbulence; vortex reconnections bear uncanny resemblance to magnetic reconnections; the anomalous behavior of intermittency is quite similar to that in classical turbulence, etc. But we should also point out that quantum turbulence shows a much richer behavior than classical turbulence because of the existence of an additional length scale due to quantum effects and, where applicable, the interaction between the normal and superfluid parts.

The term quantum turbulence was first introduced in 1982 by Barenghi (1982) in his Ph.D. thesis, and was disseminated widely by Donnelly and Swanson (1986) after the symposium dedicated to the memory of the great classical fluid dynamicist G. I. Taylor. Its use came into vogue slowly but has accelerated quite significantly in recent years. This increase in use is a recognition that quantum turbulence is a field of some importance spanning quantum fluids, on the one hand, and classical turbulence, on the other. This book identifies similarities and differences between quantum turbulence and its more widely known counterpart of classical turbulence, and bridges low-temperature physics, condensed-matter physics, fluid dynamics, and atomic physics. One of our aims is to bring these communities together. We have kept in mind not only experts who are constantly enlarging the field but also beginners with some exposure to basic physics and fluid dynamics, whether they are theoretically inclined or wish to focus on simulations or experiments. Indeed, we shall draw our account from experiments, theory, and numerical simulations to create a self-contained description of the subject that not only consolidates recent advances but also points to further open problems.

We are grateful to a number of colleagues with whom we have had fruitful interactions over the years, too numerous to list here. We should, however, mention Joe Vinen (Barenghi *et al.*, 2022), who did pioneering measurements well ahead of the recent crop, and Russ Donnelly (Hammer *et al.*, 2015; Niemela and Sreenivasan, 2022), who promoted the field through his own work and that of his students. Alas, neither is with us today. We are greatly honored to dedicate this book to their memories.

1
Introduction

Turbulence is a state of fluid motion that is spatially and temporally complex. One might say that its study began about five centuries ago, when Leonardo da Vinci observed that water falling into a pond creates coherent eddies or whirls, which he compared to the curls of a woman's braided hair; see Fig. 1.1. The insight that eddies are the building blocks of turbulence was confirmed in 1839 by Hagen, who visualized the vortical motion of water with sawdust (Eckert, 2019), and by Reynolds (1883) in his pioneering studies of turbulent pipe flow. Reynolds visualized the flow of water by injecting a narrow jet of dye into the center of the flow along a larger glass pipe. He found that the dyed jet remains narrow and distinct when the water velocity is small; the dye breaks up and diffuses rapidly when the velocity is increased, filling the cross-section of the glass pipe at some point that marks the appearance of turbulence. A short-exposure photograph taken with an electric spark reveals that, instantaneously, the turbulent water consists of vortices; see Fig. 1.2. Turbulence as a subject of study has grown substantially since then because of its importance to a variety of fields such as astrophysics, geophysics, and various engineering applications including such large-scale projects as controlled fusion; what we have in front of us is the result of a rich combination of efforts of theorists, experimentalists, and simulation experts alike.

While Leonardo understood that the flow of water shapes the natural landscape, today we appreciate that turbulence also shapes processes that surround us (from jet engines to interstellar medium) or are contained inside us (from the blood flow in the aorta to airflow in lungs). Its various facets have continually been understood better over the last few decades, but it remains an unfinished problem. The subject is moving forward constantly, yet some of its essential problems have remained as current as ever. For example, the existence or breakdown of the smoothness of the Navier–Stokes equation has remained an unclaimed prize among the Millennium Problems of the Clay Mathematics Institute. And, as an illustration on the practical side, one cannot yet predict the pressure drop in a pipe using the Navier–Stokes

Figure 1.1 Leonardo da Vinci's studies of water (circa 1507), pen and ink on paper. (Top) Turbulent water in a pond. (Bottom) Leonardo's analogy between turbulent water flow and braided hair. Leonardo da Vinci, Public domain, via Wikimedia Commons.

equations alone. Because of its intellectual challenges and practical importance, it continues to challenge the attention of physicists, mathematicians, and engineers. One reason why so many scientists from various disciplines have to deal with one aspect or another of fluid turbulence is that the phenomenon spans an enormous range of scales – more than thirty orders of magnitude – from astronomical, when turbulent description is useful for explaining the shapes of spiral galaxies, the

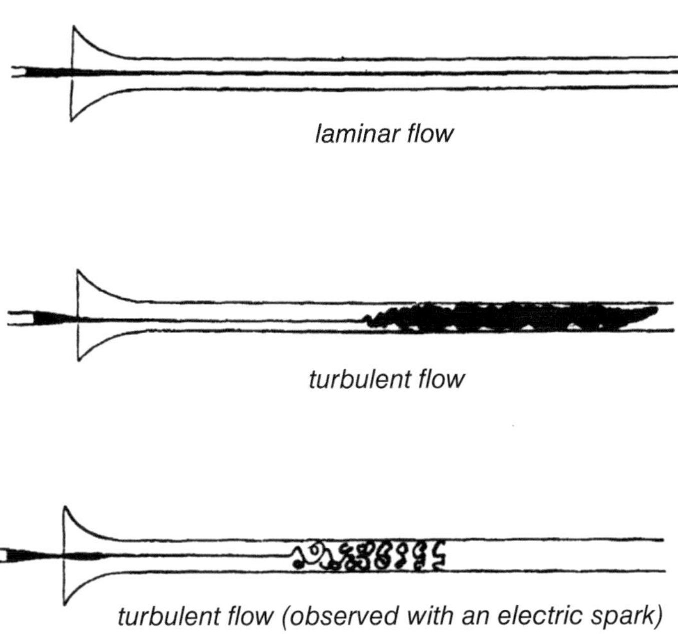

Figure 1.2 The experiment of Osborne Reynolds (1883). A dye, injected into the center of the glass pipe, visualizes the flow of water from left to right. (Top) At small velocity, the flow is laminar and the jet of dye remains narrow. (Middle) At large velocities, the flow becomes turbulent and the dye spreads into the entire pipe cross-section. (Bottom) The short exposure photograph obtained using an electric spark reveals that turbulence consists of coherent eddies.

behavior of interstellar wind, the stream of interstellar gas and dust that is moving past the solar system, on the one hand, to the tiny core of a quantized vortex in superfluid helium on the other. Moreover, phenomena occurring at scales differing by many orders of magnitude can be linked together by the same underlying physics, such as the theoretically predicted Kibble–Zurek mechanism for the generation and decay of cosmic strings and quantized vortices in helium superfluids.

This book is devoted to the study of turbulence in quantum fluids displaying superfluidity, mainly in superfluid ^{4}He and ^{3}He-B. In general, superfluid helium possesses many extraordinary physical properties. The fundamental laws of physics become manifest when the temperature of the quantum liquid decreases, allowing us to use a helium droplet as a condensed matter model system for the early universe (Volovik, 2007). Turbulence in quantum fluids, which we shall refer to as *quantum turbulence*, is a relatively new area of research. A number of reviews are available on this subject, starting from the pioneering article by Vinen and Niemela (2002), to the works of Nemirovskii (2012); Skrbek and Sreenivasan (2012); Barenghi *et al.* (2014a); Tsubota *et al.* (2017), and to the more recent perspective of Skrbek *et al.* (2021). The aim of this book is to provide a coherent account of recent developments and attempt to draw qualitative and quantitative connections between quantum turbulence and turbulence in ordinary fluids, or the *classical turbulence*.

As stated in the Preface, the term *quantum turbulence* was first introduced in 1982 in the Ph.D. thesis of Barenghi (1982) and popularized by Donnelly and Swanson (1986). We shall base this book on quantum turbulence on the current work by distinct scientific communities: low-temperature physicists, condensed-matter physicists, fluid dynamicists, and atomic physicists. One of our aims is indeed to bring these communities together. We shall take the evidence and draw inferences from experiments, theory, and numerical simulations.

Quantum fluids are so called because their physical properties depend on quantum physics, classical physics being insufficient to describe them. The most studied quantum fluids are the low temperature phases of liquid helium $\left({}^4\text{He and }{}^3\text{He}\right)$ and, more recently, ultracold atomic gases. We shall concentrate particularly on liquid helium, for which more experimental information is available, using results from the study of atomic gases only when relevant to the general problem of turbulence in quantum fluids.

From the standpoint of fluid dynamics, quantum fluids differ from classical fluids (such as air or water) in three respects:

(i) at nonzero temperature (or in the presence of impurities) they exhibit a two–fluid behavior;
(ii) under suitable conditions they can flow freely, without the dissipative effect of viscous forces, hence the term *superfluidity*;
(iii) their local rotation is constrained to thin *vortex lines* (also called *quantized vortices* or *superfluid vortices*);[1] unlike eddies in ordinary fluids, which are continuous and can have arbitrary size and strength, vortex lines are discrete, individual structures whose circulation and core thickness are determined by quantum mechanical constraints.

From the point of view of turbulence the fundamental property is quantized vorticity, which is an extraordinary manifestation of quantum mechanics at macroscopic length scales. This property arises because of the existence of a macroscopic complex wavefunction (sometimes called an order parameter), which governs the dynamics of the system and endows quantum turbulence with some attractive properties.

How these three properties of quantum fluids endow flows with certain distinguishing characteristics, how these characteristics are similar to, and different from, those of classical fluids are the topics that we take up in succeeding chapters. We expect our account to serve as both introductory material for graduate students and

[1] Exceptions exist, such as the superfluid ^{3}He-A phase where continuous vortices or vortex sheets can be formed.

a reference material for expert researchers. After reading the first three chapters, the remaining chapters can be read reasonably independently, but a chronological reading is recommended for a systematic understanding of the subject.

One further point is useful. Superfluid ^{4}He, ^{3}He, and atomic condensates exist only at very low temperatures (of the order of a kelvin, a millikelvin and a microkelvin, respectively). However, very small absolute temperatures are not necessary for the existence of quantum fluids: for example, exciton–polaritons condensates are superfluid at room temperature, and the interior of neutron stars is extraordinarily hot by normal standards. Quantum mechanics is required when the temperature is significantly lower than some characteristic temperature of the system, such as the *Fermi temperature* T_F in a gas of fermions.

2
Quantum Fluids

Quantum fluids are so called because some of their macroscopic properties depend on quantum physics. In this book we shall mainly restrict our attention to the two most studied quantum fluids: the superfluid phase of ^{4}He and the superfluid B-phase of ^{3}He. Their main fluid dynamic properties are described in the later part of this chapter, but it is useful to discuss the elements of the atomic structure here. Quantum turbulence occurs (or may occur) in other superfluid systems as well: one-component and two-component atomic Bose–Einstein condensates, exciton–polariton condensates, bipolar gases, spinor condensates, the interior of neutron stars, etc. The existence of a superfluid mixture of ^{4}He and ^{3}He was predicted long ago but experimentally has not (yet) been realized. The reason we focus on quantum turbulence in ^{4}He and ^{3}He is that they have been studied for many years, unlike some of the recently discovered superfluids listed above, which are current research subjects. We shall include brief remarks on them as well.

2.1 Liquid Helium

The element helium has two nonradioactive isotopes, ^{4}He (which is common) and ^{3}He (which is rare). The superscript 4 refers to the fact that the nucleus of ^{4}He contains two protons and two neutrons (the ^{4}He nucleus is also known as an alpha particle); the superscript 3 means that this lighter isotope consists of two protons and only one neutron. In ^{4}He, the two protons, two neutrons, and two electrons (which make a fully populated s-orbital) constitute a spherically symmetric atom with zero total spin; ^{4}He atom is therefore a boson, a general term used for particles with integer spin. In ^{3}He, the missing neutron is crucial: with only one neutron in its nucleus, the ^{3}He atom is a fermion of spin $1/2$.

In spite of this difference in atomic structure between ^{4}He and ^{3}He, there are no appreciable differences in their physical properties at room temperature and ambient pressure: both isotopes behave essentially as ideal gases. If cooled, the two gases

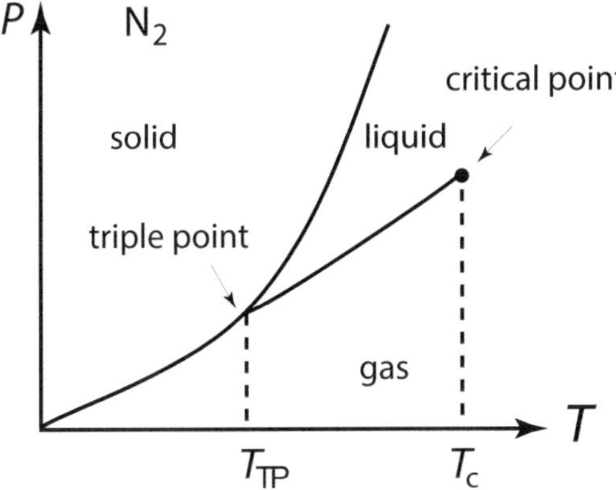

Figure 2.1 Phase diagrams of nitrogen (N_2). Note the triple point at $T_{TP} = 63$ K (where gaseous, liquid, and solid phases coexist) and the critical point at $T_c = 126$ K (where the equilibrium saturated vapour pressure (SVP) line terminates).

liquify (at normal pressure) at similar temperatures: 4.2 K for ^{4}He and 3.2 K for ^{3}He. Indeed, it was the liquefaction of ^{4}He at such low temperatures (much lower than the temperature needed to liquify other known gases like nitrogen, oxygen, and hydrogen), first achieved by Heike Kamerlingh Onnes in Leiden on July 10, 1908, which marked the beginning of modern low-temperature physics.

On the other hand, at low temperatures the thermal disorder of the atoms is reduced and the fundamental properties of matter become more apparent. In particular, a gas of ^{4}He atoms is governed by Bose–Einstein statistics, and a gas of ^{3}He atoms by Fermi–Dirac statistics. Therefore, it is at low temperatures of the liquid phases that the fundamental differences between ^{4}He and ^{3}He become manifest; their properties are rather different from each other, and, of course, very different from those of ordinary classical liquids.

Figures 2.1–2.3 compare the low-temperature region of the equilibrium pressure–temperature phase diagram of a typical substance (here, for example, nitrogen as in Fig. 2.1) to the phase diagrams of the two isotopes ^{4}He (Fig. 2.2) and ^{3}He (Fig. 2.3). The most striking feature of both isotopes of helium is the absence of the triple point. In order to solidify either isotope, external pressures exceeding 25–33 bar must be applied to the liquid at very low temperatures. One can interpret this property in terms of melting of the solid at ambient pressure due to the quantum mechanical zero-point motion (zero-point energy is indeed the lowest possible energy of a system in the quantum mechanical ground state). In contrast, in classical systems all motions vanish at 0 K.

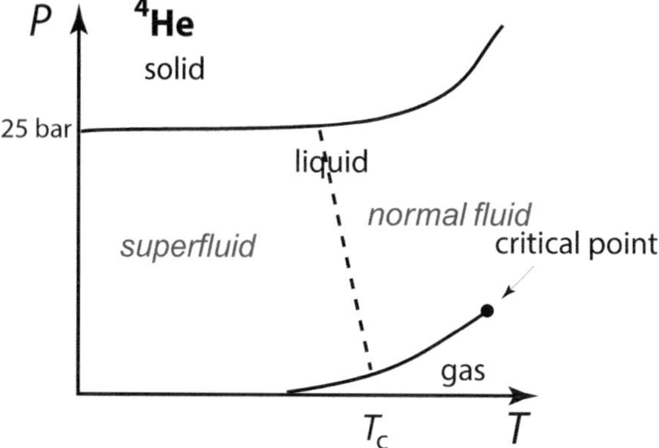

Figure 2.2 Phase diagram of ^{4}He. Note the absence of a triple point, which means that helium remains liquid down to $T = 0$ unless a large pressure is applied, guaranteeing the existence of two liquid phases: He I (which is a normal liquid) and He II (which is a super fluid), separated by the so-called *lambda-line*, shown by the dashed line. The critical temperature T_c in He II is also called the *lambda temperature*, $T_\lambda \approx 2.17$ K, which marks the superfluid transition on the SVP line.

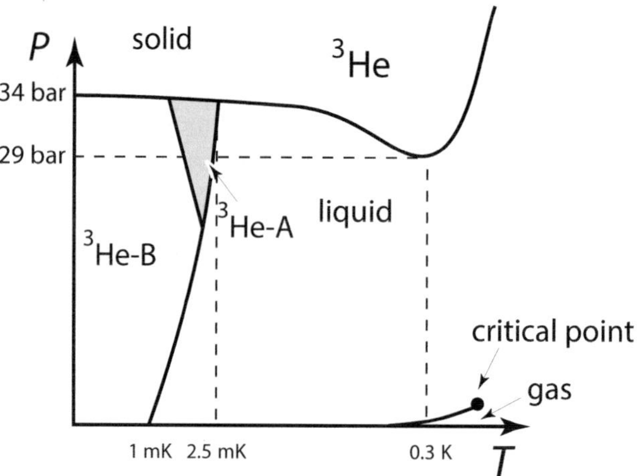

Figure 2.3 Phase diagram of ^{3}He. Unlike ^{4}He, there are several superfluid phases; the two most common bulk phases are the A-phase (the small shaded region) and the B-phase. Note that the superfluid transition at SVP is at a much lower temperature than that in ^{4}He, typically around 1 mK.

2.2 He I and He II

Let us first examine ^{4}He (Fig. 2.2). Below critical temperature, at pressures exceeding the corresponding pressure at the equilibrium saturated vapour pressure

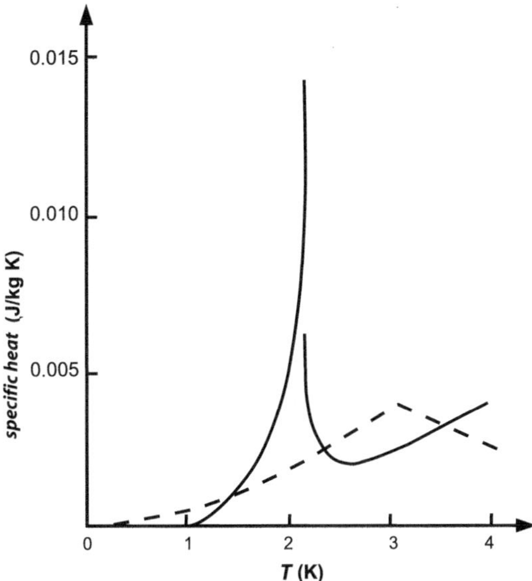

Figure 2.4 Temperature dependence of the specific heat of liquid ^{4}He (solid line) and of an ideal Bose gas (dashed line) at saturated vapour pressure.

(SVP) curve, liquid ^{4}He behaves as an ordinary fluid and is only remarkable for its extraordinarily small (kinematic) viscosity (two orders of magnitude smaller than the viscosity of water, three orders of magnitude lower than that of air). When measuring the physical properties of liquid ^{4}He (density, specific heat, thermal conductivity, etc.), the early experimentalists soon discovered strange effects if the temperature T is lowered below the critical value $T_c = 2.17$ K along the SVP curve. These effects were so unusual that they gave the historical name *helium II* (He II) to liquid ^{4}He below this critical temperature in order to distinguish it from *helium I* (He I) above it. The critical temperature T_c was called T_λ because of the shape of the specific heat dependence, C_V vs. T, which looks like the Greek letter lambda; the specific heat appears to diverge as $T \to T_\lambda$, see Fig. 2.4, which also compares it to the theoretical specific heat of an ideal Bose gas of helium atoms of the same density (London, 1938). The clear difference arises due to strong interactions between helium atoms in the liquid. On the modern temperature scale (T90), the accepted value (Donnelly and Barenghi, 1998) of the critical temperature is $T_\lambda = 2.1768$ K at the SVP curve. Notice the slope of the (dashed) lambda-line in Fig. 2.2: T_λ becomes slightly reduced if the pressure is increased.

Besides the unusual behavior of the specific heat, the most noticeable effects found in the helium II region $T < T_\lambda$ are the very high thermal conductivity (more than 10^6 times that of copper) and the extremely low viscosity, which, in some experiments, was found to be effectively zero. It was this second property – the ability of helium II to flow, under certain experimental conditions, through narrow gaps and capillaries without any viscous friction, discovered by Kapitza (1938) and Allen and Misener (1938) – that resulted in it being named a *superfluid* by the former. This term was introduced in analogy with "superconductivity," the property (discovered by Kamerlingh Onnes in 1911) by which, in some materials, electrical currents can flow without electrical resistance. The physical reason for superfluidity is the bosonic character of interacting ^{4}He atoms that can undergo quantum mechanical Bose condensation in three-dimensional (3D) momentum space.

In this book we are not concerned with the history of the discovery of superfluidity and the controversies associated with it: the race between Kapitza (1938), on the one hand, and Allen and Misener (1938), on the other; the separate development of the two-fluid model by Tisza (1938) and by Landau (1941, 1947); the long time that was necessary to fully appreciate the link between superfluidity and Bose–Einstein condensation (BEC), which was theoretically predicted in 1924 (Bose, 1924; Einstein, 1925) but experimentally realized only in 1995 by Ketterle, Cornell, and Wieman in conditions that were originally envisaged by Bose and Einstein, which, despite the early insights of London (1938), were not fully appreciated during his time. These topics go beyond the scope of this book and have been already described elsewhere in detail by, for example, Balibar (2007). The relation between superfluidity, Bose–Einstein condensation, and superconductivity is also beyond the scope of this book; on these topics, we point the interested reader to the books of Annet (2004), Tilley and Tilley (1986), and the introductory textbook of Barenghi and Parker (2016).

2.3 The Two-Fluid Model

As we already briefly mentioned above, Tisza was the first to introduce a two-fluid model, extending the early microscopic attempt to describe the phenomenon of superfluidity by London (1938). Below the Bose condensation temperature, Tisza (1938) associated the superfluid component with the condensate and the normal component, or the normal fluid, with the rest of particles constituting the ideal Bose gas of this model, which qualitatively agrees with and explains some of the extraordinary properties of He II. However, as interactions are critically important for superfluidity – without which the critical superfluid velocity of the Bose gas would be zero (see Section 3.12) – Tisza's picture could not explain the underlying physics with sufficient detail and disagreed with several experimental facts. It was therefore heavily criticized by Landau (1941, 1947) who offered his own phenomenological model for the two-fluid behavior. Despite the same name, this model differs in many aspects from Tisza's. The important idea here is the quantum superposition of states, since it is not possible to take one helium atom

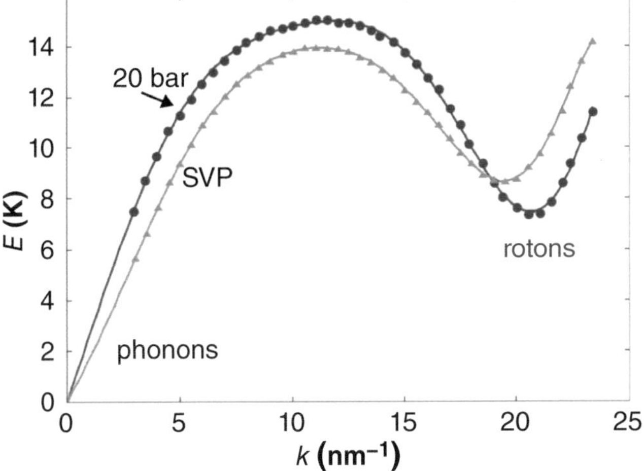

Figure 2.5 Dispersion relation in superfluid helium II at SVP and at 20 bar. E is the energy of excitations of wave number k. The data are taken from the neutron scattering experiments (Gibbs *et al.*, 1999). The speed of first sound (the slope of the curve $E(k)$ vs. k at small k) increases with pressure while the roton energy gap decreases and disappears at about 25 bar when helium solidifies.

out of He II and decide whether it is a part of the normal or superfluid component. Thanks to his phenomenal intuition, Landau guessed the form of dispersion relation in He II, which is of the form shown in Fig. 2.5. To explain the early specific heat data of He II, Landau postulated the existence of two types of elementary excitations, which he called phonons and rotons. As is customary in condensed-matter physics, phonons are (longitudinal) sound waves. The physics of rotons was poorly understood for a long time, but with the availability of better data it became clear that both phonons and rotons belong to the same branch of dispersion relation, with rotons representing quasiparticles possessing higher momenta. In Landau's two-fluid model, the normal fluid consists of these elementary excitations, the phonons and the rotons. Although Landau's original two-fluid model without modifications does not account for quantum turbulence, it is essential to understand the motion of He II and accurately describe its flow at small velocities preceding the appearance of quantum vortices (see Chapter 3).

The main features of Landau's two-fluid model are the following. For temperature $T < T_\lambda$, liquid ^{4}He (helium II) is described as a mixture of two fluid components, the superfluid and the normal fluid. The viscous normal fluid consists of a "gas" of thermal excitations – phonons and rotons – that carry the entire entropy content of the liquid. At relatively "high" temperatures, approximately above 1 K, the mean free path is small, and the thermal excitations can be described hydrodynamically. This normal fluid component coexists with the inviscid superfluid component (related,

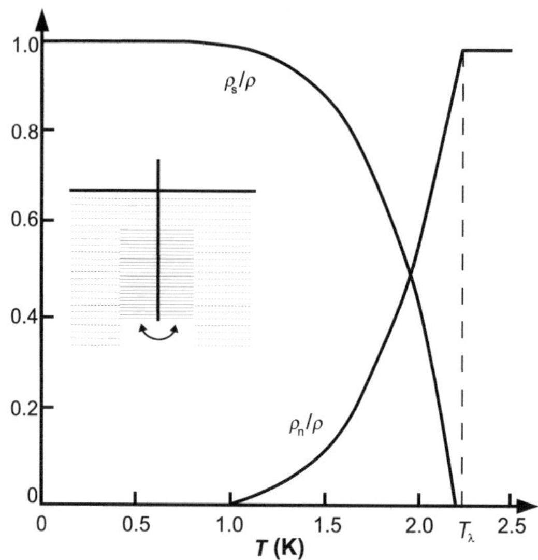

Figure 2.6 Superfluid (ρ_s/ρ) and normal fluid (ρ_n/ρ) fractions vs. temperature. The inset shows schematics of the Andronikashvili experiment: the stack of closely spaced thin discs torsionally oscillating on a fiber in a He II bath.

but not equal to the condensate fraction) carrying no entropy. The density ρ of helium II is nearly temperature independent, of value $\rho \approx 145\,\text{kg/m}^3$, and can be decomposed into two parts:

$$\rho = \rho_n + \rho_s. \tag{2.1}$$

The normal fluid and superfluid densities, ρ_n and ρ_s, are both strongly temperature dependent, as shown in Fig. 2.6. In the low-temperature limit ($T \to 0$), helium II is entirely superfluid ($\rho_s/\rho \to 1$, $\rho_n/\rho \to 0$), while in the high-temperature limit ($T \to T_\lambda$), superfluidity vanishes ($\rho_s/\rho \to 0$, $\rho_n/\rho \to 1$). In practice, in the absence of impurities ^{4}He can be considered a pure superfluid at temperatures below approximately 1 K (where $\rho_n/\rho \cong 0.07$).

Experimental confirmation of the temperature dependence of normal and superfluid densities was provided by the famous experiment of Andronikashvili (1946, 1948). The experiment was based on a torsional pendulum. An oscillating pile of closely spaced discs hung on a fiber; the distance between the discs was smaller than the viscous penetration depth $\sqrt{2\nu_n/\omega}$ of the normal fluid, where ω is the angular frequency of oscillation and ν_n the kinematic viscosity of the normal fluid. The experimental conditions were chosen such that the normal fluid was entrained within the discs by viscous forces, moved with them, and contributed to the angular momentum of the pendulum, while the superfluid was not moving with them and

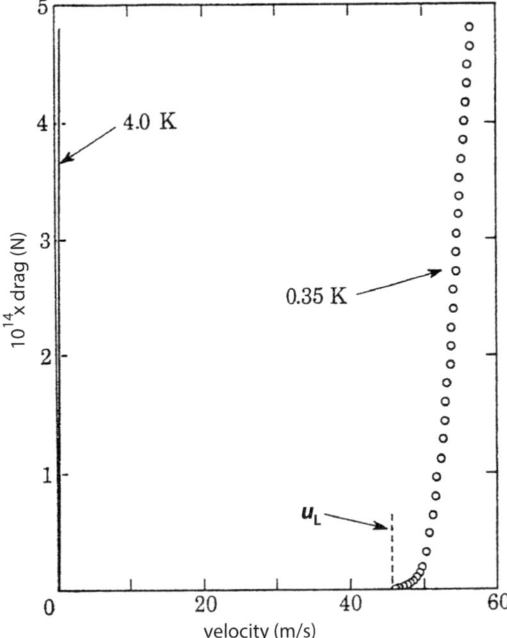

Figure 2.7 The drag on an ion (circles) moving through superfluid ^{4}He at 0.35 K as a function of the average ion velocity. For comparison, the equivalent plot for an ion moving through normal (non-superfluid) ^{4}He at 4.0 K is also shown (solid line), emphasizing the qualitative difference that exists between the two cases. It is clear that drag in the superfluid sets in abruptly at a critical velocity that is very close to the critical velocity for roton creation, as predicted by Landau. Used with permission of The Royal Society (U.K.) from Allum *et al.* (1976): permission conveyed through Copyright Clearance Center, Inc.

stayed at rest. Normal and superfluid densities could then be deduced from the observed temperature dependence of the oscillation frequency of the pendulum.

If no quantized vortices (Chapter 3) are present, normal and superfluid components support two independent velocity fields, $\boldsymbol{u}_n$ and $\boldsymbol{u}_s$, respectively. The two fluids interpenetrate, flowing freely through each other. On the basis of physical considerations, Landau argued that the superfluid component is irrotational, $\nabla\times\boldsymbol{u}_s = \mathbf{0}$. From the form of the dispersion relation, using the laws of conservation of energy and momentum (Barenghi and Parker, 2016), Landau predicted that if an object moves faster than a critical velocity $u_c \approx 60$ m/s (at low pressure) in the superfluid at rest (or, vice versa, if the superfluid flows at velocity faster than u_c with respect to the container's walls), then thermal excitations (rotons) are emitted and the motion of helium II is frictionless no longer.

The existence of this critical velocity is illustrated in Fig. 2.7, which compares the drag offered by liquid He I and He II to a moving negative ion, measured by

Allum *et al.* (1976) at elevated pressure, where the Landau critical velocity due to roton creation drops to about 45 m/s.

It has to be remarked that, in a more general sense, Landau's criterion applies to any superfluid: on exceeding a certain critical velocity u_c (which in fermionic superfluids is called the *Landau's pair-breaking velocity*), it becomes energetically favorable to generate quasiparticles, which means the onset of dissipation, hence the disappearance of superfluidity.

Neglecting viscous effects and terms that are quadratic in the velocities, the two-fluid equations can be written as

$$\frac{\partial \rho}{\partial t} + \nabla \cdot (\rho_s \boldsymbol{u}_s + \rho_n \boldsymbol{u}_n) = 0, \tag{2.2}$$

$$\frac{\partial (\rho S)}{\partial t} + \nabla \cdot (\rho S \boldsymbol{u}_n) = 0, \tag{2.3}$$

$$\rho_s \frac{\partial \boldsymbol{u}_s}{\partial t} = -\frac{\rho_s}{\rho} \nabla P + \rho_s S \nabla T, \tag{2.4}$$

$$\rho_n \frac{\partial \boldsymbol{u}_n}{\partial t} = -\frac{\rho_n}{\rho} \nabla P - \rho_s S \nabla T, \tag{2.5}$$

where S is the entropy per unit mass and P is the pressure. Notice that Eq. (2.2) ensures conservation of mass, and (having neglected viscous dissipation) Eq. (2.3) guarantees conservation of entropy.

2.3.1 First and Second Sound

The existence of two distinct interpenetrating fluids in helium II has physical consequences for the propagation of sound and heat, which are both relevant to quantum turbulence.

The first consequence is the so-called second sound. As we shall see in Chapter 3, second sound is particularly important in the study of quantum turbulence because it is used to detect quantized vortices, characterizing the turbulence intensity in the superfluid component. Small amplitude perturbations of the uniform rest state of Eqs. (2.2–2.5) propagate in the form of two modes. The first mode (analogous to sound in ordinary fluids, hence called the *first sound*) is a wave in which P and ρ oscillate with time, S and T remain essentially constant, and $\boldsymbol{u}_s$ and $\boldsymbol{u}_n$ move in phase with each other. First sound propagates with the speed

$$c_1 = \left(\frac{\partial P}{\partial \rho} \right)^{1/2}, \tag{2.6}$$

which is approximately temperature independent, as shown in the left panel of Fig. 2.8, reaching $c_1 \approx 240$ m/s at very low temperatures (nominally below about

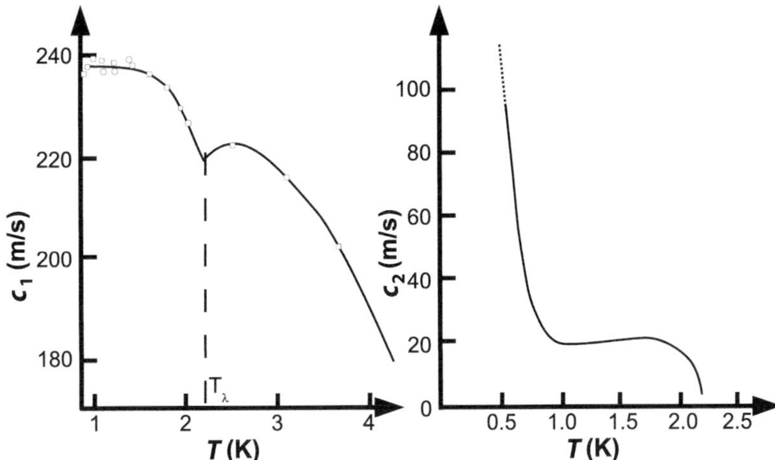

Figure 2.8 Temperature dependence of the (left) first and (right) second sound speed in liquid ^{4}He. First sound (normal and superfluid components oscillating in phase in He II) becomes the usual longitudinal sound (i.e., pressure) wave in He I.

1.5 K). The second mode, already introduced as the *second sound*, is a wave in which T and S oscillate with time. Pressure P and density ρ remain essentially constant, and $\boldsymbol{u}_s$ and $\boldsymbol{u}_n$ move antiphase to each other. Second sound represents a temperature or entropy wave rather than the usual density wave in classical fluids, and propagates with the speed

$$c_2 = \left(\frac{\rho_s T S^2}{\rho_n C_V} \right)^{1/2}, \tag{2.7}$$

where C_V is the specific heat at constant volume. The second sound speed is about one order of magnitude lower than the first sound speed, $c_2 \approx 20$ m/s, and tends to zero as $T \to T_\lambda$, as seen in the right panel of Fig. 2.8.

2.3.2 Coflow and Thermal Counterflow

Because of the existence of two independent fluid components, the superfluid and the normal fluid, the flow of helium II can take various forms. The closest analogue to classical viscous flows is *coflow*, obtained by pushing the normal and superfluid components in the same direction with the same average velocity so that they move together; this is achieved by simple mechanical forcing, such as bellows or propellers. It is also possible to make the normal fluid and the superfluid components flow with different average velocities, a situation generally called *counterflow*.

A special case of counterflow is the *thermal counterflow*. The prototype configuration is a channel that is closed at one end and open to the helium bath at the other

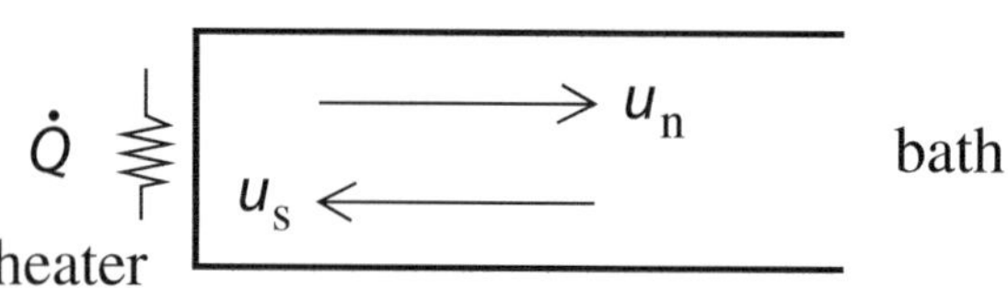

Figure 2.9 Schematic of a counterflow channel. The heater (an electrical resistor) introduces a known heat flux (energy per unit time per unit area of the channel), $\dot{q} = \dot{Q}/A$, where A is the channel cross section. The heat is carried away by the normal fluid toward the bath. Superfluid flows in the opposite direction to conserve mass.

end, as in Fig. 2.9. For simplicity, we assume a channel of constant cross section, A, steady state conditions, and (idealized, as we shall see in Chapter 6) uniform superfluid and normal fluid profiles, u_s and u_n. At the closed end of the channel, an electrical resistor dissipates a known power $\dot{Q}$. This heat is carried away by the normal fluid,

$$\dot{q} = \frac{\dot{Q}}{A} = \rho S T u_n, \tag{2.8}$$

and the superfluid flows in the opposite direction toward the resistor,

$$u_s = -\frac{\rho_n}{\rho_s} u_n, \tag{2.9}$$

so that the net mass flux is zero (the channel is blocked at the side of the resistor), so that

$$\rho_s u_s + \rho_n u_n = 0. \tag{2.10}$$

In this way, a counterflow of the two fluids, $u_{ns} = |u_n - u_s|$, is set up in opposite directions, which is proportional to the applied heat flux:

$$u_{ns} = \frac{\dot{q}}{\rho_s S T}. \tag{2.11}$$

This effect explains the exceptional heat conducting ability of He II, which is exploited in cryogenic engineering applications (Van Sciver, 1986): the internal counterflow of the two fluids prevents temperature gradients being set up. However, as we shall discuss in Section 4.5, if the heat flux exceeds a certain critical value $\dot{q}$, a tangle of superfluid vortices appears, inducing a temperature gradient along the channel thus limiting the extraordinary heat transfer properties of He II.

2.4 Superfluid ^{3}He

We have already mentioned that the fermionic and bosonic structures of ^{3}He and ^{4}He – leading, respectively, to Fermi–Dirac and Bose–Einstein quantum statistics

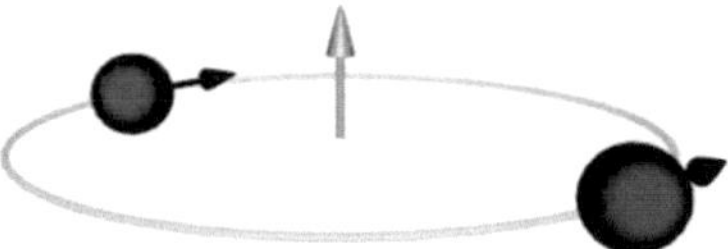

Figure 2.10 Schematic view of a Cooper pair in superfluid ^{3}He. The pair consists of two ^{3}He atoms, each with spin 1/2, circling around their center of mass. In contrast with most superconductors, the total spin and the total angular momentum of ^{3}He Cooper pairs are 1, due to p-wave pairing.

– result in great differences in their properties at very low temperatures. Indeed, although ^{3}He also undergoes the transition to a superfluid state at about $T_c \cong 1$ mK, the underlying physics is different and more similar to the superconducting transition than the lambda transition in ^{4}He. Superconductivity can be viewed as superfluidity of charged electrons in a crystal lattice occurring due to the so-called Cooper pairing of ^{3}He atoms, schematically shown in Fig. 2.10. Several phases of superfluid ^{3}He have been theoretically predicted and experimentally observed. Generally, the superfluid properties of ^{3}He are much more complex than those of He II, and only a short description is provided below and in Chapter 8. Their full description falls outside the scope of this book, but a more detailed description can be found in Tilley and Tilley (1986) and Vollhardt and Wolfle (2013).

Briefly, the phase diagram in Fig. 2.3 shows the two most common superfluid phases, A and B. Various types of quantized vortex structures exist in the strongly anisotropic phase A, including vortex sheets or the so-called continuous vortices possessing no hard cores where superfluidity would be suppressed. The hydrodynamic properties of the A-phase are very interesting and differ in many important respects from those of He II. We shall, however, not consider them here, as quantum turbulence in the A-phase, if it exists, has not yet been seriously investigated.

Quantum turbulence in liquid ^{3}He has so far been experimentally studied in its superfluid B-phase, which in many ways has similar properties to He II. There are, however, important differences. Contrary to liquid He I, which has an extremely low kinematic viscosity above the lambda temperature, the normal liquid ^{3}He just above its superfluid transition temperature represents a rather thick liquid with viscosity that is comparable to that of air or light industrial oil. This is a consequence of the fermionic nature of ^{3}He atoms. The physical properties of normal liquid ^{3}He are described in the frame of the so-called Fermi liquid theory developed by Landau (Landau and Lifshitz, 1987); one of its particular features is a steep increase of viscosity with falling temperature, $\propto 1/T^2$. As a consequence, within the framework of the two-fluid model, the high kinematic viscosity of the normal fluid leads, in most experiments, to the normal fluid being effectively at rest in the container. What

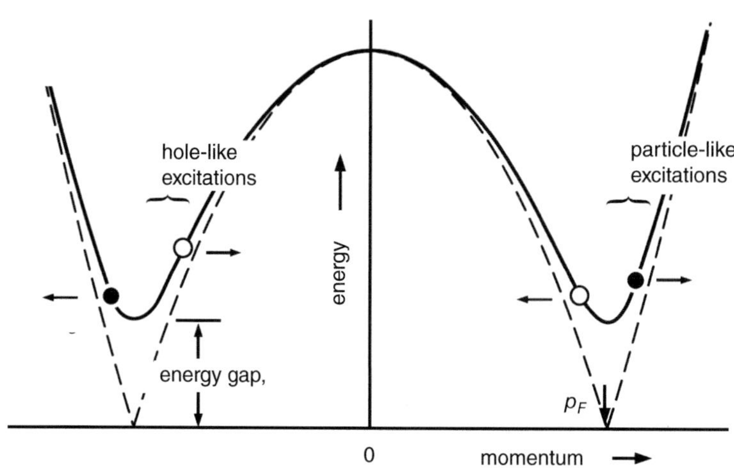

Figure 2.11 (Solid line) The dispersion curve for the quasiparticle excitations in superfluid ^{3}He. There are two classes of excitations: quasiparticles (filled circles) with the group velocity parallel to the momentum, and quasiholes (open circles) with group velocity antiparallel to the momentum. The direction of the momentum is indicated by horizontal arrows.

immediately follows is that second sound, although detected in superfluid ^{3}He by Lu and Kojima (1985), is heavily damped by the thick normal component.

Quasiparticles and quasiholes are always simultaneously present in ^{3}He. The dispersion relation, shown in Fig. 2.11, includes an energy gap and differs from that in bosonic ^{4}He, as discussed earlier in this chapter. In fermionic superfluids, the density of quasiparticles falls off exponentially (faster than in He II) with decreasing temperature; in ^{3}He-B one deals with an essentially pure superfluid at temperatures lower than about $0.2T_c$. When the superfluid fraction of ^{3}He-B moves, the dispersion curve is tipped by the Galilean transformation (similar to the case of ^{4}He); the Landau critical velocity corresponds to the situation when the energy gap disappears.

2.5 Summary

Let us summarize and compare the main flow properties of He II and superfluid ^{3}He-B – our working quantum fluids – from the point of view of fluid dynamics.

The conceptually simplest (though experimentally the most difficult) case is the zero-temperature limit, which, in practice, is below about 0.3 K for He II, and below about $0.2T_c$ for ^{3}He-B, where the critical temperature T_c varies between about 1–2 mK depending on the pressure. In this low-temperature limit, both He II and ^{3}He-B can be treated in most cases as pure superfluids.

At higher temperatures (but below T_λ), both He II and ^{3}He-B display two-fluid behaviors and most of their physical properties can be usefully described by the two-fluid model postulating the existence of an inviscid superfluid component and a viscous normal component, the latter carrying all the entropy content of the liquid. At these higher temperatures, one might therefore expect more or less similar properties in both He II and ^{3}He-B. There are, however, important differences, and the presence of bosonic thermal excitations (phonons and rotons) in He II and of fermionic thermal excitations in ^{3}He-B changes the physical properties of the quantum fluid in the following way. In ^{3}He-B, given typical geometrical and experimental restrictions, the normal fluid hardly moves, being almost clamped to the container walls because of its relatively large viscosity. This difference manifests in fundamentally different macroscopic properties of flows of these two superfluids, as we shall see in subsequent chapters.

3
Quantized Vortices

What makes superfluid hydrodynamics particularly interesting is the quantum mechanical constraint that the circulation around vortices is quantized. This property is central to the subject of this book. The idea of quantized circulation belongs to Onsager (1949). It was further developed by Feynman (1955) and experimentally confirmed by Vinen (1961).

Quantized circulation manifests itself differently in various superfluids and can take the form of rather exotic structures, such as vortex sheets, spin-mass vortices, doubly quantized continuous vortices, or even half-quantum vortices. In this chapter, we shall mainly discuss basic properties of the simplest line defects, or vortices, in 3D flows of He II and ^{3}He-B superfluids. We start with line vortices in He II that are singly quantized; i.e., the circulation around them is a single quantum unit, as we shall see below.

3.1 Quantized Circulation

The quantization of circulation is a consequence of the existence of a macroscopic wavefunction $\Psi(\boldsymbol{x}, t)$ where $\boldsymbol{x}$ is the position and t is time. Because $\Psi(\boldsymbol{x}, t)$ is complex, we can write it in terms of its real and imaginary parts as $\Psi(\boldsymbol{x}, t) = \mathrm{Re}(\Psi(\boldsymbol{x}, t)) + i\,\mathrm{Im}(\Psi(\boldsymbol{x}, t))$. However, to show the connection between quantum mechanics and fluid dynamics, it is more convenient to write $\Psi(\boldsymbol{x}, t)$ in terms of its magnitude, $|\Psi(\boldsymbol{x}, t)|$, and phase, $\phi(\boldsymbol{x}, t)$, as

$$\Psi(\boldsymbol{x}, t) = |\Psi(\boldsymbol{x}, t)|\, e^{i\phi(\boldsymbol{x}, t)}. \tag{3.1}$$

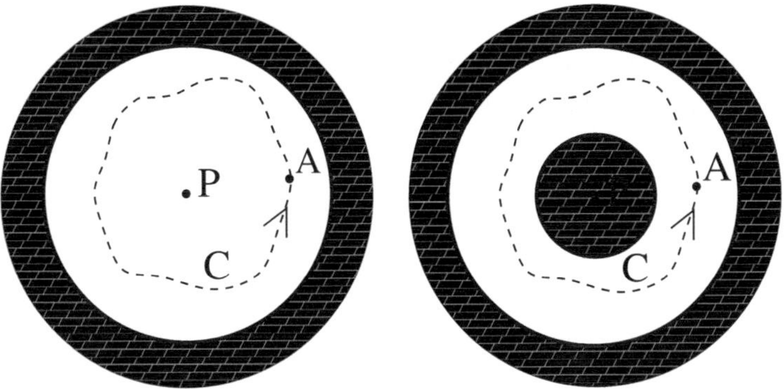

Figure 3.1 Path C around a point P within (Left) a simply connected fluid region and (Right) a doubly connected fluid region.

By applying standard quantum mechanical prescriptions to He II, we define the fluid density and velocity[1] in terms of the magnitude and phase of Ψ, as

$$\rho_s(\boldsymbol{x}, t) = m|\Psi(\boldsymbol{x}, t)|^2, \tag{3.2}$$

$$\boldsymbol{u}_s(\boldsymbol{x}, t) = \frac{\hbar}{m}\nabla\phi(\boldsymbol{x}, t), \tag{3.3}$$

where $\hbar = h/(2\pi)$ is the (reduced) Planck constant and $m = 6.646 \times 10^{-27}$ kg is the mass of the helium atom. Now, consider a region containing He II. The circulation Γ is defined as the integral along any simple (i.e., not self-intersecting) closed path C around a point P within the fluid,

$$\Gamma = \oint_C \boldsymbol{u}_s \cdot d\boldsymbol{r}. \tag{3.4}$$

As illustrated in Fig. 3.1, the fluid region itself can be either simply connected or multiply connected. Suppose that the fluid region is simply connected; then (by definition) we can continuously shrink the curve C to the point P without C leaving the region; in this case Stokes' theorem applies (the line integral of $\boldsymbol{u}_s$ along C becoming the integral of $\nabla \times \boldsymbol{u}_s$ over the surface S bounded by C), and, by using Eq. (3.3) and the fact that the curl of a gradient is always zero, we conclude that the circulation is zero. That is,

$$\Gamma = \int_S (\nabla \times \boldsymbol{u}_s) \cdot d\boldsymbol{S} = 0. \tag{3.5}$$

Now suppose that the fluid region is multiply connected (doubly connected in Fig. 3.1). In this case Stokes, theorem does not apply. We use Eqs. (3.3) and (3.4),

[1] For simplicity, at this stage we neglect the distinction between the superfluid component and the condensate. We shall take up this topic in Chapter 5.

and note that the phase difference between the starting point (A) along the path C and the final point (A again) must be $2\pi q$, where q (called the winding number or charge) must assume integer values $0, 1, 2, \ldots$ for the wavefunction to be single valued. We obtain

$$\Gamma = \frac{\hbar}{m} \oint_{\mathrm{C}} \nabla\phi \cdot d\boldsymbol{r} = q\kappa. \tag{3.6}$$

Since, as we shall see later in this section, multiply quantized ($q > 1$) vortices are unstable in He II, the circulation around a vortex line is restricted in practice to $q = 1$, and we have

$$\kappa = \frac{h}{m} = 9.97 \times 10^{-8}\,\mathrm{m^2/s}, \tag{3.7}$$

where κ is called the *quantum of circulation*. Equation (3.7) shows that the angular momentum per atom is quantized in units of the Planck constant (Onsager, 1949; Landau and Lifshitz, 1987).

An example of a multiply connected region of helium is a cylindrical container with a wire stretched from top to bottom along the axis. This configuration was used by Vinen (1961) to measure κ for the first time.

The transformation from simply connected to multiply connected region can occur spontaneously if the superfluid becomes threaded by vortex lines, which are essentially tiny tubular holes of atomic size. The appearance of vortex lines takes place in an ordered way when helium rotates or in a disordered way when the helium flow becomes turbulent.

A *vortex line* is defined as a 3D space-curve along which $\Psi = 0$; that is, both $\mathrm{Re}(\Psi) = 0$ and $\mathrm{Im}(\Psi) = 0$. Since along this curve the phase

$$\phi = \arctan\left(\frac{\mathrm{Im}(\Psi)}{\mathrm{Re}(\Psi)}\right) \tag{3.8}$$

is undefined, we say that a vortex line is a topological defect of the phase. The presence of a vortex line makes He II multiply connected because the superfluid density $\rho_s = m|\Psi|^2$ vanishes on the centerline (axis) of the vortex, creating a hollow tube of radius $a_0 \approx 10^{-10}$ m, as estimated by experiments with vortex rings (Rayfield and Reif, 1964).

Consider a straight singly charged ($q = 1$) vortex line. If we assume cylindrical coordinates with the z-axis aligned along the vortex and evaluate the circulation using Eq. (3.6) along a circle C of radius $r \gg a$, we find that the superfluid velocity has the form of an azimuthal flow around the axis of magnitude

$$u_s = \frac{\kappa}{2\pi r}\,. \tag{3.9}$$

Note that, although $u_s \to \infty$ as $r \to 0$, the momentum is finite as $\rho \to 0$ (the vortex core is hollow). What is known of the nature of the vortex core will be discussed in Section 3.2.

In summary, a vortex line in helium II is a hollow, Angström-size tube with fixed circulation κ and fixed core size. The velocity field around the hole, Eq. (3.9), corresponds to a change of 2π of the quantum mechanical phase around the vortex line. Unlike a classical vortex, which decays due to viscous forces, the superflow around an isolated vortex line is persistent. It never stops because the superfluid has zero viscosity, and cannot become weaker because the circulation is fixed: the vortex is topologically protected by the quantization of the circulation. The classical Kelvin theorem applies, so the circulation is conserved and vortex lines form either closed vortex loops or terminate at solid boundaries or free surfaces. Another feature that distinguishes a vortex line from a classical vortex is that, since the vortex core radius a_0 is fixed, we cannot increase the vorticity by stretching the vortex.

Finally, we show why multiply charged He II vortices are energetically unstable. (In Chapter 8 we shall discuss the more complex situation in superfluid ^{3}He.) Consider a straight vortex of circulation $q\kappa$ in a cylinder of radius R and height h. Its total kinetic energy and angular momentum are, respectively,

$$\begin{aligned} E_v &= \int_V \rho_s \frac{u_s^2}{2}\, dV = \frac{\rho_s q^2 \kappa^2 h}{4\pi} \ln\left(\frac{R}{a_0}\right), \\ L_v &= \int_V \rho_s r u_s dV \approx \frac{\rho_s q \kappa h}{2} R^2, \end{aligned} \tag{3.10}$$

where we have used $u_s = q\kappa/(2\pi r)$, written the volume element in cylindrical coordinates $dV = dz\, d\theta\, r\, dr$, introduced the lower cutoff at $r = a_0$ to account for the vanishing density in the core region, and neglected a_0^2 compared to R^2. Equation (3.10) shows that, given the angular momentum L_v, there is less energy E_v in many singly charged vortices than in a single multiply charged vortex. In the presence of any dissipation, therefore, a multiply charged ($q > 1$) vortex will split into q singly charged ($q = 1$) vortices. The instability of multiply charged vortices is not only energetic but also dynamic (that is to say, it occurs even without a mechanism for decreasing the energy of the system); according to the recent work by Patrick *et al.* (2022), the dynamical instability results from a super-radiant bound state inside the vortex core.

3.2 Nature of the Vortex Core

The precise nature of the vortex core in He II is still not well understood. There is little doubt, however, that the vortex core radius is approximately $a_0 \approx 10^{-10}$ m in He II and about 100 times larger (the precise value depending on pressure) in

^{3}He-B. The main evidence comes from time-of-flight experiments of charged He II vortex rings (Rayfield and Reif, 1964) and the classical Hamiltonian relations that relate energy E_R, velocity u_R, and impulse p_R of a hollow-core circular vortex ring of radius $R \gg a_0$ (Barenghi and Donnelly, 2009), as

$$u_R = \frac{\kappa}{4\pi R}\left(\ln\frac{8R}{a_0} - \frac{1}{2}\right), \quad E_R = \frac{\rho_s \kappa^2 R}{2}\left(\ln\frac{8R}{a_0} - 2\right), \quad p_R = \rho_s \kappa \pi R^2. \tag{3.11}$$

Unfortunately, no flow visualization technique yet exists that directly reveals the nature of the vortex core at a length scale of the order of $a_0 \approx 10^{-10}$ m. Lacking this information precisely, it is often assumed in the literature that a vortex line in He II is either hollow (with the superfluid density equal to zero for $r < a_0$ and jumping to the bulk value ρ_s for $r > a_0$) as for Eq. (3.11), or similar to the vortex solution of the Gross–Pitaevskii equation (GPE) (see Section 5.1.1 and Fig. 5.1), with density changing smoothly over the characteristic distance a_0, from ρ_s at infinity to zero on the vortex axis. What happens exactly inside the vortex core ($r < a_0$) does not seem to be directly relevant to most experimental results of quantum turbulence. Indeed, in the popular vortex filament model (VFM) used to model turbulence in the superfluid component, the vortex core size is assumed to be infinitesimally small compared to all other length scales.

There are, however, important features of quantum turbulence for which a better understanding of the vortex core would be useful, for example, vortex reconnections. The best microscopic description currently available is the following (Galli *et al.*, 2014; Amelio *et al.*, 2018): The N-body wavefunction $\Psi(\boldsymbol{R})$ (where $\boldsymbol{R} = (\boldsymbol{r}_1, \boldsymbol{r}_2, \ldots, \boldsymbol{r}_N)$ are the coordinates of N atoms) is an eigenstate of the angular momentum operator with eigenvalues that are integer multiples of $N\hbar$. Therefore, $\Psi(\boldsymbol{R})$ must be of the complex form $\Psi(\boldsymbol{R}) = \Psi_0(\boldsymbol{R})e^{i\Omega(\boldsymbol{R})}$ where $\Omega(\boldsymbol{R})$ is the phase. In the *fixed phase approximation*, one chooses $\Omega(\boldsymbol{R})$ and solves the resulting Schrödinger equation for $\Psi_0(\boldsymbol{R})$ allowing interatomic correlations at short distances. The simplest choice is the *Onsager–Feynman phase* $\Omega(\boldsymbol{R}) = \sum_{j=1}^{N} \theta_j$ where θ_j is the azimuthal angle of atom j with respect to the vortex axis; this choice yields the characteristic velocity field $u_\theta = \kappa/(2\pi r)$, which diverges as $r \to 0$, and the density profile that vanishes at the axis. Unlike the GPE case, this solution displays steady density oscillations near the edge of the core of wavenumbers typical of rotons. Indeed, it is a major improvement of this approach over the GPE that the spectrum of elementary excitations contains rotons. Such oscillations also appear if one solves the *nonlocal* GPE (Berloff and Roberts, 1999), which also yields rotons. Better agreement with experiments is obtained if the Onsager–Feynman assumption is used only as an initial guess and interparticle correlations are also taken into account in determining the phase (Ortiz & Ceperley, 1995). The improved model yields features (Ortiz & Ceperley, 1995; Sadd *et al.*, 1997) that are schematically

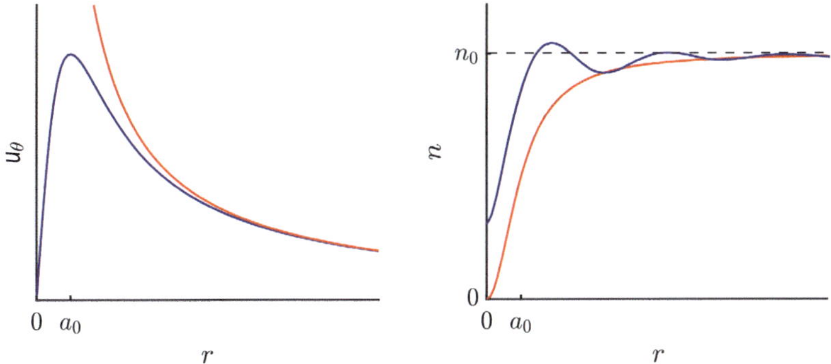

Figure 3.2 (Left) Azimuthal velocity profile u_θ vs. r and (Right) schematic number density profile $n(r)$ vs. r in the GPE model (red curves) and N-body quantum mechanics (blue curves). Data from Galli *et al.* (2014).

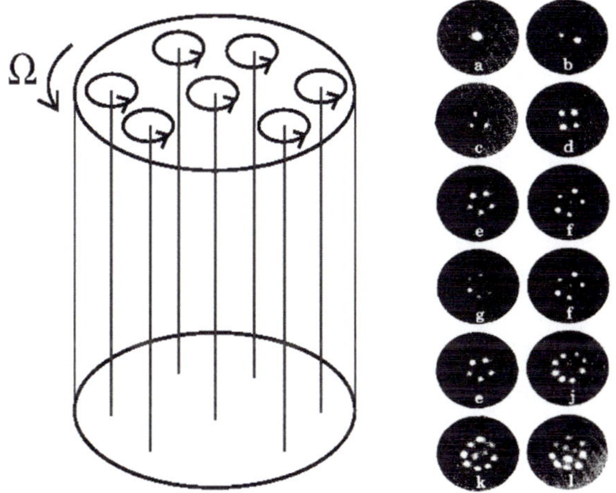

Figure 3.3 (Left) Schematic superfluid vortex lattice in a vessel rotating at constant angular velocity Ω. (Right) Vortex lattices in ^{4}He at increasing angular frequency of rotation, experimentally visualized by Packard and collaborators by trapping electrons in the vortex lines; each dot is the tip of a vortex, looking along the axis of rotation. Reprinted figure with permission from Yarmchuk *et al.* (1979). Copyright 1979 by the American Physical Society.

summarized in Fig. 3.2: the density drops in the core (as for the GPE model) but remains nonzero on the vortex axis (unlike the GPE), and the azimuthal velocity near the axis takes the form of a Rankine vortex with a crossover from $u_\theta \sim r$ behavior to $u_\theta \sim 1/r$ behavior at $r \approx a_0$; this second feature means that the vorticity is approximately constant inside the core.

3.3 Vortex Lattice

According to classical fluid mechanics, an ordinary viscous fluid filled in a container that rotates at constant angular velocity Ω about the vertical axis has a height profile (neglecting capillary phenomena) given by $z = (\Omega r)^2/(2g)$, where r is the radial distance from the axis and g is the acceleration due to gravity. The fluid velocity profile $\boldsymbol{u} = \Omega r \hat{\boldsymbol{\phi}}$ matches that of solid body rotation, hence the vorticity is $\boldsymbol{\omega} = \nabla \times \boldsymbol{u} = 2\Omega \hat{\boldsymbol{z}}$, where $\hat{\boldsymbol{\phi}}$ and $\hat{\boldsymbol{z}}$ are the unit vectors along the azimuthal and axial directions, respectively.

In He II, according to the two-fluid model, the superfluid component is irrotational ($\nabla \times \boldsymbol{u}_{\mathrm{s}} = \boldsymbol{0}$) but the normal fluid is not subject to this restriction, so at finite temperature one might expect a modified height profile $z = (\rho_{\mathrm{n}}/\rho)\ (\Omega r)^2/(2g)$ controlled by the normal fluid fraction ρ_{n}/ρ. However, experiments by Osborne (1950) determined that the height profile of He II obeys the classical formula $z = (\Omega r)^2/(2g)$ independently of temperature. Following the work of Onsager and Feynman, it became apparent that the solution of this puzzle is that the superfluid component becomes threaded by a hexagonal lattice of rectilinear singly quantized vortex lines (see Fig. 3.3) with areal density (number of vortex lines per unit area) obeying the so-called *Feynman rule*,

$$n_{\mathrm{v}} = \frac{2\Omega}{\kappa}. \tag{3.12}$$

The superfluid vortex lattice can be related thus to classical rotation. A classical liquid rotating at angular velocity Ω has vorticity 2Ω in the direction of the rotation. Interpreting Eq. (3.12), we note that, by forming n_{v} vortices per unit area, with each vortex carrying the circulation κ, the superfluid mimics the vorticity of a classical fluid that rotates as a solid body, so that for characteristic length scales much greater than the distance between quantized vortices (which we estimate as $\ell \approx n_{\mathrm{v}}^{-1/2}$ on dimensional grounds) we have $\langle \omega_{\mathrm{s}} \rangle = \langle \nabla \times \boldsymbol{u}_{\mathrm{s}} \rangle = 2\Omega$, where the symbol $\langle \cdot \rangle$ denotes averaging over said length scales. This result will become useful when trying to understand the underlying physics of quantum turbulence.

Packard and collaborators (Yarmchuk *et al.*, 1979) used photographic techniques to mark the points where the vortex lines intersect the free surface (see Fig. 3.3, right) thus providing indirect evidence of rotating vortex lattices in rotating He II (see Fig. 3.3, left). Direct in situ visualization in three dimensions of the vortex lattice in helium eluded physicists for a long time. Better images of the vortex lattice were instead obtained by imagining related atomic Bose–Einstein condensates (Abo-Shaeer *et al.*, 2001; Coddington *et al.*, 2004), see Fig. 3.4. It is only more recently (Bewley *et al.*, 2006) that classical visualization methods such as Eulerian particle image velocimetry (PIV) or Lagrangian particle tracking velocimetry (PTV), which we shall discuss in chapters to follow, have been successfully applied in the difficult

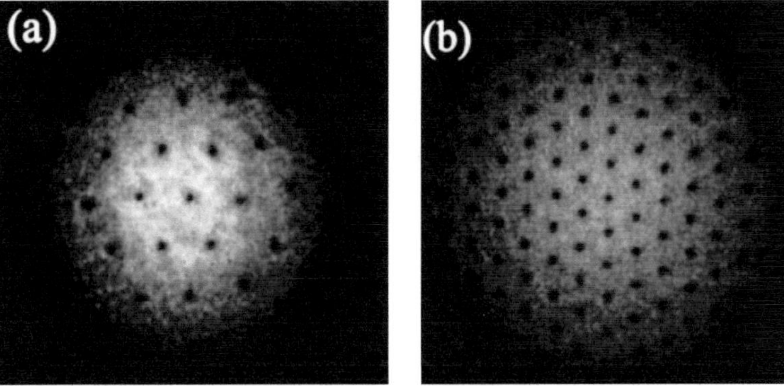

Figure 3.4 Observation of vortex lattices in rotating atomic Bose–Einstein condensates at (a) slow and (b) fast angular velocity. Reprinted figure with permission from Coddington *et al.* (2004). Copyright 2014 by the American Physical Society.

low-temperature environment of liquid helium. In particular, Bewley (2006) directly confirmed Feynman's relation.

It is easy to assess the condition for the first vortex to appear in the cylindrical vessel of radius R by minimizing the expression $F' = F - \boldsymbol{L}_{\mathrm{v}} \cdot \boldsymbol{\Omega}$, where F is the free energy and $\boldsymbol{L}_{\mathrm{v}}$ is the angular momentum of the fluid. This so-called lower critical velocity is

$$\boldsymbol{\Omega}_{\mathrm{c}}^{1} = \frac{\kappa}{R^2} \ln\left(\frac{R}{a_0}\right), \tag{3.13}$$

corresponding to what is known as the Feynman criterion (Feynman, 1955), derived on the basis of similar energy considerations for vortex rings. This result does not mean, however, that upon exceeding this critical velocity Ω_{c}^{1} vortex lines necessarily appear in the flow.

3.4 Vortex Nucleation

Nucleation of quantized vortices is a complex problem, a thorough discussion of which we will not provide in this book. Briefly, it is easy to show that the creation of a length of vortex line must be opposed by a potential barrier. Consider a uniform flow of a superfluid, with density ρ_{s} and velocity u_{s}, parallel to a plane solid boundary, as in Fig. 3.5. A stationary vortex of the appropriate sign, at a distance x from the wall and normal to the flow experiences two Magnus forces: one of magnitude $\rho_{\mathrm{s}}\kappa v_{\mathrm{s}}$ away from the wall, and one of magnitude $\rho_{\mathrm{s}}\kappa^2/4\pi x$ towards the wall due to the image of the vortex in the wall. The resulting potential energy, U, has a maximum value approximately equal to $\left(\rho_{\mathrm{s}}\kappa^2/4\pi\right)\ln(x/a_0)$ at a distance from the wall equal to $\kappa/4\pi u_{\mathrm{s}}$. Except at temperatures very close to the λ-transition

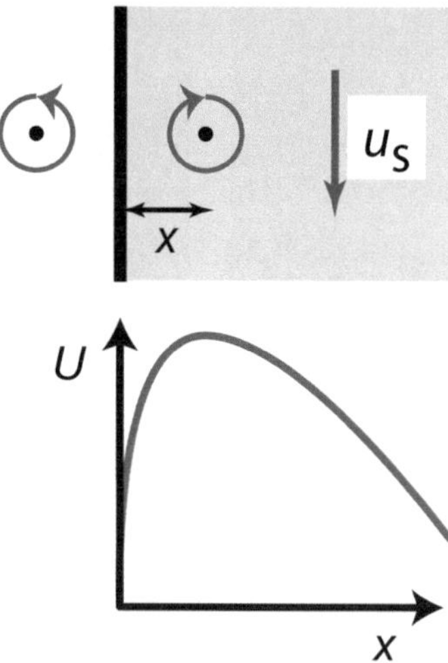

Figure 3.5 Illustrating the potential barrier opposing vortex nucleation. The energy barrier exists due to interaction of a vortex in the superfluid (shaded area) with the wall, which is the same as the interaction with the image vortex of opposite circulation behind the wall. As a result, the vortex-image vortex pair moves parallel with the wall in each other's velocity field that is a constant, u_s.

or at velocities much larger than those characteristic of quantum turbulence, this potential barrier cannot be overcome, either thermally or by quantum tunneling. Thus, the superflow ought to remain frictionless up to an intrinsic critical velocity that is very much larger than is usually observed in experiments. In practice, this so called *intrinsic* vortex nucleation in He II requires $u_c \approx 10\,\text{m/s}$, large enough to make it unlikely, unless induced by a fast-moving ion. In ^{3}He-B or in Bose–Einstein condensates (BECs) the situation is different, as we shall discuss in Section 3.9.

It has long been recognized that, in practice, frictionless superflow usually breaks down by an *extrinsic* process (Schwarz, 1990) in which an existing small length of vortex line expands under the influence of the superflow. There must still be some effective barrier because there would otherwise be no frictionless flow, but it must be small. In any macroscopic He II sample, the seeds are always present in practice. These *remanent vortices* might have been left from an earlier experiment, or might have been formed by the Kibble–Zurek mechanism (Zurek, 1985) when helium is cooled through the second-order λ-transition. The basic idea of this nucleation mechanism is that, when cooling through the second-order phase transition, the phase of the macroscopic wavefunction, unable to adjust everywhere at the same time, leaves vortex lines as topological defects.

Vortex lines thus appear spontaneously while cooling helium through the critical temperature, but they decay and disappear at lower temperatures. However, the

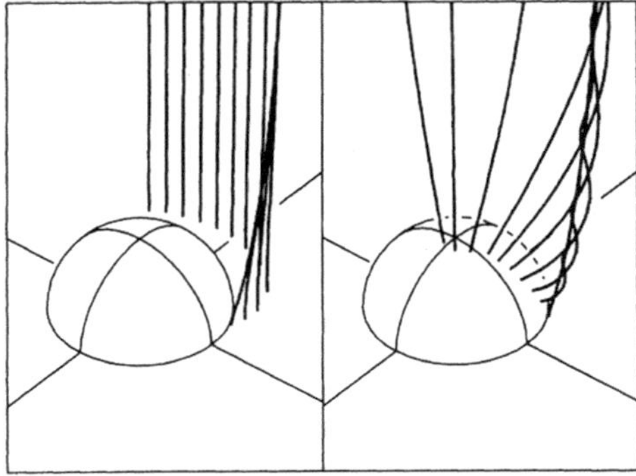

Figure 3.6 (Left) A vortex line approaches a pinning site, modelled as a smooth hemispherical bump, and (Right) comes to rest on top of it. Reprinted figure with permission from Schwarz (1985). Copyright 1985 by the American Physical Society.

walls of the vessel containing He II are rough in comparison with the vortex core size, providing suitable pinning centers to anchor remanent vortices. Figure 3.6 shows a numerical simulation by Schwarz (1985) of the pinning process. In fact, it is hardly possible to create any macroscopic He II sample that is entirely free of remanent vortices. This is why the critical velocities for vortex generation in He II are observed to be low in practice, of order a few cm/s, rather than of the order of 10 m/s, as would be required by the intrinsic vortex nucleation mechanism.

Note in passing that, due to the much bigger size of vortex cores in the case of ^{3}He-B or Bose condensates, the situation is very different, as will be discussed separately in Section 3.9 and Chapter 8, in connection with the transition to quantum turbulence in these systems.

3.5 Mutual Friction

Three properties of vortex lines are important for quantum turbulence. The first is the *mutual friction* (Vinen, 1957; Barenghi *et al.*, 1983) that couples the superfluid and the normal fluid. In fact, vortex lines interact with phonons and rotons that make up the normal fluid; these quasiparticles are scattered off the vortex cores, giving rise to a friction force acting on the vortex lines as they move with respect to the normal fluid. As a result, the normal and the superfluid velocity fields are no longer independent, but couple together.

Within the framework of the phenomenological two-fluid model, Eqs. (2.4) and (2.5) can be modified in order to take into account the presence of vortex lines via additional mutual friction coupling terms. Assuming small velocities, we neglect the nonlinear terms on the left-hand side; we also assume that normal fluid and superfluid densities are constant, and obtain

$$\rho_s \frac{\partial \boldsymbol{u}_s}{\partial t} = -\frac{\rho_s}{\rho} \nabla P + \rho_s S \nabla T - \boldsymbol{F}_{ns}, \tag{3.14}$$

$$\rho_n \frac{\partial \boldsymbol{u}_n}{\partial t} = -\frac{\rho_n}{\rho} \nabla P - \rho_s S \nabla T + \boldsymbol{F}_{ns} + \eta \nabla^2 \boldsymbol{u}_n, \tag{3.15}$$

with $\nabla \cdot \boldsymbol{u}_s = \nabla \cdot \boldsymbol{u}_n = 0$, where η is the dynamic viscosity of the normal fluid and $\boldsymbol{F}_{ns}$ is the mutual friction force (per unit volume) that couples the normal and superfluid velocity fields in the presence of vortex lines.

The term $\rho_s S \Delta T$ is responsible for the so-called *fountain effect*. Indeed, in steady flows and in the absence of vortex lines, the superfluid equation, Eq. (3.14), reduces to the well-known London equation for fountain effect,

$$\frac{\Delta P}{\Delta T} = \rho S, \tag{3.16}$$

first derived by F. London on the basis of different considerations. This equation, which has been verified in many experiments, means that in He II a gradient of temperature leads to a gradient of pressure (and vice versa). The normal fluid equation, Eq. (3.15), together with the no-slip boundary condition for $\boldsymbol{u}_n$, relates the profile of $\boldsymbol{u}_n$ to the applied pressure gradient.

Mutual friction is best described via its action on second sound, the special mode of oscillation that occurs in He II due to its two-fluid nature. Second sound can be generated easily in He II by applying an AC voltage to a resistive heater immersed in He II and can be detected by a sensitive thermometer. In the second sound wave, the normal fluid and the superfluid oscillate antiphase to each other so that the total density and pressure remain constant. The second sound wave is attenuated by vortex lines because the quasiparticles (phonons and rotons), which are the constituents of the normal fluid, are scattered off the cores of the vortex lines.

Hall and Vinen (1957) studied second sound in rotating helium, and found that the second sound wave is unaltered (to a first approximation) if the wave propagates in the direction along the vortex lines; however, if the wave propagates across the vortex lines, the amplitude of the wave is attenuated. These experimental results are consistent with a mutual friction force (per unit volume) of the form

$$\boldsymbol{F}_{ns} = B \frac{\rho_s \rho_n}{\rho} \hat{\boldsymbol{\Omega}} \times [\boldsymbol{\Omega} \times (\overline{\boldsymbol{u}_s} - \overline{\boldsymbol{u}_n})] + B' \frac{\rho_s \rho_n}{\rho} \boldsymbol{\Omega} \times (\overline{\boldsymbol{u}_s} - \overline{\boldsymbol{u}_n}), \tag{3.17}$$

where $\overline{\boldsymbol{u}_s}$ and $\overline{\boldsymbol{u}_n}$ are velocity fields averaged over fluid parcels containing many vortex lines. The experimentally observed quantities B and B' are weakly dependent on the wave frequency, but their values are well known and tabulated (Donnelly and Barenghi, 1998); together with other main physical properties of ^{4}He along the SVP curve, the values of B and B' are available on an interacting website,[2] which computes cubic spline interpolations at any desired temperature.

Mutual friction obviously disappears in the zero-temperature limit, where the normal fluid is absent. We shall discuss this important limit when appropriate in the rest of the book.

3.6 Kelvin Waves

The second property that is important for quantum turbulence is that vortex lines support *Kelvin waves*, propagating helical displacements of the vortex core. Kelvin waves arise from the tension of the vortex lines (the kinetic energy of a circulating superfluid per unit length of vortex line, Eq. (3.10)). One can show that a small-amplitude deformation of wavelength $\lambda = 2\pi/k$, where k is the wavenumber, propagates along the vortex line in the form of a circularly polarized wave, which, in the long wavelength limit, has the dispersion relation given by

$$\omega \approx \frac{\kappa k^2}{4\pi} \ln\left(\frac{1}{k a_0}\right), \tag{3.18}$$

where ω is the angular frequency of the wave. This dispersion relation was first derived by Thomson (1880) for waves on a classical thin-cored vortex filament in an inviscid fluid. A well-known example of classical Kelvin waves is often visible in the sky: the tips of the wings of an aircraft generate trailing vortices of opposite circulation that, downstream, become unstable (Crow instability) and oscillate. The difference between classical Kelvin waves and Kelvin waves in He II is that in He II the circulation κ is not arbitrary but prescribed. Individual Kelvin waves resulting from the relaxation of reconnecting vortex lines have been directly observed in experiments by Fonda *et al.* (2014). Kelvin waves on a vortex lattice (Henderson & Barenghi, 2004) extend beyond the cutoff frequency of the classical spectrum of inertial waves in a rotating liquid.

At finite temperatures, Kelvin waves in superfluid helium are damped by mutual friction, but propagate almost undamped below 1 K. The nonlinear interactions of Kelvin waves shift their energy to larger and larger wavevectors, a process known as the *Kelvin wave cascade*. At large values of k, Kelvin waves rotate rapidly and decay due to acoustic emission. As we shall discuss in Chapter 10, acoustic emission provides an important mechanism for the decay of quantum turbulence in He II in the zero-temperature limit.

At large amplitude, Kelvin waves affect the motion of vortex lines; for example, Kelvin waves along a vortex ring slow down the translational velocity of the ring

[2] www.mas.ncl.ac.uk/helium/

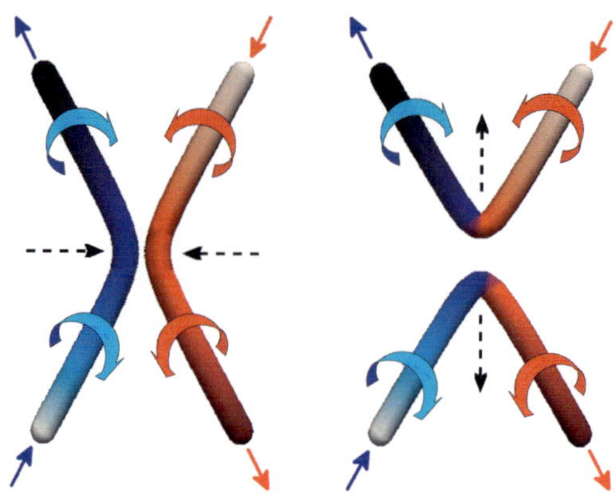

Figure 3.7 Schematics of vortex reconnection. (Left) Before reconnection; (Right) after reconnection. Color gradients along the vortices and blue/red arrows represent vorticity directions along the vortices and directions of flow velocity around them. Dashed black arrows indicate vortex motion, first toward each other and then away from each other.

(Barenghi *et al.*, 2006). Vortex lines also sustain other nonlinear waves such as breathers and solitons (Salman, 2013).

3.7 Vortex Reconnections

The third crucial property of vortex lines is that, when they collide, they split and reconnect, changing their topology. The reconnection process is schematically illustrated in Fig. 3.7: two vortex lines (the direction of the flow around each line being represented by the shaded blue/red colors and by the blue/red arrows) approach each other increasingly rapidly, exchange heads and tails, and move away from each other as they slow down.

Reconnections of streamlines and vortex lines occur naturally in classical viscous flows governed by the Navier–Stokes equation. Indeed, without reconnections, almost none of the flows that we observe in ordinary life – particularly turbulent flows – would exist. Reconnections of thin, concentrated vortices in ordinary fluids are similar to reconnections of vortex lines in quantum fluids. An example of such reconnection is sometimes visible in the sky, when wingtip vortices that trail a flying aircraft develop unstable Kelvin waves that grow in amplitude, as mentioned in the previous section; when this amplitude becomes of the order of the separation between the two vortices, reconnections take place, creating beautiful vortex rings (see, for example, Van Dyke 1982, figure 116). Similar reconnections of magnetic field lines occur in plasmas (e.g., the solar corona, the Earth's magnetosphere).

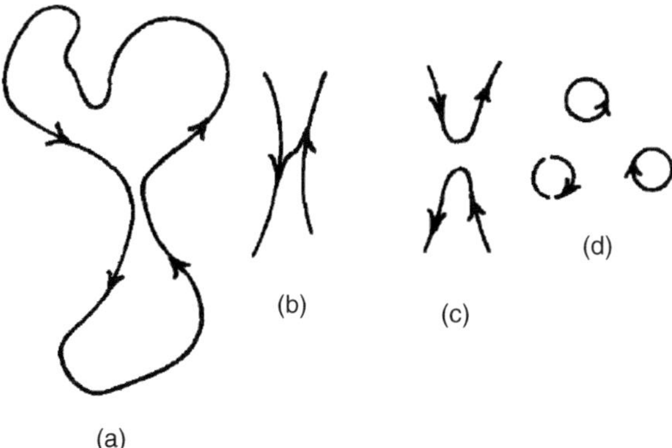

Figure 3.8 Vortex reconnections as envisaged by Feynman. Reprinted from Feynman (1955) with permission from Elsevier.

The possibility of superfluid vortex reconnections was first raised by Feynman (1955), who argued that distorted vortex loops can break up and reconnect, forming smaller loops, as shown in Fig. 3.8.

The necessity of vortex reconnections for turbulence was appreciated by Schwarz (1988) when developing the VFM. Lacking any experimental evidence of reconnections, Schwarz clearly stated that reconnections are an independent *ansatz*. In the VFM, vortex reconnections are computed algorithmically when two vortex lines become sufficiently close to each other (compared to the typical discretization distance along the vortex lines; see Chapter 5).

Later, by solving the Gross–Pitaevskii equation numerically, Koplik and Levine (1993) demonstrated that quantized vortex lines do indeed reconnect when they collide. An exact analytical solution of reconnecting vortex lines near the reconnection point in the planar approximation was obtained by Nazarenko and West (2003). Direct experimental visualization of vortex reconnections was developed first by Bewley *et al.* (2008b) in superfluid helium using tracer particles – see Fig. 3.9 – and by Serafini *et al.* (2015, 2017) in atomic BECs using a stroboscopic technique.

Classical reconnections depend on viscous or resistive dissipation mechanisms that turn energy into heat; in the inviscid limit (zero viscosity, or infinite electrical conductivity) these mechanisms cannot operate. In fact, reconnections of vortex lines occur in quantum fluids at almost constant energy, with only a small fraction of the kinetic energy of the vortex configuration turning into acoustic energy (sound waves). The effect was demonstrated by Leadbeater *et al.* (2001) by numerically solving the Gross–Pitaevskii equation. Figures 3.10 and 3.11 show a reconnection event and the corresponding sound emission. It must be stressed that there is no direct experimental evidence yet for this sound emission: this is not surprising as the direct visualization would require one to resolve a density wavepacket of the order

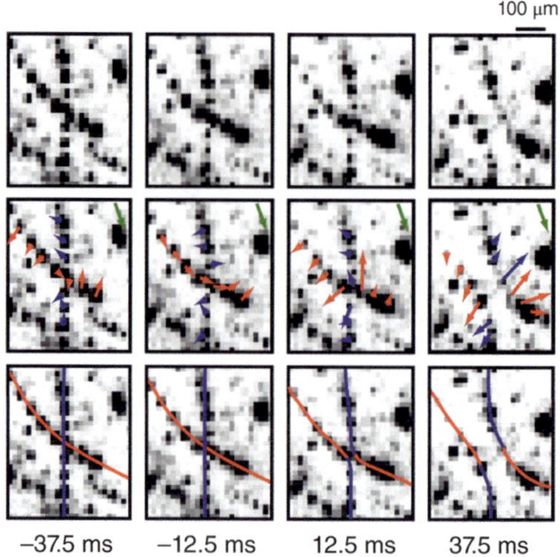

Figure 3.9 Experimental images of (Top) particles trapped on reconnecting vortices, (Middle) the corresponding velocity vectors, and (Bottom) the pre/post interpretations of the vortex configurations. The positive/negative time intervals shown at the bottom refer to the instant of reconnection. Reprinted from Paoletti *et al.* (2010) with permission from Elsevier.

of only ten healing lengths (i.e., about $10\,a_0$). However, since vortex reconnections are frequent events in a turbulent vortex tangle, the loss of kinetic energy and the corresponding sound generation are significant (by vortices that reconnect or, in general, undergo rapid accelerations), and were invoked to explain the observed decay of turbulence at very low temperatures (Davis *et al.*, 2000).

The possibility that vortex reconnections obey universal rules independently of the initial conditions has been considered in the literature. De Waele and Aarts (1994) suggested that the angle of the reconnecting vortex cusps is universal, but Tebbs *et al.* (2011) found counterexamples (collisions between vortex rings and vortex lines). A cascade of vortex rings (Kerr, 2011; Kursa *et al.*, 2011) is another feature that was found to be common in antiparallel reconnections but not universally so. In seeking universality, a more promising feature is the temporal scaling of the minimum distance δ between the two reconnecting vortex strands, shown in Fig. 3.12 before and after the reconnection. The scaling is analyzed by fitting the expressions

$$\delta(t) = A^-(t_0 - t)^{\alpha^-} \quad \text{for } t < t_0, \qquad \delta(t) = A^+(t - t_0)^{\alpha^+} \quad \text{for } t \geq t_0, \tag{3.19}$$

and examining the prefactors $A^\pm$ and the exponents $\alpha^\pm$, where t_0 is the time of the reconnection and the symbols $\pm$ refer to pre/post reconnection dynamics.

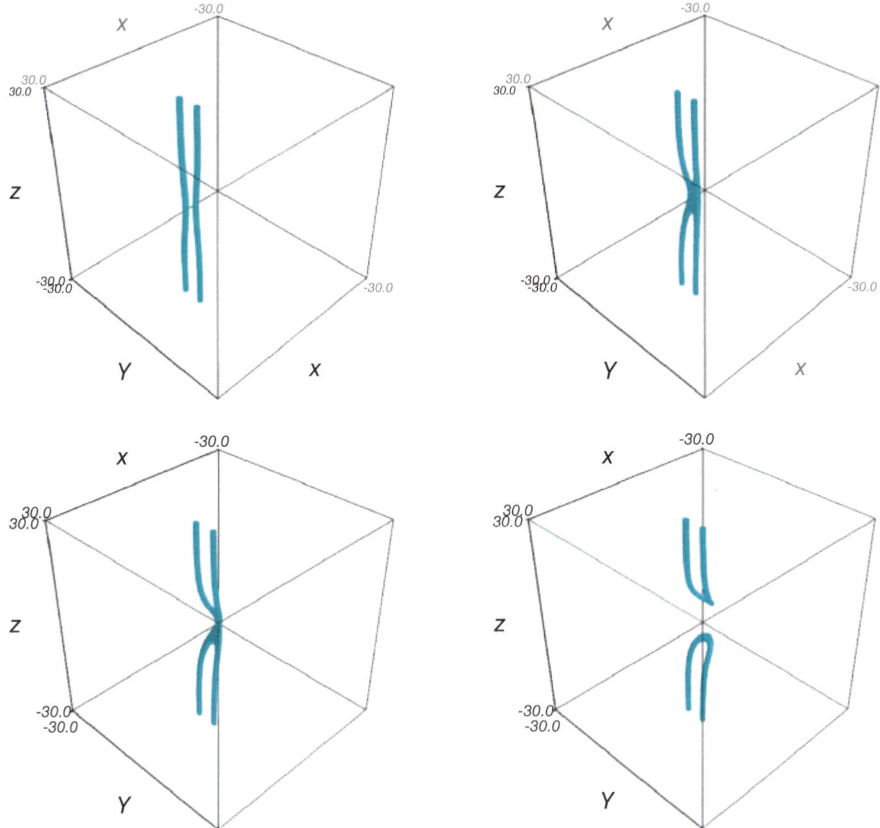

Figure 3.10 Reconnection of two antiparallel vortex lines obtained by solving the Gross–Pitaevskii equation numerically. Plotted are the density isosurface around the axes of the vortex lines. Reprinted from Zuccher *et al.* (2012) with the permission of AIP Publishing.

The experiments of Paoletti *et al.* (2010), later confirmed more convincingly by Fonda *et al.* (2019), found $\alpha^{\pm} = 0.5$ in agreement with dimensional analysis. Numerical investigations, however, gave some conflicting results (Zuccher *et al.*, 2012; Rorai *et al.*, 2016; Villois *et al.*, 2016). More recently, Galantucci *et al.* (2019) applied dimensional analysis that included the effects of the local vortex curvature and (for atomic condensates) density gradients and the presence of boundaries (represented by image vortices), and predicted two limiting scaling regimes: $\alpha^{\pm} = 1/2$ and $\alpha^{\pm} = 1$. Galantucci *et al.* (2019) identified both regimes in numerical simulations of reconnections in superfluid helium and trapped atomic Bose–Einstein condensates over a wider range of initial conditions than in previous literature, and showed that the crossover between the two regimes is determined by the balance between interaction-dominated motion and individually driven dynamics. Finally,

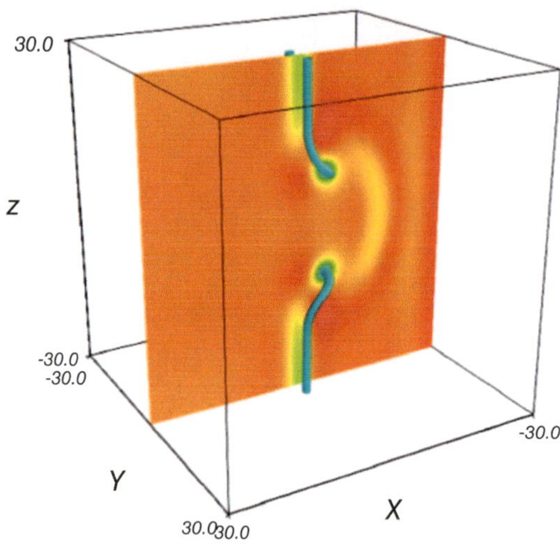

Figure 3.11 The density of the fluid on the plane between the two reconnecting vortex lines shown in Fig. 3.10. Note the rarefaction pulse that is ejected from the reconnection event. Reprinted from Zuccher *et al.* (2012) with the permission of AIP Publishing.

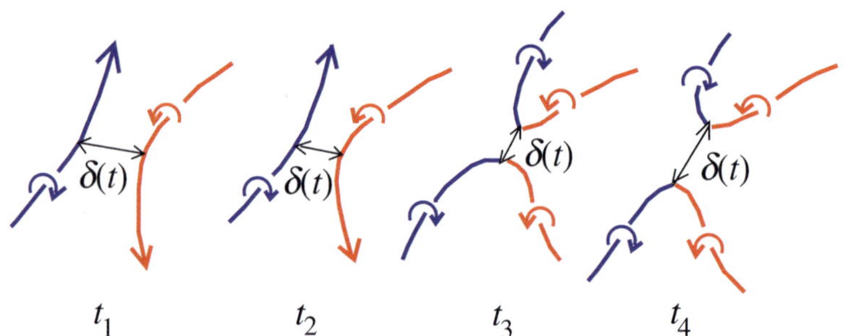

Figure 3.12 Schematic reconnection of two vortex strands to illustrate the minimum distance $\delta(t)$ between the two reconnecting vortices. The times t_1 and t_2 and the times t_3 and t_4 correspond to before and after reconnection, respectively.

they found that at short distances (of the order of magnitude of the superfluid healing length), $\delta(t) \sim |t - t_0|^{1/2}$ before and after the reconnection, as predicted by Nazarenko and West (2003) by Taylor-expanding the solution ψ of the Gross–Pitaevskii equation (5.7) around the reconnection point where $\psi = 0$. At such short distances, the density $|\psi|^2$ becomes small because the two vortex tubes start merging with each other, thus depleting the cubic nonlinearity, and we are indeed in the regime identified by Nazarenko and West (2003). The schematic scaling behavior of $\delta(t)$ found by Galantucci *et al.* (2019) is illustrated in Fig. 3.13.

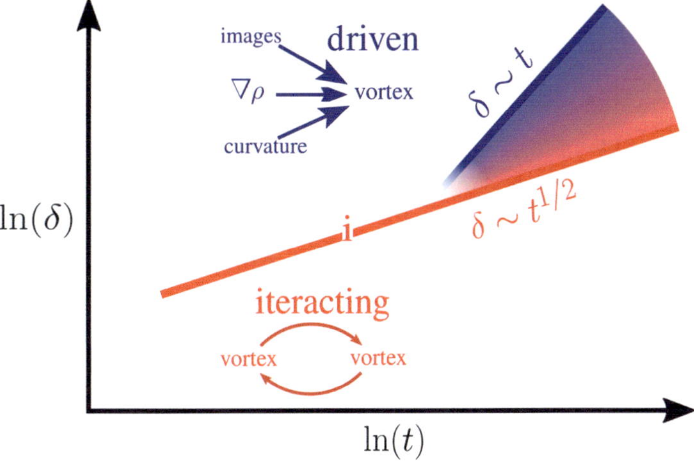

Figure 3.13 Routes to the reconnection of two vortex lines. According to Galantucci *et al.* (2019), two limiting scaling laws exist. In the first limiting behavior, $\delta(t) \sim |t - t_0|^{1/2}$, the minimum vortex separation δ between the vortex lines is determined by the mutual interaction of the two reconnecting vortex strands. In the second limiting behavior, $\delta(t) \sim |t - t_0|$, the separate motion of each vortex, hence δ is set by the background density gradient, or by the presence of images, or by the curvature of the vortex lines. In some vortex configurations, a crossover occurs between these two regimes. When δ becomes of the order of the vortex core radius or less, the scaling becomes $\delta(t) \sim |t - t_0|^{1/2}$, as predicted by Nazarenko and West (2003). From Galantucci *et al.* (2019).

The natural question is whether vortex reconnections in ordinary viscous fluid obey the same scaling laws of quantized vortices revealed by Galantucci *et al.* (2019). Recently, Yao and Hussain (2020) computed reconnections of slender vortex tubes using the Navier–Stokes equation, and confirmed both scalings, $\delta(t) \sim |t - t_0|^{1/2}$ and $\delta(t) \sim |t - t_0|$, and the expected crossover.

3.8 Donnelly–Glaberson Instability

We have seen that, in the presence of vortex lines, normal fluid and superfluid components are coupled by a mutual friction. It is instructive to understand how friction affects the motion of the simplest nontrivial vortex structure: the vortex ring. It is well known (Barenghi and Donnelly, 2009) that a thin-core inviscid vortex ring of circulation κ, radius R, and core radius $a_0 \ll R$ moves with constant speed u_{R} and energy E_{R} given by Eq. (3.11). Note that if we increase (decrease) the radius R of the ring, the ring travels slower (faster) and has more (less) energy. In the simple case of a ring traveling in the positive z-direction in the presence of uniform superfluid and normal fluid velocities u_{s} and u_{n} also aligned in the positive z-direction, the ring's radius obeys the equation (Barenghi *et al.*, 1983)

$$\frac{dR}{dt} = \frac{\gamma}{\rho_s \kappa}(u_n - u_s - u_R), \tag{3.20}$$

where γ is a friction coefficient. As expected, since $\gamma \to 0$ in the $T \to 0$ limit, the ring's radius and energy remain constant. If $T > 0$ but $u_n = u_s = 0$, then $dR/dt < 0$ and the ring shrinks, losing energy by friction to the background normal fluid at rest. If $u_n - u_s \neq 0$, then the ring will shrink (grow) gaining (losing) energy from (to) the normal fluid depending on whether $u_n - u_s - u_R > 0$ or $u_n - u_s - u_R < 0$. In particular, note that, if $u_s = 0$, the ring expands if there is normal fluid flowing in the same direction as the ring.

A similar effect, the *Donnelly–Glaberson instability*, was discovered experimentally by Cheng *et al.* (1973) and then explained by Glaberson *et al.* (1974) and Ostermeier and Glaberson (1975) in the more controlled configuration of a vortex lattice with a heat flow applied in the direction of rotation. Consider a bucket of He II rotating at angular velocity Ω about the z-direction; the resulting vortex lattice has a real density $n_v = 2\Omega/\kappa$. If a heat flow is applied in the z-direction, the vortex lines will become unstable to the exponential growth of Kelvin waves of initially infinitesimal amplitude. Figure 3.14 shows the instability simulated by Tsubota *et al.* (2003a) using the VFM. Detailed analysis by Tsubota *et al.* (2004) shows that the growth rate of the Kelvin waves is given by $\sigma = \alpha\left(ku_{ns} - \beta k^2\right)$ with $\beta = (\kappa/(4\pi))\ln(1/(ka_0))$. The largest growth rate thus occurs for wavenumbers $k = u_{ns}/(2\beta)$, at timescale $\tau = 1/\sigma = 4\beta/\left(\alpha u_{ns}^2\right)$, as verified by the numerics.

3.9 Vortex Lines in ^{3}He-B

An important difference between ^{4}He and ^{3}He arises from the fact that the elementary superfluid particle is not a ^{3}He atom, but a Cooper pair of two ^{3}He atoms moving about their center of mass. As a consequence, the elementary quantum of circulation is

$$\kappa = \frac{h}{2m_3} \cong 0.664 \times 10^{-7}\ \mathrm{m^2/s}, \tag{3.21}$$

where m_3 denotes the mass of the ^{3}He atom. This quantum of circulation is of the same order of magnitude as for He II but only slightly smaller.

The commonly created vortices in the ^{3}He-B phase are similar to He II vortex lines in that they are singular (that is, they possess cores that differ from the bulk B-phase) and singly quantized. The core size, as already mentioned in Chapter 2, is about two orders of magnitude larger than that in He II, of the order of the *coherence length* in ^{3}He-B, which, depending on the pressure, ranges from about 10 to 70 nm.

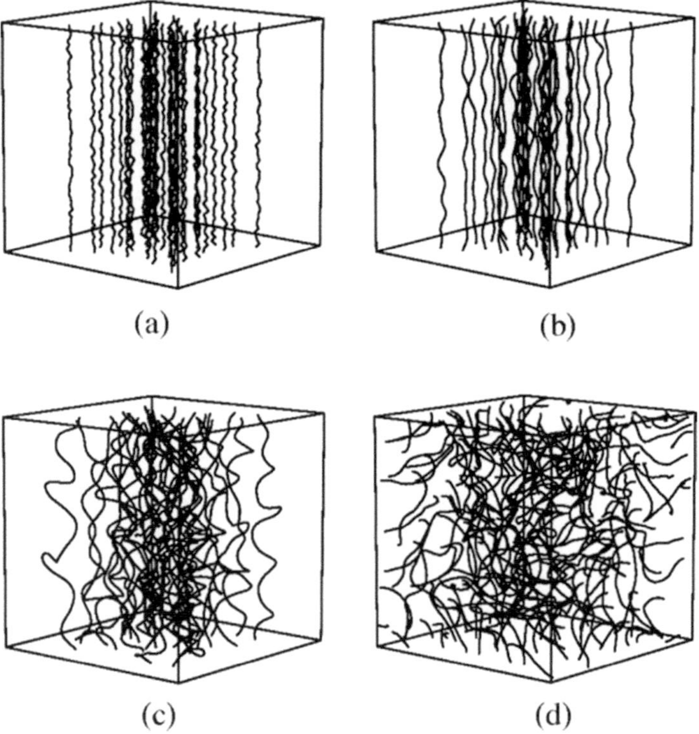

Figure 3.14 Donnelly–Glaberson instability as seen in the rotating frame. Panel (a) shows vortex lines in the frame rotating at angular velocity Ω; the lines have been seeded with small random wiggles to speed up the instability. Panel (b): a counterflow $u_n - u_s$ is applied in the (vertical) direction of rotation and Kelvin waves along the vortex lines grow until (c) their amplitude exceeds the intervortex distance and vortex reconnections scramble the lattice, resulting in a polarized rotating turbulence (d). Reprinted figure with permission from Tsubota *et al.* (2003a). Copyright 2003 by the American Physical Society.

Thus the vortex core is a macroscopic object where phase transitions can occur, as was indeed experimentally observed. Because of the large core, vortex pinning is less likely to occur, especially in experimental apparatus made up of smooth (e.g., quartz) material. This allows both extrinsic and intrinsic vortex nucleation to occur.

3.10 Quantum Turbulence (QT)

Here, we briefly introduce the primary topic of this book.

Disordered vortex configurations are created by stirring superfluid helium by a thermal or mechanical means. This is quantum turbulence: a time-dependent, apparently random *tangle* of vortex lines that interact with each other, and, at finite temperatures, with the normal fluid via the mutual friction force. The best

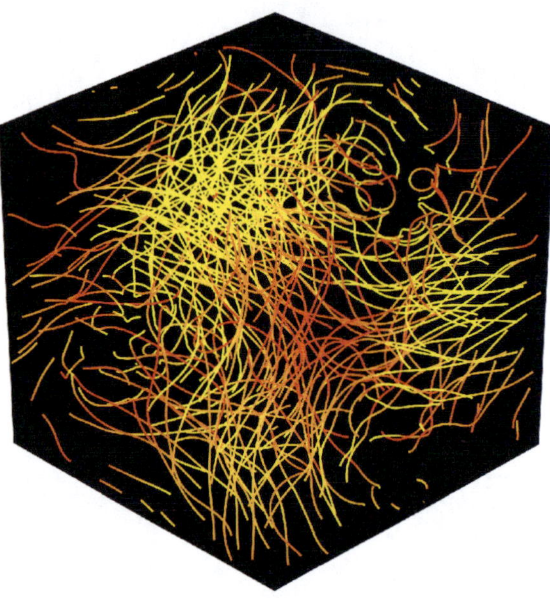

Figure 3.15 Snapshot of a vortex tangle computed numerically using the vortex filament method in a periodic domain. The tangle is driven by a normal fluid ABC flow and displays the Kolmogorov $k^{-5/3}$ scaling (see Chapter 2). The vortex lines are color-coded based on the helicity H (see Section 5.4) to show the amount of nonlocal interaction (yellow is strong, corresponding to regions of polarized vortex lines, and red is weak). (Credit A. W. Baggaley.)

visualization of a vortex tangle is still provided by numerical simulations. Figure 3.15 shows the instantaneous snapshot of a typical tangle of quantized vortices in He II computed using the VFM, which will be discussed in Chapter 5. In the figure, some vortex lines appear to terminate at a boundary. This is an effect of the periodic boundary conditions used in this particular simulation; as we already discussed, vortex lines are either closed loops or terminate at the walls or at the free surface.

The flow represented in Fig. 3.15 (and in Fig. 3.14d) is very complex. To interpret the figure, one must bear in mind that the figure only represents the location of the axes of the vortex lines; the superfluid flows around each vortex line with fixed speed $u_s = \kappa/(2\pi r)$, where r is the distance to the vortex axis. Plotting the vortex lines without any information about their orientation, as done here, means that it is impossible to infer the actual velocity field; one can only guess from the figure that it is very complex.

This complexity in itself does not guarantee any correspondence with classical turbulence, as can be seen from figures in the classical turbulence literature that aim to reveal coherent vortex structures by plotting isosurfaces of the vorticity magnitude, $|\boldsymbol{\omega}|$, at some arbitrary prescribed level, as done by Farge *et al.* (2003) in Fig. 3.16, by Farge *et al.* (2001), and by Ishihara *et al.* (2009) at a higher resolution.

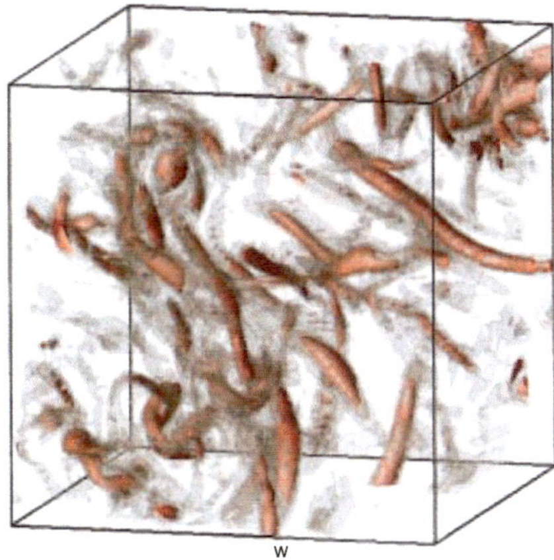

Figure 3.16 Snapshot of homogeneous isotropic turbulence computed by directly solving the Navier–Stokes equation in a periodic domain until a statistically steady state is reached. What is plotted are contours of the magnitude of the vorticity, $|\omega|$, at some arbitrary levels. Reprinted from Farge *et al.* (2003) with the permission of AIP Publishing.

It is worth pointing out that there are a few key differences between Fig. 3.15 and Fig. 3.16. First, the length and the shape of the vortex tubes shown in Fig. 3.16 are arbitrary because they depend on the chosen isosurface; the vortex lines shown in Fig. 3.15 represent the actual topological defects of the system (there are no more defects, nor fewer). Second, outside the vortex lines, the flow depicted in Fig. 3.15 is inviscid and irrotational (essentially the solution to the Laplace equation), whereas the flow around the vortex tubes of Fig. 3.16 is viscous and vortical. Despite these differences, there exist many similarities as explored in the rest of the book.

To further clarify the terminology: By *quantum turbulence* we mean the turbulence of any quantum fluid displaying quantized vorticity and superfluidity, hence two-fluid behavior at finite temperature. By *superfluid turbulence* we mean the turbulence of the superfluid component of a quantum fluid.

The simplest case – sometimes called the prototype turbulence, though this characterization is naive as we shall describe later – occurs in the zero-temperature limit, where there is no normal fluid and both quantum and superfluid turbulence mean an apparently random tangle of vortex lines. At finite temperature, the situation is more complex. The superfluid turbulence (i.e., the vortex tangle in the superfluid component) interacts via the mutual friction force with the normal component that itself may or may not be turbulent. Quantum turbulence thus can be seen to be simpler (under some conditions in the zero-temperature limit; see later) or more complex (at finite temperature with the coupled turbulent velocity fields) than its classical counterpart.

The simplest and most convenient way to characterize the intensity of the vortex tangle is to measure (numerically or experimentally) the total length Λ of superfluid vortex lines contained in the region of interest. In a cube of size D, the quantity

$$L = \frac{\Lambda}{D^3} \tag{3.22}$$

defines the *vortex line density*. From the vortex line density we infer that the typical distance between the vortex lines, called the *quantum length scale*, is

$$\ell \approx L^{-1/2}. \tag{3.23}$$

Obviously, this length scale characterizes superfluid turbulence more completely than it does quantum turbulence at finite temperatures (Skrbek *et al.*, 2021).

3.11 Ions and Neutral Excimer Molecules in Liquid Helium

Positive and negative helium ions exist in liquid helium (in both He II and superfluid ^{3}He-A and -B phases) and can be used for the generation and, especially, for probing quantum turbulence.

Positive ions are effectively solid snowballs made of about 20 helium atoms, while negative helium ions effectively represent an electron in an otherwise empty bubble of nanometre size, possessing a rather large hydrodynamic added mass of the order of 200 bare helium atoms. The motion of both ion species, which, for most purposes, can be thought of as charged spheres, can be controlled by an externally applied electric field and is affected by the presence of vortex lines. Due to the superfluid pressure gradient arising from the circulation, an attractive interaction exists between the ions and the vortex lines, allowing for the trapping of the ions on vortex cores.

Helium excimer molecules He_2^* (a bound state of two ^{4}He atoms with one atom in an excited state) can be produced in liquid ^{4}He following the ionization or excitation of ground state helium atoms. The singlet state molecules radiatively decay in a few nanoseconds, but the triplet state molecules are metastable with a radiative lifetime of about 13 s. Similar to negative ions, these triplet molecules form bubbles in liquid helium with a radius of about 6 Å, and can be used as tracers for probing quantum turbulence.

3.12 Atomic Bose–Einstein Condensates

Vortex lines also exist in trapped atomic BECs (Barenghi and Parker, 2016). BECs are small ($\approx$100 μm) clouds of ultracold, dilute gases that, unlike helium, almost achieve the ideal conditions under which Bose–Einstein condensation takes place, as first envisaged by S. N. Bose and A. Einstein. The first successful experiments used rubidium $\left({}^{87}\mathrm{Rb}\right)$ and sodium $\left({}^{23}\mathrm{Na}\right)$ atoms cooled below the critical temperature ($T_\mathrm{c} \approx 100\,\mathrm{nK}$) by combining laser and magnetic evaporative cooling. Similar to helium, BECs are effectively pure superfluids at sufficiently low temperatures below the critical temperature. Unlike helium, however, the BEC clouds can be shaped by external confining potentials, with typical geometries ranging from balls to long cigars to flat pills. This feature allows the creation of effectively 1D, 2D, and 3D systems (vortex lines and vortex points exist only in 3D and 2D systems, while in 1D topological defects take the form of bright and dark solitons). Vortex points in 2D BECs are particularly interesting, as the best physical realizations of 2D inviscid flows are soap films, which, unlike a BEC, are affected by surface tension, viscosity, and friction with respect to the surrounding air. In helium, the quantum of circulation κ and the vortex core radius a_0 are fixed by nature; in BECs, all physical properties can be engineered, including the strength and the sign of interactions between the bosons (e.g., bright solitons). Finally, it is relatively easy to directly image vortex lines in 2D BECs. For the above reasons, BECs have proved to be ideal objects to study 2D vortex dynamics in great detail. For example, the existence of a quantum *von Kármán vortex street* past a moving cylindrical obstacle has been predicted by Sasaki *et al.* (2010) and experimentally observed by Kwan *et al.* (2016); another example is the emergence of large-scale vortex structures via an inverse energy cascade in 2D turbulence (Groszek *et al.*, 2016), as predicted by Onsager (1949).

From the point of view of this book, which is devoted to 3D quantum turbulence, the study of turbulence in BECs is relatively new in comparison with that of superfluid helium. The difficulties are two-fold: the limited range of length scales experimentally available, and the complexity of flow visualization.

To understand these difficulties, we remark that the vortex core radius in He II is $a_0 \approx 10^{-10}\,\mathrm{m}$ (and is about 100 times larger in ^{3}He-B). The average distance between vortex lines in turbulence experiments varies greatly depending on parameters, but typical values are $\ell \approx 10^{-4}$–$10^{-6}\,\mathrm{m}$. The size of the helium sample varies from thin capillary tubes $\left(\text{radius } D \approx 10^{-3}\,\mathrm{m}\right)$ to counterflow channels $\left(D \approx 10^{-2}\,\mathrm{cm}\right)$ to the SHREK experiment (Saint-Michel *et al.*, 2014) ($D \approx 0.8\,\mathrm{m}$), giving typical length scale ratios as large as $D/a_0 \approx 10^9$ and $\ell/a_0 \approx 10^6$.

In contrast, the vortex core radius in a typical BEC is often only one order of magnitude smaller than the size of the entire BEC. Figure 3.17 shows vortex lines in a spherically trapped atomic condensate computed by solving the Gross–Pitaevskii equation (an excellent quantitative model of BECs at temperatures $T/T_\mathrm{c} \ll 1$). The figure corresponds to experimentally realistic parameters. What is plotted is a surface of constant density, which captures the spherical edge of the condensate and the hollow vortex cores. The smallness of the vortex configuration is quite apparent.

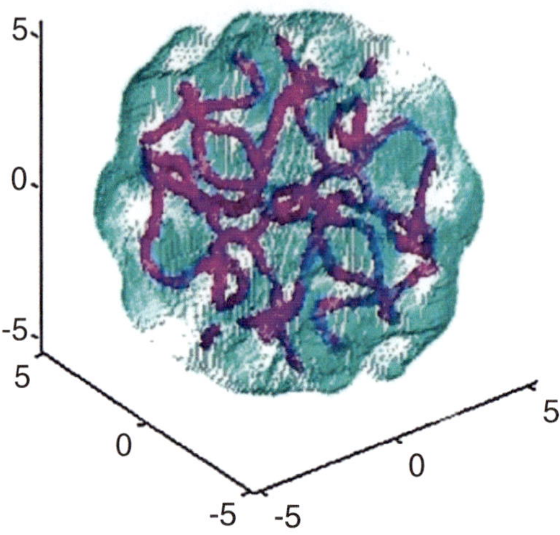

Figure 3.17 Computed turbulent tangle of vortex lines in an atomic BEC. Plotted is a surface of constant density; the light blue curve shows the outer edge of the condensate, and the purple tubes are the cores of the quantized vortices. Reprinted figure with permission from White *et al.* (2010). Copyright 2010 by the American Physical Society.

We shall see in subsequent chapters that, when describing quantum turbulence and attempting to make comparisons with classical turbulence, the issue of available k-space is crucial for verifying the existence of concepts such as scaling and energy cascade. It would be desirable but difficult to have larger BECs to make direct connections with turbulence in helium and in ordinary fluids. Despite the difficulty, 3D turbulence has been generated in atomic BECs (Henn *et al.*, 2009), and there is now experimental evidence of an energy cascade (Navon *et al.*, 2016; Garcia-Orozco *et al.*, 2020) despite the limited k-space. It is likely that BECs will be ideal systems to bridge the gap between chaos and turbulence.

The second difficulty that has limited the progress in BEC as a quantum turbulence system is the absence of an established technique for measuring the vortex line density, although good visualization is available for 2D BECs. This feature limits direct comparison with quantum turbulence in helium.

For these two reasons, the experimental investigation of 3D quantum turbulence in BEC, reviewed by Tsatsos *et al.* (2016), is still in its infancy, but we expect rapid progress to occur in the near future. The reasons for our optimism are two recent noteworthy experimental developments: a stroboscopic imaging method that has been developed and applied to the interaction and reconnection of individual vortex lines by Serafini *et al.* (2017), and a technique to confine cold gases in *box traps* developed by Gaunt *et al.* (2013) such that the resulting condensate density is uniform, making better contact with helium research.

3.13 Superconductivity

Although this book is mainly concerned with quantum turbulence in helium, which includes superfluid properties, it is useful to make a brief reference to superconductors. Indeed, the difference between the ideal fluid and the superfluid can be better appreciated by noticing the link between superfluidity and superconductivity, and the more commonly known relation between an ideal conductor and a superconductor. A superconductor (of the first kind or the second kind below the lower critical magnetic field) behaves as an ideal diamagnetic substance that expels, below certain critical conditions, an externally applied magnetic field from its interior, a property known as the *Meissner effect*. The Meissner effect is caused by a supercurrent circulating along the perimeter of the specimen; this supercurrent induces inside the specimen a magnetic field with the same magnitude but the opposite orientation to the externally applied magnetic field, thus cancelling it out. In superfluidity, the corresponding feature is that the superflow is always curl free, or potential, independently of whether a rotating or a quiescent sample is cooled through the superfluid transition. Superconductors placed in a subcritical magnetic field are thus analogous to vortex-free superfluids. The analogy between superfluidity and electromagnetism is indeed very deep: an irrotational and incompressible superfluid velocity field satisfies

$$\nabla \times \boldsymbol{u}_{\mathrm{s}} = 0, \nabla \cdot \boldsymbol{u}_{\mathrm{s}} = 0, \tag{3.24}$$

from which one sees a striking similarity with Maxwell's equations. In vacuum, the magnetic induction obeys the same equations,

$$\nabla \times \boldsymbol{B} = 0, \nabla \cdot \boldsymbol{B} = 0. \tag{3.25}$$

This analogy will be used later while calculating the superfluid velocity field from the Biot–Savart law in the presence of vortex lines.

It is known that vortex lines exist in superconductors of the second kind and may reconnect. The physical quantity that is quantized here is the magnetic flux in units of $2\pi\hbar/(2e)$, where $2e$ is the charge of two electrons constituting a Cooper pair. Due to screening effects, the circulating supercurrent around the vortex core in superconductors decreases with the distance r much faster than the $1/r$ characteristic of superfluids, and the motions of a vortex line and a flux tubes are not the same. If displaced, the vortex line moves (almost) along the binormal and the flux tube (almost) along the normal direction (Chapman, 1995), which most likely explains the absence of quantum turbulence in superconductors.

3.14 Summary

The quantum mechanical constraint that the circulation is quantized in the superfluid components of quantum fluids leads to the existence of quantized vortices. Singly quantized vortices, commonly called vortex lines, are the stable structures in He II and in superfluid ^{3}He-B. Upon exceeding some critical flow velocity, these vortex lines can be created either intrinsically or extrinsically by reproduction on the basis of existing remanent vortices. Under special circumstances vortex lines can form a lattice, but in a general case they create an apparently disordered tangle, primarily through reconnections, whose dynamics constitutes an essential ingredient of quantum turbulence.

The intellectually simplest case (though experimentally the most difficult one to attain) is the zero-temperature limit, which, in practice, is below about 0.3 K for He II and below about $0.2\,T_c$ for ^{3}He-B, where the critical temperature T_c varies around 1–2 mK depending on the pressure. In this low temperature limit both He II and ^{3}He-B can be treated as pure superfluids, and quantum turbulence occurs in its simplest form, sometimes called the prototype of turbulence, representing just a tangle of vortex lines, this being the *pure superfluid turbulence*. It must be noted, however, that the mechanisms to dissipate kinetic energy in decaying He II and ^{3}He-B turbulence are different, due to the different (bosonic versus fermionic) natures of superfluidity in these two quantum fluids, and due to the different (subatomic versus mesoscopic) size of the vortex cores.

At nonzero temperatures, vortex lines couple the otherwise independent normal and superfluid velocity fields via the action of the mutual friction force. In He II, both the normal fluid and superfluid can easily become highly turbulent, creating an unusual double-turbulent fluid system that has no direct analogy in classical fluid dynamics. In ^{3}He-B, given typical geometrical and experimental restrictions, the normal fluid hardly moves because of its relatively large viscosity and offers substantial resistance to the motion of vortex lines via the large mutual friction force.

In this book we concentrate our attention on superfluid He II and superfluid ^{3}He-B, the two quantum fluids where most turbulence experiments and simulations have been performed and that allow us to draw comparison between quantum turbulence and classical turbulence. However, we shall make reference when appropriate to turbulence in atomic Bose–Einstein condensates – systems that also display superfluidity and quantized vorticity.

4

Experimental Methods

4.1 General Remarks

As discussed in the previous chapter, the physical properties of quantum fluids where quantum turbulence has been experimentally studied – superfluid ^{4}He, superfluid ^{3}He-B, and BECs – are very different (though the behavior of quantum turbulence is similar). This implies that it is possible to take advantage of special properties of these quantum fluids to develop a plethora of experimental techniques for generating and probing turbulent flows.[1] While some of these techniques are the same or similar to those commonly used in classical experimental turbulence research, others are unique to a particular quantum fluid and to the particular experiment at hand. In this chapter, we shall describe various ways of generating and detecting flows of He II and ^{3}He-B, which, as has been stated in earlier chapters, are the working fluids that have so far provided most of the quantitative and semiquantitative experimental data.

It is worth mentioning that liquid helium can be prepared with high levels of purity, much higher than in other areas of condensed-matter physics, because all impurity atoms (except the other helium isotope) freeze out at the low temperatures of interest. Isotopically pure ^{4}He (with ^{3}He content below 10^{-13}) can be prepared using a thermal counterflow technique (Hendry and McClintock, 1987). In ^{3}He at very low temperatures of the order of mK, an efficient self-cleaning process removes all remaining impurities (which adhere to the walls) leaving, with very high probability, the entire bulk sample completely free of any impurity including ^{4}He atoms; ^{3}He thus represents the cleanest substance to which we have access. The physical properties of ^{4}He and ^{3}He are well known and tabulated (Arp and McCarty, 1998; Donnelly and Barenghi, 1998; McCarty, 1972). In experiments, these properties can be varied over several decades in situ simply by controlling the

[1] We do not describe the experimental techniques for BECs, based on optics and lasers. These are, of course, significantly different from the general cryogenics techniques shared by researchers working in He II and ^{3}He turbulence.

temperature and pressure. From a practical point of view, an important restriction for studies on quantum turbulence is the energy dissipation rate, which is experimentally limited to about 1 W in He II experiments and to about 1 nW in ^{3}He-B experiments.

All experimental studies must be performed at cryogenic conditions in specifically designed and highly sophisticated cryostats containing special cryogenic inserts. The temperature ranges from above 2 K down to absolute zero (in practice, to anything below about 100 mK in He II and about 200 μK in ^{3}He-B). It is relatively easy to generate a turbulent flow of He II thermally by applying heat flux, or, in both He II and ^{3}He-B, by some mechanical means similar to those used to generate classical turbulence at ambient temperatures.

We have to bear in mind that even normal liquid He I above T_λ has a kinematic viscosity (Arp and McCarty, 1998; Donnelly and Barenghi, 1998) that is two orders of magnitude smaller than that of water (in fact, He I is the classical fluid with the smallest kinematic viscosity); thus, obtaining a sizeable sample of initially quiescent He I is not straightforward and requires special care. Below the λ-point the situation is more complex. Above approximately 1 K, we have the normal fluid component, which has kinematic viscosity similar to He I, and, in addition, the inviscid superfluid component. It is practically impossible to prepare a macroscopic sample of He II free of vortex lines. A sample of He II, even if left undisturbed to relax for a very long time, will always contain remanent vortex lines. These vortex lines may be left over from an earlier experiment, or may have been formed by the Kibble–Zurek mechanism (Zurek, 1985) when the helium was cooled through the λ-transition. Remanent vortex lines become pinned to the nooks and crannies of the container walls, which must always be considered rough when compared to the microscopic size of the vortex core $\left(10^{-10}\,\mathrm{m}\right)$. The density of remanent vortex lines depends on the geometry and can be estimated within reason, thanks to the experimental work of Awschalom and Schwarz (1984). The situation is different in ^{3}He-B due to the much larger vortex core size; and, as we already stated, with special care, it is possible to prepare a macroscopic sample of ^{3}He-B containing no vortex lines.

Following a brief history of experimental research in quantum turbulence, we describe various methods of generating and detecting turbulence in He II and ^{3}He-B flows.

4.2 Generation of Steady and Decaying Coflows

We begin with mechanical methods of generation, resulting in the so-called *coflows*: Normal and superfluid components are forced together simultaneously and are driven in the same overall direction, so that the normal fluid and superfluid velocity fields are almost identical at large scales (though not close to the walls where different boundary conditions apply for the normal and superfluid components). For smaller length scales, of the order of magnitude of the average distance between vortex lines (quantum length scale, see Section 3.10) and less, the two velocity fields are always different due to fundamental physical reasons. Again, the situation in ^{3}He-B is different and will be considered separately.

4.2.1 Steady Rotation, Spin-Up/Down

Rotating cryostats have always played an important role in experimental investigations of ^{4}He and ^{3}He. The seminal example is solid body rotation in a rotating container described in Chapter 2. Just as a lattice of rectilinear vortex lines in the rotating container mimics solid body rotation, a partial polarization of vortex lines (a bundle of lines oriented parallel to each other, coexisting with other random lines) may, under certain conditions, mimic turbulent eddies of classical turbulence. As in classical viscous fluids, a sudden spin-up or spin-down of a container filled with He II or ^{3}He-B results in a turbulent flow.

From the point of view of quantum turbulence, spin-down experiments are particularly interesting. The procedure starts with a steadily rotating bucket filled with helium, which is either suddenly stopped or brought to lower angular velocity, as illustrated in the top panel of Fig. 4.1. We shall later discuss in more detail the underlying physics of generation and decay of quantum turbulence in the He II spin-down experiments led by Walmsley, Golov, and collaborators (Walmsley *et al.*, 2007, 2008, 2014; Walmsley and Golov, 2008; Zmeev *et al.*, 2015; Walmsley and Golov, 2017), as well as their techniques based on negative ions, as shown in the bottom panel of Fig. 4.1. As for quantum turbulence in ^{3}He-B, a crucial role has been played over decades by the ROTA rotating cryostat group in Helsinki, led by Lounasmaa, Krusius, and, more recently, Eltsov (Finne *et al.*, 2003; Eltsov *et al.*, 2009, 2014).

4.2.2 Von Kármán Flows

Von Kármán flows can be easily generated in He II above 1 K, driven by counter-rotating discs and/or propellers of various shapes using cryogenic (Maurer and Tabeling, 1998) or room temperature (Schmoranzer *et al.*, 2009) motors.

An important and informative experiment was performed by Maurer and Tabeling (1998), who measured the frequency spectrum of local pressure fluctuations in liquid helium using a total-head pressure tube (see Section 4.7) in both classical and quantum turbulence generated between counter-rotating discs with blades, using He I and He II. The apparatus is shown schematically in the left panel of Fig. 4.2.

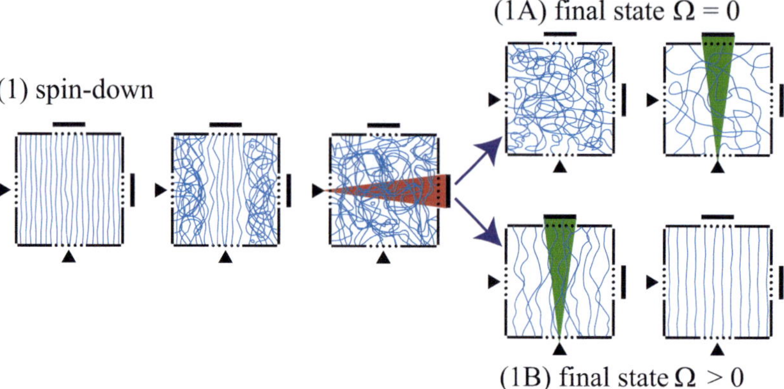

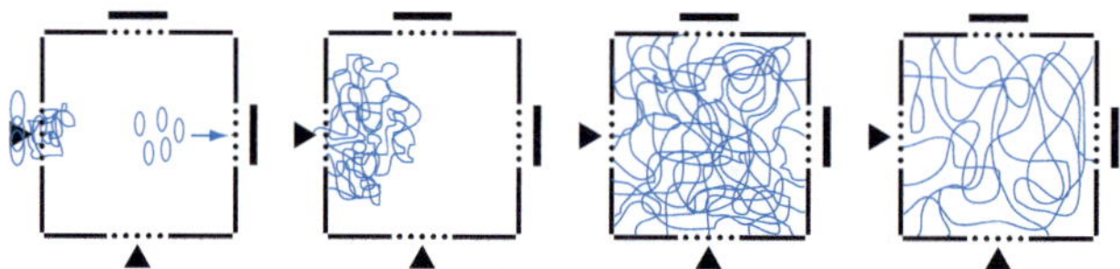

Figure 4.1 Cartoon of two different techniques for creating quantum turbulence: (1) spin-down to (1A) rest and (1B) to finite angular velocity; and (2) colliding charged vortex rings. For each type, the time evolution proceeds from left to right. In the early stages, the turbulence can be strongly inhomogeneous, but eventually fills the entire container homogeneously. The red and green shaded areas indicate how L can be probed in the horizontal and vertical directions separately, using a short pulse of injected charged vortex rings. Reproduced with permission from Walmsley *et al.* (2014).

A large turbulent facility of this kind named SHREK (see Fig. 4.3) has been completed at Grenoble (Rousset *et al.*, 2014). Turbulent von Kármán flows are generated by counter-rotating plates of various shapes; SHREK has the possibility of controlling the rotation of each plate independently. Turbulence can be created in different working fluids, including gaseous, normal, and superfluid liquid ^{4}He. This large and technically impressive apparatus, connected to a powerful 400 Watt/1.8K refrigerator, achieves very high Reynolds numbers in the 10^7 range in a cryogenic environment. It is equipped with various up-to-date detection sensors that allow fundamental and large-scale experiments in classical and quantum von Kármán flows.

Figure 4.2 also shows schematically the Grenoble cryogenic helium wind tunnels (Roche *et al.*, 2007; Salort *et al.*, 2012a). Here, flows are driven by the centrifugal force generated by propellers, without relying on viscous or thermal effects. This type of forcing is well suited for liquid helium, irrespective of its superfluid density

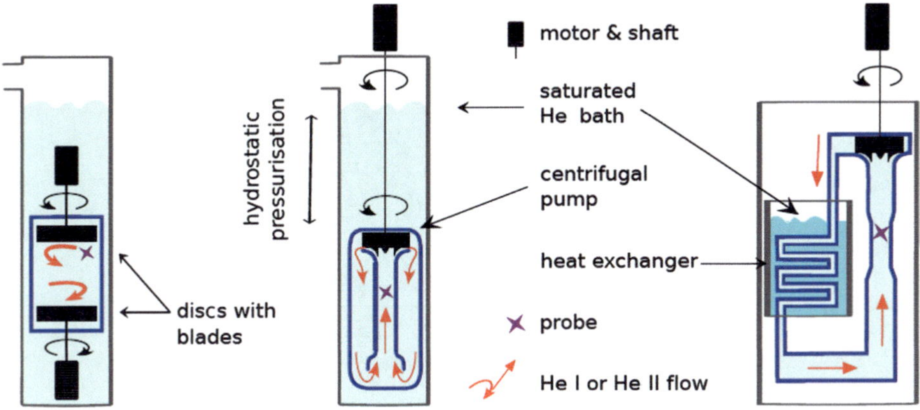

Figure 4.2 (Left to Right) von Kármán flows (Maurer and Tabeling, 1998), wind-tunnels (Roche *et al.*, 2007; Salort *et al.*, 2012a), and circulator cooled through a heat exchanger (Salort *et al.*, 2010). From Barenghi *et al.* (2014a).

fraction (the density of liquid He I and the total density of He II depend only weakly on the temperature). To ensure cavitation-free operation in He I, these flows can be pressurized hydrostatically by a column of liquid ^{4}He above; in He II, cavitation is strongly suppressed by the high thermal conductivity. Without pressurization, bubbles will form in He I preventing the comparison of turbulence above and below T_λ in the same apparatus. The configuration shown in the right panel of Fig. 4.2 consists of a pressurized helium loop cooled by a heat exchanger (Salort *et al.*, 2010).

4.2.3 Superfluid Jets

As in classical fluid dynamics, jets represent an experimental flow configuration suitable for studying turbulence. For example, Duri *et al.* (2011) in Grenoble used a centrifugally driven cryogenic helium wind tunnel to generate a liquid helium jet. A stable round jet with a Reynolds number of 4×10^6 could be sustained in both He I and He II down to a minimum temperature of 1.7 K.

Additionally, He II jets can be generated by the fountain pump. This simple device operates on the basis of the fountain effect (see Eq. (3.16)), discovered by Allen and Jones (1938), with no analogy in classical hydrodynamics. The fountain pump is basically a closed volume containing a resistive heater, one side of which is connected to the helium bath via a superleak, usually in the form of a porous pill of either a thin powder such as a jeweler's rouge or a sintered silver. The narrow pores of the superleak inhibit the flow of the normal component of He II, while

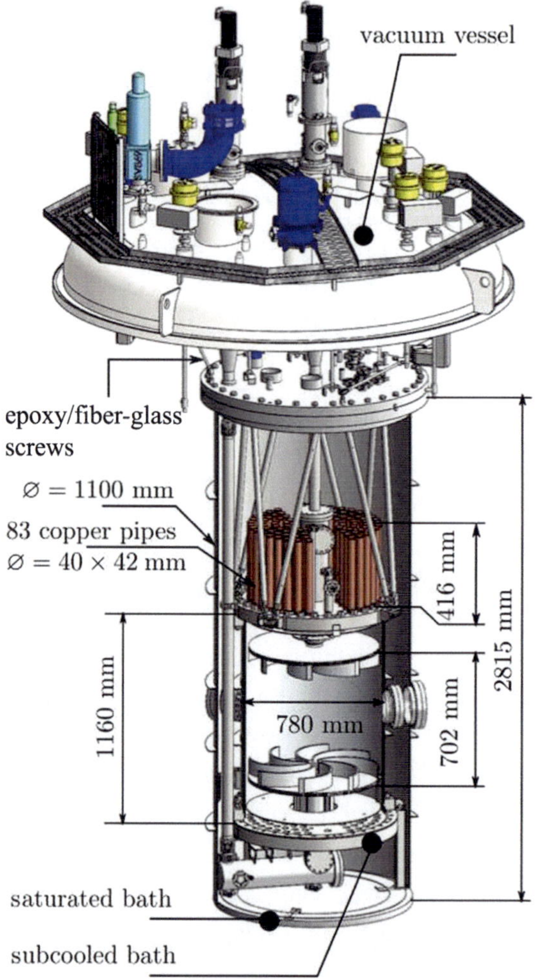

Figure 4.3 Sketch of the SHREK experiment with the von Kármán cell (780 mm in diameter), its 83 copper pipes providing heat exchange between the von Kármán cell and the saturated bath, which is enclosed in a large tank (1.1 m diameter, 2.8 m height). The whole assembly is hung on a flange (top of the figure), which is the upper flange of the "multipurpose cryostat." Reprinted from Rousset *et al.* (2014) with the permission of AIP Publishing.

the superfluid component passes through from the bath freely, responding to the applied heat inside the pump, thus increasing the pressure inside the pump. This results in generation of either a free or a submerged jet from a capillary mounted on the opposite side of the fountain pump.

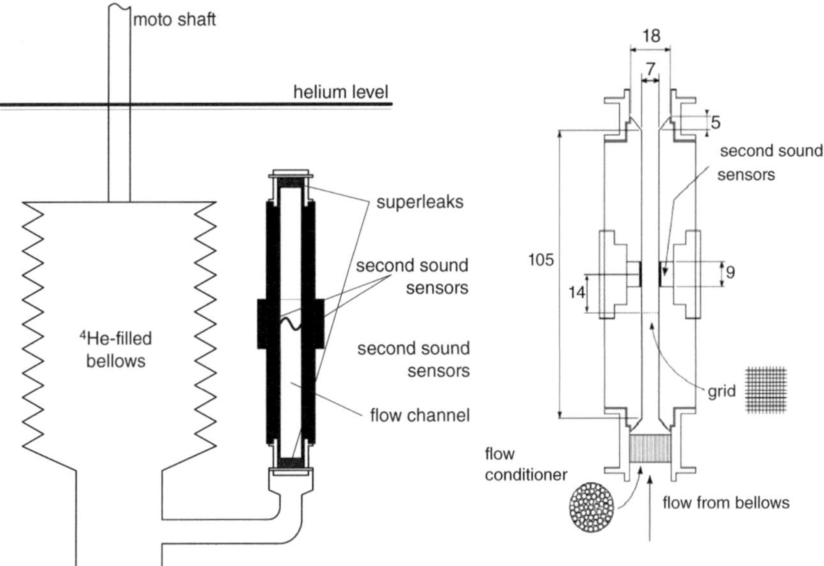

Figure 4.4 Schematic view of bellows-induced He II flow apparatus. (Left) Generation of pure superflow through a channel of square cross section with the use of a pair of sintered silver superleaks. (Right) Generation of coflow with an additional option of creating cryogenic wind tunnel grid turbulence. (Left panel) Reprinted figure with permission from Babuin *et al.* (2012). Copyright 2012 by the American Physical Society. (Right panel) Reprinted from Varga *et al.* (2015) with the permission of AIP Publishing.

4.2.4 Compressible Cryogenic Bellows

Another way of generating steady or decaying coflows in He II (as well as in complementary classical flows of He I/cryogenic He gas) is to squeeze or expand (via an external computer-controlled linear motor) a compressible cryogenic bellows (Babuin *et al.*, 2012) or a pair of them (Xu and Van Sciver, 2007). A schematic view of such a compact arrangement, widely used by the Prague group, is shown in Fig. 4.4. Bellows are generic, convenient drivers of He II flows and can be further modified by adding superleaks (Babuin *et al.*, 2012), flow conditioners, or grids (Babuin *et al.*, 2014a,a) in order to obtain steady or decaying pipe flows, pure superflows, or grid turbulence (see Section 4.4). The disadvantage is that the steady flow can be sustained for only limited durations, typically 10–100 seconds, depending on the volume of the bellows and the size of the channel. Moving bellows can be stopped abruptly, providing the possibility of studying the temporal decay of quantum turbulence.

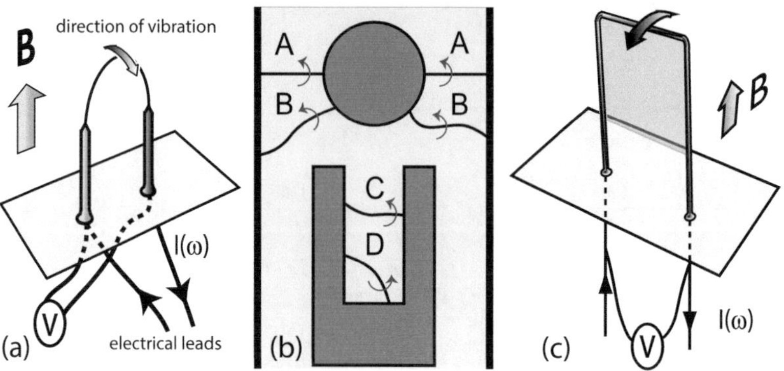

Figure 4.5 Schematic view of (a) an oscillating wire and (c) a fine grid; **B** represents the magnetic field applied in the plane of the loop. Panel (b) illustrates the likely position of remanent vortices A, B, C, and D attached to the oscillating objects. From Vinen and Skrbek (2014).

4.3 Flows around Oscillatory Objects

Various types of oscillatory flows have been successfully used for probing the hydrodynamic properties of helium flows since the discovery of superfluidity of He II. The motivation behind oscillating flows is obvious: In the confined geometry of a cryostat it is technically much simpler to generate flows due to oscillating objects than by using the more complex external drives needed for classical pipe and channel flows.

The seminal experiment of Andronikashvili (1946, 1948) – the measurement of torsional oscillations of a pile of closely spaced discs (see Fig. 2.6) – was the first direct determination of the densities of normal and superfluid fractions in He II, while measurements with torsionally oscillating discs (Hollis-Hallet, 1952) and spheres, capillary flows, or U-tube oscillations (Donnelly and Penrose, 1956) have been employed to determine *critical velocities* above which the ideal two-fluid picture does not hold. Indeed, Onsager's idea of quantized circulation (Onsager, 1949) was experimentally confirmed by Vinen (1961) using an oscillating cylinder in the form of a straight vibrating wire.

Oscillating cylinders used in the most recent experiments have actually been vibrating wires, usually bent in the shape of a loop, as shown in the left panel of Fig. 4.5; this shape avoids closely spaced resonant modes that are associated with a straight wire. If a magnetic field is applied in the plane of the loop, resonant vibrations can be excited by passing an alternating current of appropriate frequency through the wire; the amplitude of these vibrations can then be measured by observing the induced voltage V across the ends of the wire. Typically, wires of diameters

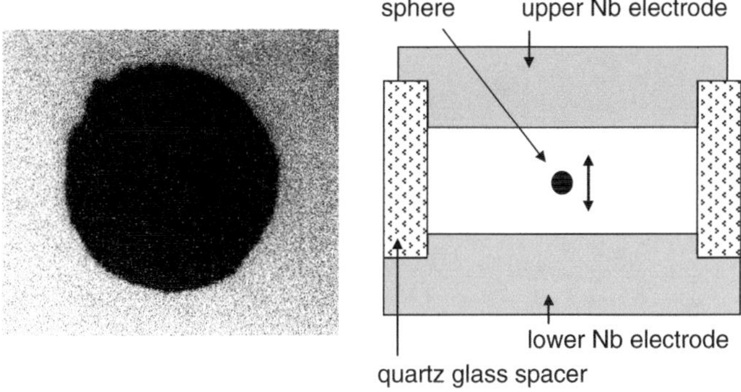

Figure 4.6 Micrograph of one of the magnetic spheres and schematic diagram of the cell used by Schoepe's group for experiments on flow due to an oscillating sphere in superfluid ^{4}He. Reprinted from Skrbek and Vinen (2009) with permission from Elsevier.

between a few to a few hundred micrometers are made of superconducting alloys (e.g., NbTi) to avoid ohmic dissipation as the current flows through them. Vibrating wires are very useful and universal cryogenic tools. They have been used over decades by many groups to investigate quantum turbulence and, in particular, the details of the transition to quantum turbulence (the underlying physics of which will be discussed in Chapter 6) in both He II and in superfluid ^{3}He-B. In the latter case, the vibrating wires suffer large damping because of the large viscosity of the normal fluid component, particularly at temperatures close to T_c. This damping disappears at lower temperatures, so that studies similar to those in He II become possible.

For the sake of comparison with classical fluid dynamics, it is desirable to consider simple oscillating objects around which the flow of classical viscous fluids is known. Torsionally or laterally oscillating discs and spheres are therefore natural choices. Experiments on the torsional oscillations of large discs and spheres in He II were reported earlier by Benson and Hollis-Hallet (1956) and very recently revisited by Schmoranzer *et al.* (2019). Pioneering systematic studies by Schoepe's group (Jager *et al.*, 1995) considered a transversally oscillating sphere. The sphere and the experimental apparatus used by them is shown schematically in Fig. 4.6. The sphere has a radius of about 100 μm, made from a ferromagnetic material ($SmCo_5$) with a very rough surface. It is placed between two horizontal niobium electrodes about 1 mm apart, so that it becomes magnetically levitated when the niobium is superconducting; the natural frequency of the vertical oscillation of the sphere is in the range 100–500 Hz. The experiment is arranged so that the sphere carries an electric charge of about 1.5 pC. Vertical resonant oscillations of the sphere can then be driven by applying an alternating voltage of the appropriate frequency between the niobium plates, and by detecting these oscillations through the induced charge

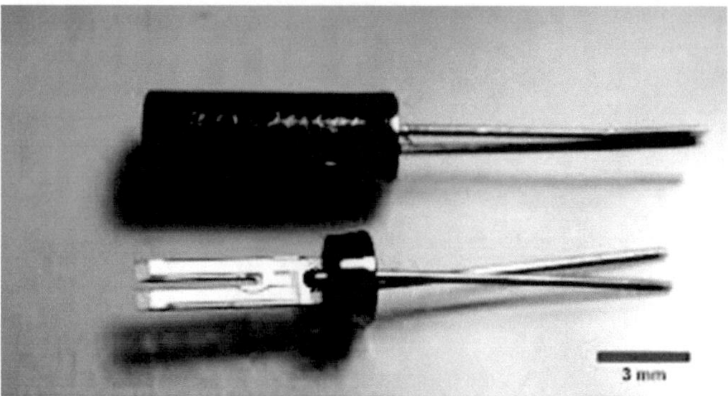

Figure 4.7 Photograph of a typical commercially produced quartz tuning fork in encapsulated and naked states.

on one plate. Experiments by Jager *et al.* (1995) were carried out in ^{4}He in the temperature range from 2.2 K down to 0.35 K.

A relatively new addition to the family of oscillating objects is the quartz tuning fork (Blaauwgeers *et al.*, 2007; Blažková *et al.*, 2008b). A quartz tuning fork is a piezoelectric resonator, which is mass produced as the frequency standard for digital watches. A typical commercially produced and encapsulated naked fork, ready to use, is shown in Fig. 4.7. It is cheap, sensitive, robust, and easy to install. Only two wires are needed to drive a fork as generator as well as sensor; no magnetic field is needed. In fact, the fork is quite insensitive to magnetic fields and thus useful as a field-independent sensor. A relatively simple electronic scheme composed of a digital oscillator, IV converter (Holt and Skyba, 2012), and a lock-in amplifier is typically used for detecting its resonant response over seven orders of magnitude of the driving force, in both viscous and quantum fluids. At large drive the prongs of the tuning fork vibrate at velocities up to 10 m/s. This set-up can be used for different types of studies, including cavitation (Blažková *et al.*, 2008a) and sound emission (Schmoranzer *et al.*, 2011; Rysti *et al.*, 2014) (these two effects must be avoided by suitable design or carefully taken into account (Bradley *et al.*, 2012) when using vibrating objects to study quantum turbulence). Above all, this set-up allows the study of the transition from laminar to turbulent state in He I and in cryogenic helium gas (Blažková *et al.*, 2007) as well as in superfluid He II and ^{3}He-B (Blažková *et al.*, 2008b), down to temperatures of the order of μK.

Modern technologies such as electron lithography allow the manufacturing of very small, micro- and nanoscale oscillating objects (Collin *et al.*, 2010; Kamppinen and Eltsov, 2018; Barquist *et al.*, 2020; Guthrie *et al.*, 2021). It is independently of fundamental physical interest to cool these objects to their ground state, in both vacuum and helium. Of interest in the present context is that these tiny devices in the form of wires or cantilevers, developed by Salort *et al.* (2012b, 2018); Barquist *et al.* (2020), can serve as very sensitive tools to probe various quantum flows. Additionally, recently developed nanofluidic Helmholtz resonators with viscously clamped normal fluid, electromechanically coupled to pure superflow inside (Varga and Davis, 2021), enable the study of crossover from 3D to 2D and of purely 2D superfluid turbulence in He II (Varga *et al.*, 2020).

4.4 Grid-Generated Quantum Turbulence

A basic and frequently explored problem in classical turbulence is homogeneous isotropic turbulence (HIT)

(i) that is held in a statistically steady state by balancing the forcing effects and dissipation in which the flow properties fluctuate about well-defined mean values; and
(ii) displays temporal decay. Both these states are presumed to be independent of initial conditions and detailed forcing (though available evidence points to a significant dependence).

Most of the relevant experimental work has occurred in wind tunnels, where the turbulence decays downstream (Comte-Bellot and Corrsin, 1966; Kistler and Vrebalovich, 1966; Bennet and Corrsin, 1978; Sreenivasan *et al.*, 1980; Sinhuber *et al.*, 2015). Turbulence without a mean flow, generated by using an oscillating grid (De Silva and Fernando, 1994) and by towing a grid through a stationary sample of fluid (van Doorn *et al.*, 1999) has also been studied. A general theory describing the decay of turbulence based on first principles has not yet been developed. However, the works of Kolmogorov (1941b), Birkhoff (1954), Comte-Bellot and Corrsin (1966), and Saffman (1967a) outline phenomenological theories relating the forms of the 3D turbulent energy spectra to the temporal decay of turbulence.

This approach was extended by Skrbek and Stalp (2000) and further developed by Skrbek and Sreenivasan (2013) by taking into account ideas of Eyink and Thomson (2000), which included the finite size of the turbulence-generating box, and intermittency and viscosity effects. Somewhat surprisingly, under certain reasonable assumptions, classical decay models describe very well a variety of experimental data on decaying vortex line density in decaying grid generated quantum turbulence, although there are important differences that we shall discuss in the following chapters. Here, we briefly describe the relevant experiments generating grid turbulence in He II.

4.4.1 Helium Wind Tunnels

Historically the first superfluid wind tunnel was constructed by Craig and Pellam (1957). They used vertical cylindrical geometry restricted by two superleaks made of semipermeable material that allowed only a net superflow, generated thermally by a heater located above the upper superleak (for thermal generation of flow, see Section 4.5). Using a propeller with mica wings, the authors demonstrated the existence of a no lift region for subcritical flow velocity, i.e., perfect potential flow.

Subsequently, a number of cryogenic wind tunnels have been constructed, operating on the same principles as those at ambient temperature, allowing the exploration of both classical cryogenic flows of He I and superflows. Two examples of such wind tunnels with flow of cryogenic helium (both He I and He II) driven by a centrifugal pump, powered via a shaft from the room temperature flange of the cryostat, are shown in Fig. 4.2. Another possibility, as already mentioned, is to use cryogenic compressible bellows driving a cryogenic helium wind with a grid added downstream; see Fig. 4.4 (right).

4.4.2 Towed Grids in He II

Experiments on decaying quantum turbulence in He II generated by a towed grid have been performed over many years in Donnelly's group at the University of Oregon, resulting in a series of papers by Smith *et al.* (1993), Stalp *et al.* (1999, 2002), Skrbek *et al.* (2000), and Niemela *et al.* (2005). The schematic of the apparatus is shown in Fig. 4.8. Turbulence is generated by towing a grid through a stationary sample of He II in a channel of square cross section (about 29 cm long and $D = 1$ cm wide, manufactured by an electroforming process). Two different grids have been used over time. The first was of unconventional design – a 65% open brass monoplanar grid of rectangular tines, 1.5 mm thick, with a mesh size M (tine spacing) of 0.167 cm; the second, consisting of 28 rectangular tines of width 0.012 cm forming 13 × 13 full meshes of approximate dimension $M = 0.064$ cm, was identical in design to that of Comte-Bellot and Corrsin (1966).

The grid was attached to a stainless steel pulling rod that exited the cryostat via a pair of tight sliding seals. The space between the seals was continually evacuated to prevent the introduction of impurities inside the cryostat during the experiment. Above the cryostat, the rod was attached to a computer-controlled linear servo motor that positioned the grid with an accuracy of about 1 mm and provided towing velocities, u_g, up to 2.5 m/s. This arrangement enabled the exploration of a wide range of mesh Reynolds numbers up to $Re_M = u_g M \rho/\mu \leq 2 \times 10^5$, where μ is the dynamic viscosity of the normal fluid and ρ the total density. The channel was

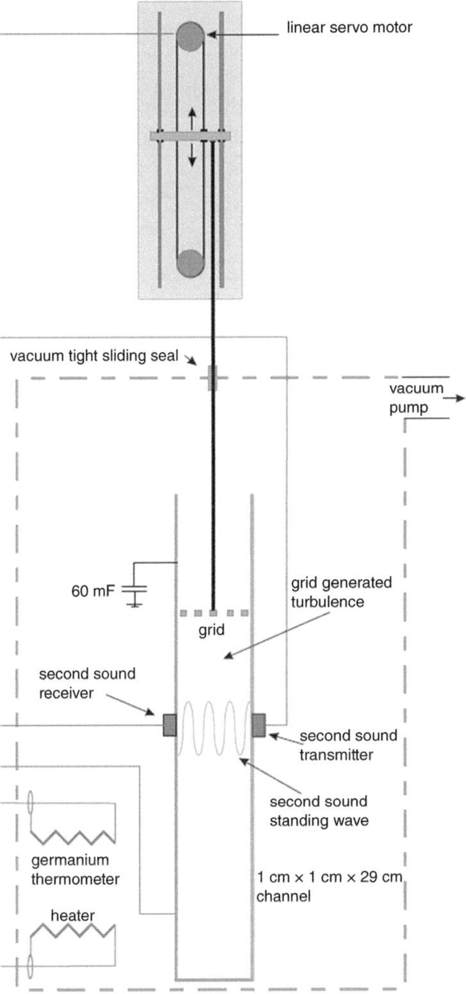

Figure 4.8 Schematic of the towed grid experimental apparatus. Reproduced with permission from Stalp (1998).

suspended vertically in the helium cryostat and totally submerged in superfluid helium during measurements.

More recently, similar towed grid experiments in normal liquid He I and in superfluid He II have been performed at the University of Florida in Gainesville (Yang and Ihas, 2018) and at Florida University in Tallahassee, USA (Gao *et al.*, 2016a).

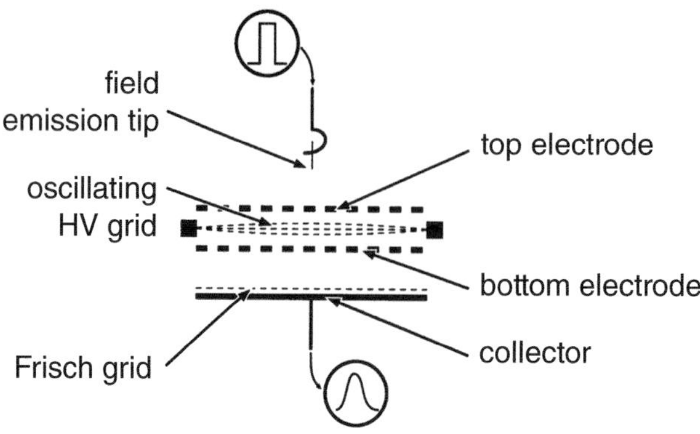

Figure 4.9 Schematic of the oscillating grid in He II used by McClintock's group in Lancaster. Reprinted from Davis *et al.* (2000) with permission from Elsevier.

4.4.3 Steadily Oscillating Grids in He II and in ^{3}He-B

The first attempts to generate quantum turbulence by an oscillating grid (and to detect it by means of ions; see Section 4.7.5) were performed by Davis *et al.* (2000) in Lancaster, using the apparatus shown schematically in Fig. 4.9. The grid is held at a high constant potential. The perforated top and bottom electrodes complete the double capacitor, and oscillations of the grid are excited by application of an alternating potential to the lower electrode, producing quantum turbulence above a certain threshold. This pioneering experiment clearly demonstrated the production and decay of quantum turbulence in the zero-temperature limit, although only qualitatively.

A pair of grids oscillating in phase, driven by a room-temperature motor via a shaft, are currently used by the Prague group to generate quantum turbulence in He II as well as classical grid turbulence in He I (Švančara and La Mantia, 2017). These flows are then studied by the particle tracking velocimetry (PTV) technique, described in Section 4.9.

The experimental arrangement for the first vibrating grid experiments in ^{3}He-B at very low temperature, achieved using the Lancaster-style nuclear cooling stage (Bradley *et al.*, 2004), is shown on the right of Fig. 4.5. The grid is made of a fine mesh of copper wires spaced 50 μm apart, leaving 40 μm square holes. Facing the grid are two vibrating wire resonators made from 2.5 mm diameter loops made of 4.5 μm NbTi wire, positioned at distances of 1 mm and 2 mm from the grid. The operational principle for the oscillating grid is the same as that for vibrating wire detectors of quantum turbulence, described in Section 4.7. An additional wire resonator is used as a background thermometer.

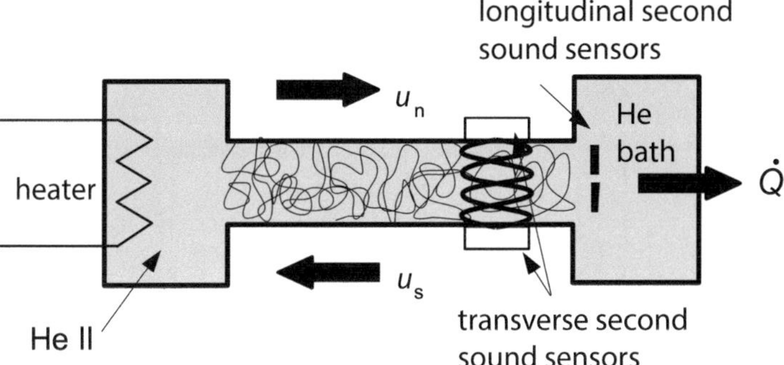

Figure 4.10 Schematic view of the generation of counterflow turbulence of He II and its detection using the transverse and longitudinal second sound attenuation method.

4.5 Generation of QT in Thermal Counterflow of He II

As already briefly explained in Chapter 2, thermal counterflow can be easily set up by applying a voltage to a resistor (heater) located at the closed end of a channel that is open to a helium bath at the other end, as illustrated in Fig. 4.10. The heat flux $\dot{Q}$ is carried away from the heater by the normal fluid alone, and, by conservation of mass, a superfluid current arises in the opposite direction; the counterflow velocity in the channel is proportional to the applied heat flux, $\dot{q}$, and takes value $u_{\mathrm{ns}} = \dot{q}/\rho_{\mathrm{s}}ST$. See Eq. (2.11).

When the heater situated at the dead end of the counterflow channel is switched on (see Fig. 4.10), it does not take long to establish a steady-state counterflow. If the applied power density is below a critical value, there are no vortex lines in the flow (except for remanent vortex lines). Quantum turbulence is generated upon increasing the heating power.

4.6 Generation of QT in He II by Sound and Ion Jets

Second sound can be seen as a time-dependent, oscillatory form of thermal counterflow. We shall use the term *ac counterflow*, as opposed to the *dc counterflow* discussed above. It is not surprising that ac counterflow can be used for the generation of quantum turbulence. Indeed, if second sound of large enough amplitude is driven, either thermally (Chagovets, 2016) or mechanically, e.g., by means of an ac pressure applied to a suitably shaped resonator via a superleak (Kotsubo and Swift, 1990), quantum turbulence is generated, first at the antinodes of the standing wave resonance; its amplitude ceases to grow with the drive upon reaching a threshold. The resulting quantum turbulence is easy to detect, either by the same second sound signal or by an additional signal (Midlik *et al.*, 2021).

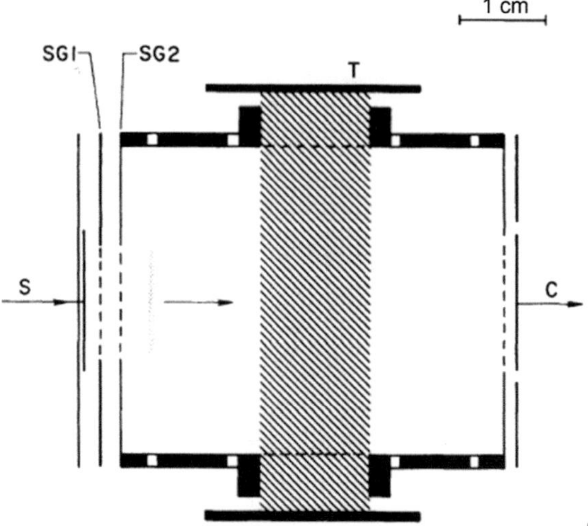

Figure 4.11 Turbulence is created by a pair of ultrasonic transducers that illuminate the hatched region. Ions are produced by the source S and manipulated by SG1 and SG2 to produce ion pulses. The drifting ion pulse can be stopped anywhere in the cell to sample vortex line density. Reprinted figure with permission from Milliken *et al.* (1982). Copyright 1982 by the American Physical Society.

First sound of high frequency and amplitude – i.e., ultrasound – can also be used to produce quantum turbulence in He II. Figure 4.11 shows the experimental arrangement of Milliken *et al.* (1982) for the generation of inhomogeneous quantum turbulence by a pair of ultrasonic transducers and turbulence detectors using negative ions (see Section 4.7). Complex physics underlies the generation of turbulence, but it is believed that turbulence is generated and maintained by acoustic streaming. The nucleation of vortex lines from the collapse of small bubbles was demonstrated by Berloff and Barenghi (2004) by solving the Gross–Pitaevskii equation numerically.

Negative ions can be used to both detect (see Section 4.7) and generate quantum turbulence in He II (Walmsley and Golov, 2017). The generation method is illustrated in the lower part of Fig. 4.1. Negative ions (electron bubbles) are injected by a sharp tungsten field-emission tip and manipulated inside the experimental cell by an applied electric field, creating a vortex tangle. This ion jets generation method creates charged vortex tangles, which opens new possibilities for studying quantum turbulence with an additional means of interesting control. For example, Coulomb repulsion among charged ions trapped on cores of quantized vortices affects the dynamics of the vortex tangle; further charging of an existing tangle would likely lead to its polarization.

4.7 Detection of Quantum Turbulence

A rich variety of experimental techniques probing turbulent flows of superfluid He II and ^{3}He-B could be, or have already been, employed. While some of these techniques are the same or similar to those commonly used in classical turbulence experiments, others are unique to the particular quantum fluid in the sense that they exploit some of the fluid's special physical properties. We begin with methods that can probe both classical and quantum flows, and are therefore generally capable of providing direct in situ comparison of turbulence of both varieties.

4.7.1 Pressure Drop and Pressure Fluctuations

Measurements of longitudinal pressure gradients along pipes and channels can be performed in both laminar and turbulent He I, He II, and cryogenic helium gas flows (Swanson *et al.*, 2002), essentially as done in ordinary fluids at ambient temperatures. For example, Donnelly and collaborators used a precise differential pressure gauge with capacitive readout, based on a flexible membrane, which was effectively one of the electrodes of a plane capacitor with its two sides connected by a capillary to the pressure ports in the pipe (Swanson *et al.*, 1998).

Pressure fluctuations in He I and He II were first measured by Maurer and Tabeling (1998) using a miniature pitot tube (the operational principle being the same as in viscous fluids at ambient temperatures) at a location where the von Kármán velocity field generated by counter-rotating discs has nonzero mean, as seen in the left panel of Fig. 4.2. The pitot tube of Fig. 4.12 thus mostly detects fluctuations of the pressure head, which in turn are present due to velocity fluctuations (for quantitative estimates in various flows of He II, see Appendix A of Salort *et al.* (2010)). The measurements can thus be related to the turbulent energy spectrum in the frequency domain; the frequency spectrum can be related, via Taylor's frozen flow hypothesis, to the energy spectral density in wavenumber space. Various types of cryogenic pitot tubes are now in use, possessing faster response time, for example in the large von Kármán flow facility SHREK (Rousset *et al.*, 2014; Salort *et al.*, 2021) (see Fig. 4.13).

4.7.2 Hot-Wire Anemometers

In classical turbulence research, one of the most commonly used sensors for fluctuation measurements is the hot-wire anemometer thanks to its high spatial and temporal resolutions. Indeed, such sensors have already proven to be very useful at cryogenic temperatures, in particular in the study of very high Reynolds number turbulence in gaseous helium, where superconducting materials have been used in

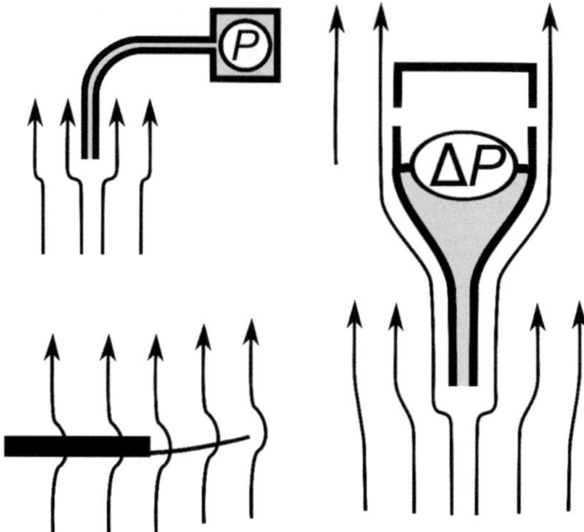

Figure 4.12 (Top left) Stagnation pressure velocity probes without static pressure reference (Maurer and Tabeling, 1998; Salort *et al.*, 2010), (Right) stagnation pressure velocity probe with a static pressure reference (right) (Roche *et al.*, 2007; Salort *et al.*, 2010), and (Bottom left) cantilever-based velocity probe (Salort *et al.*, 2012b). In the arrangement depicted on the right, the measurement of differential pressure (denoted as ΔP) removes the static pressure variation of the flow associated with turbulent pressure fluctuations and acoustic background noise. The streamline patterns shown should not be taken literally. From Barenghi *et al.* (2014a).

their design (Castaing *et al.*, 1992). As for their use in superfluids, it is often assumed that hot-wires cannot work for the following reason. In classical fluids, their operation is based on the fact that at length scales typical of hydrodynamic experiments, forced convection is far more efficient than natural convection and molecular diffusion. In superfluid helium, a counterflow between the normal and superfluid components exists, providing another thermal transfer mechanism, whose efficiency is limited by the generation of a vortex tangle sustained by the counterflow itself. Even for large heat fluxes, the counterflow mechanism is very efficient and forced convection does not improve the heat transfer significantly. Despite these odds, Duri and coworkers designed and tested a hot-wire anemometer in coflowing He II with very promising results (Duri *et al.*, 2015). The good correlation of the hot-wire signal with a validated local velocity sensor, as well as the observed Kolmogorov inertial range scaling of the power spectra, are indications that the response of hot wires in He II is mainly related to the local velocity of the coflowing He II; thus it appears that the very high effective conductivity of He II does not diminish the utility of the sensor. Although further studies are needed to understand the detailed

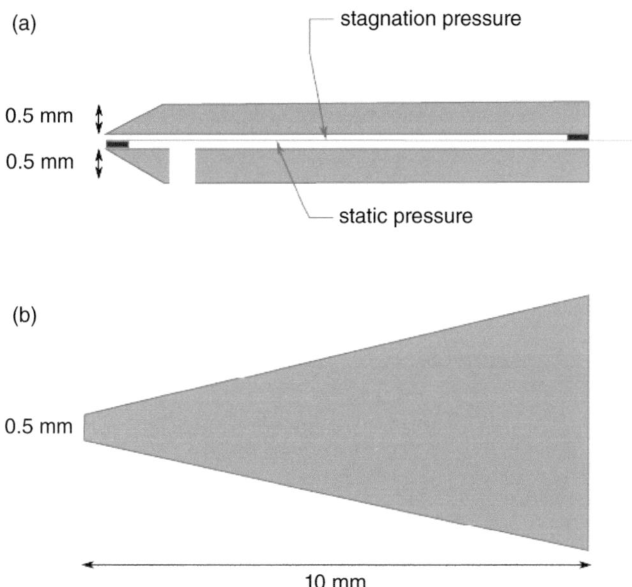

Figure 4.13 Schematic miniature pitot sensor. (a) Side view. (b) Top view. The gap between the upper and lower parts is 20 μm. Reproduced from Salort *et al.* (2021), used under CC BY 4.0 (https://creativecommons.org/licenses/by/4.0).

underlying physics within the thin shell surrounding the hot-wire (Rickinson *et al.*, 2020), and the results obtained by this means ought to be treated with care, the hot-wire anemometer can be considered a sensitive local velocity sensor in (coflowing) He II (Diribarne *et al.*, 2021).

4.7.3 Mechanical Probes

We have seen that oscillating structures such as cylinders, wires, spheres, and tuning forks can be used to generate turbulent flow. The same oscillating structures can also be used as detectors of such flows. For a review, see Skrbek and Vinen (2009). Almost all the experiments therefore involve measuring the force that liquid helium exerts on the structure (neglecting the internal damping and/or energy losses of the structure itself) as a function of the amplitude of oscillations. This force can be divided into a part that varies quadratically with the velocity, which can be interpreted as a change in the hydrodynamic effective mass of the structure, and a drag part that is in phase with the velocity and therefore dissipative. The amplitude of the drag force is usually expressed in the form

$$F_{\mathrm{d}} = \frac{1}{2} C_{\mathrm{d}} A \rho u^2, \tag{4.1}$$

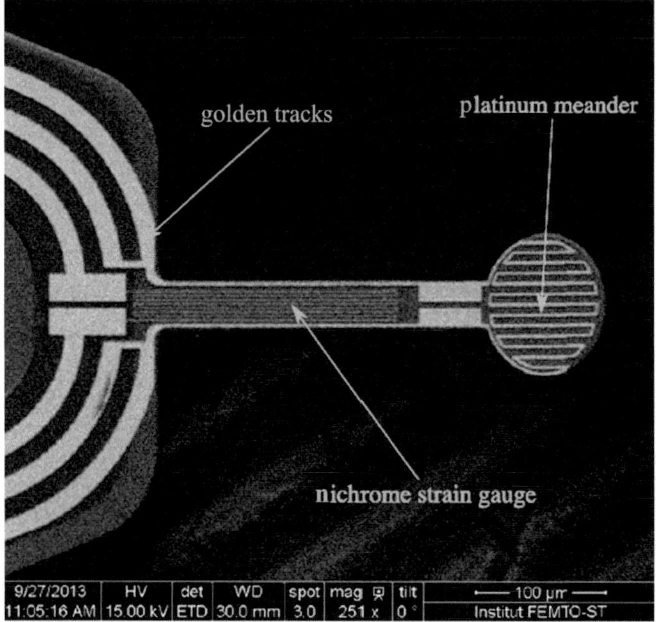

Figure 4.14 Scanning electron microscope picture of the cantilever anemometer used in the SHREK facility. Reproduced from Salort *et al.* (2021), used under CC BY 4.0 (https://creativecommons.org/licenses/by/4.0).

where C_d is a dimensionless drag coefficient, A is the projected area of the structure in the direction perpendicular to the velocity u, and ρ is the total density of the fluid. The variation of C_d with u is similar for all oscillating structures studied, and so one can obtain useful information about the transition from laminar flow to quantum turbulence. The underlying physics of this transition will be discussed in Chapter 6.

We already mentioned that modern electron lithography techniques allow the manufacturing of very small oscillating objects that can serve as sensitive tools to probe quantum flows (Collin *et al.*, 2010; Kamppinen and Eltsov, 2018; Barquist *et al.*, 2020; Guthrie *et al.*, 2021). One example is the cantilever designed by Salort *et al.* (2012b), shown in Fig. 4.12, bottom left. The actual version of a miniature cantilever used in the SHREK facility is shown in Fig. 4.14.

4.7.4 Second Sound Attenuation

Second sound attenuation is the powerful and historically oldest technique that directly measures the vortex line density, L, in He II. The technique, suitable for a variety of experiments in superfluid ^{4}He, was developed by Vinen (1957) in his pioneering experiments on thermal counterflow. Since second sound, briefly introduced in Chapter 2, can be thought to consist of antiphase oscillations of normal and superfluid components, this technique cannot be used in ^{4}He below about 1 K, where the amount of normal fluid is negligible (see Fig. 2.6). Another difficult region for the second sound technique is the temperature range just below the λ-point, where the second sound velocity is small and strongly temperature dependent, as shown in the right panel of Fig. 2.8. Further, second sound attenuation cannot be used in ^{3}He at any temperature due to the large kinematic viscosity of the normal fluid.

To understand the use of second sound in detecting the vortex line density L, we consider the seminal work of Hall and Vinen (1957). In experiments with a rotating container of He II, Hall observed an *excess attenuation* of second sound travelling in the direction perpendicular to the rotation axis, which was attributed to the presence of a lattice of rectilinear vortex lines. That is,

$$\alpha_L = \frac{B\Omega}{2u_2}, \tag{4.2}$$

where Ω is the angular velocity of the container, u_2 is the second sound velocity, and B is a (dimensionless) temperature-dependent (and weakly frequency dependent) mutual friction coefficient (Barenghi *et al.*, 1983; Donnelly and Barenghi, 1998). By excess attenuation we mean an attenuation effect additional to the bulk attenuation of second sound when helium is at rest (and vortex lines, except remanent ones, are absent). No excess attenuation is detected if the second sound wave travels along the rotation axis of the vortex lines. The extra sound attenuation arises from the scattering of the elementary excitations that make up the normal fluid.

By considering a second sound resonance as an infinite series of reflected waves in a rotating cavity, the extra attenuation (in the limit of small attenuation) becomes

$$\alpha_L \approx \frac{B\kappa L}{4u_2} = \frac{\pi\Delta_0}{u_2}\left(\frac{a_0}{a} - 1\right), \tag{4.3}$$

where a and a_0 are the amplitudes of the second sound standing wave resonance with and without vortex lines, respectively, and Δ_0 denotes its width in the absence of vortex lines.

What is detected by the second sound sensors in the experiment, however, is neither the total vortex length, nor the individual projected lengths. The attenuation of a second sound wave of angular frequency ω in the presence of a straight vortex line is $\sin^2(\theta)\omega B/(2\Omega)$ where θ is the angle between the direction of the vortex line and the direction of second sound propagation. This "sine squared law" has been confirmed experimentally by measuring second sound signals in a container of helium held at tilted angles with respect to the axis of rotation of the cryostat (Snyder and Putney, 1966; Mathieu *et al.*, 1984). If the vortex tangle is assumed to be isotropic, since $\langle \sin^2(\gamma) \rangle = 2/3$, where $\langle\ \rangle$ denotes the average over the unit

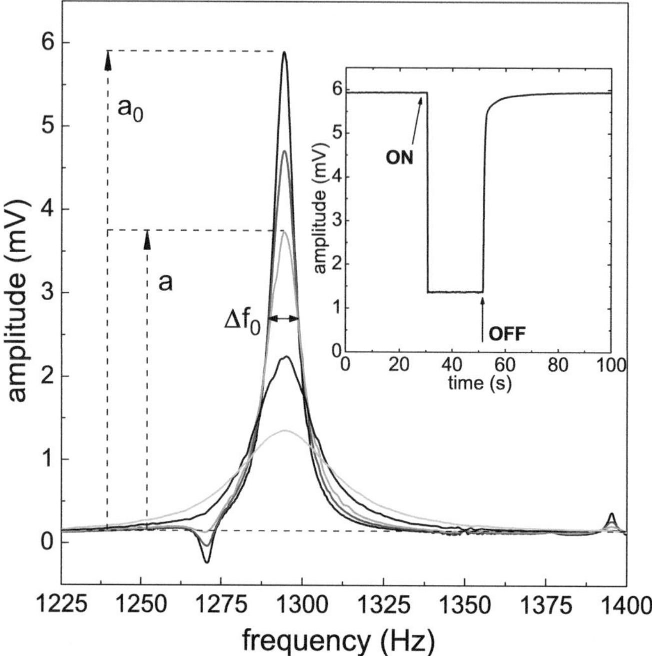

Figure 4.15 (Main panel) Second sound resonance curves (fundamental mode) for different steady-state flow velocities in the 7 mm wide channel at T=1.35 K. The tallest curve, a_0, corresponds to the case of no flow in the channel, while others correspond to flow velocities, from top to bottom, of 1.73, 2.60, 4.33, and 6.07 cm/s, in order. (Inset) The second sound amplitude at resonance monitored in time as a steady flow with a velocity of 6.07 cm/s is switched on and off (same flow velocity as the most attenuated curve in the main plot). Reprinted figure with permission from Babuin *et al.* (2012). Copyright 2012 by the American Physical Society.

sphere, the second sound sensors detect $L_{\text{eff}} = 2L/3$, where L is the actual vortex line density.

Equation (4.3) can be extended to the case of a homogeneous vortex tangle, taking into account that vortices oriented parallel to the second sound propagation do not contribute to the excess attenuation. Then we have

$$L = \frac{6\pi\Delta_0}{B\kappa}\left(\frac{a_0}{a} - 1\right). \tag{4.4}$$

Let $p = a_0/a$ and $P = 1 - \cos(2\pi d\Delta_0/u_2)$; then, for small $d\Delta_0/u_2$, where d is the length of the resonator, it can be shown that a more precise formula is

$$L = \frac{3u_2}{B\kappa d}\ln\left[\frac{1 + p^2P + \sqrt{2p^2P + p^4P^2}}{1 + P + \sqrt{2P + P^2}}\right]. \tag{4.5}$$

Parenthetically, Eqs. (4.4) and (4.5) differ from that used in analyzing, for example, the Oregon towed grid experiment and the early Prague thermal counterflow experiments (up to 2005), where the "sine squared law" was not taken into account. Values of L presented in these older papers must be multiplied by a factor of $3\pi/8 \approx 1.2$. This is only a minor correction as far as decaying turbulence is concerned, but it affects the value of the effective kinematic viscosity derived from the decay data, whose reported values are too low by the factor $(3\pi/8)^2 \approx 1.4$, as will be discussed in Chapter 9.

An example in which second sound frequency sweeps across a standing wave resonant frequency, measured by Babuin *et al.* (2012) in a flow channel with and without quantized vortices present, is shown in Fig. 4.15. For a full, comprehensive description of the second sound method and a step-by-step derivation of the relevant quantitative experimental relationships starting from the two-fluid equations for He II, we direct the reader to Babuin *et al.* (2012) and Varga *et al.* (2019).

4.7.5 Negative Ions

Helium ions have been successfully used by Milliken *et al.* (1982) to investigate the decay of inhomogeneous quantum turbulence created by ultrasonic transducers in He II at about 1.5 K, as schematically shown in Fig. 4.11. Ions are produced by the radioactive source S and manipulated by applying a voltage to grids SG1 and SG2 to produce ion pulses. The drifting narrow ion pulse can be stopped anywhere in the apparatus. Some ions are trapped into the cores of vortex lines; since vortex lines trap ions at a known rate, by collecting these trapped electrical charges, the vortex line density L can be determined by an interrupted flight pulse-ion technique. By repeating the experiment many times with different time delays after stopping the turbulence-producing transducers, it was possible to determine the decay law of vortex line density in inhomogeneous, essentially unbounded, quantum turbulence.

Awschalom *et al.* (1984) utilized the ion pulse technique to probe the local properties of counterflow turbulence in a channel of $1 \times 2.3\,\mathrm{cm}^2$ rectangular cross section, as shown in Fig. 4.16. These authors used a pulsed-ion technique based on a tritium source to obtain spatially resolved measurements of the vortex line density and the normal fluid velocity profile in turbulent counterflow. Again, a narrow pulse of negative ions was gated into the channel and allowed to propagate to a particular position under the action of an applied electric field. The field was then switched off, allowing the pulse to remain at this position for as long as several seconds. Some of the ions in the pulse were trapped by vortex lines; the rest drifted along with the local normal fluid velocity. Later, the electric field was turned back on and the surviving pulse was measured in real time by means of one or more electrometers as it arrived at one of three collectors on the opposite side of the

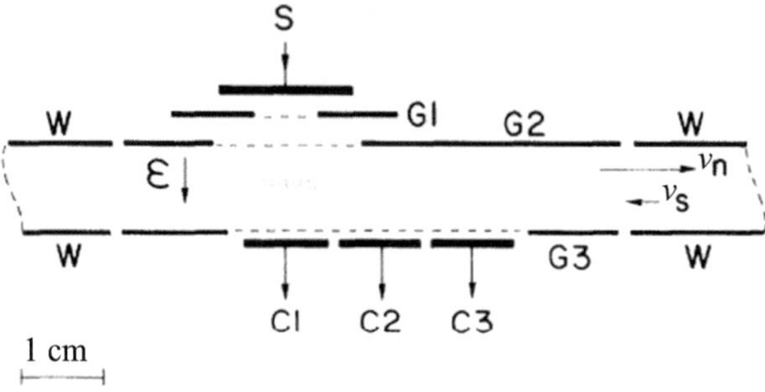

Figure 4.16 Experimental arrangement of Awschalom *et al.* (1984). S denotes the tritium ion source, G1, G2, and G3 are pulsed grids, C1, C2, and C3 are collectors, W denotes the channel wall, the superfluid and normal fluid velocities in the counterflow are indicated by arrows. The dotted region indicates a typical ion pulse; the ions are moved by an electric field ε. Reprinted figure with permission from Awschalom *et al.* (1984). Copyright 1984 by the American Physical Society.

channel, positioned along the channel axis. From the amplitude of the observed pulse and its position on the collector, the local values of L and u_n were determined, with a spatial accuracy of better than 1 mm, providing information on the local structure of counterflow He II turbulence. The remarkable result was that, within the experimental uncertainty of up to a few percent, L (rather low) was found to be constant over the entire cross section of the channel, except for a small region of width about 1 mm from the walls where the technique fails.

For measurements of the decaying quantum turbulence in He II below 1 K, an improved technique based on negative ions was introduced by Golov and coworkers in Manchester (Walmsley *et al.*, 2007; Walmsley and Golov, 2008; Walmsley *et al.*, 2008; Walmsley and Golov, 2017). Negative ions (electron bubbles) were injected by a sharp field-emission tip and manipulated by an applied electric field. Bare ions dominate for $T > 0.8$ K while, for $T < 0.7$ K, they become self-trapped on the core of vortex rings of about 1 μm diameter (which are generated spontaneously by the accelerating electron bubble in the applied electric field). Short pulses of charged vortex rings were sent across the experimental cell. The relative reduction in their amplitude at the collector on the opposite side of the helium cell, due to the interaction of the charged vortex rings with the vortex tangle in the cell, was converted into vortex line density L. An accurate in situ calibration was used as these experiments were performed on a rotating cryostat, which provided a known density of rectilinear vortex lines, $L = 2\Omega/\kappa$, corresponding to the rotation rate of the superfluid.

An interesting update of the ion detection technique combined with a towed grid has been more recently developed in Manchester by Zmeev (2014). The device has

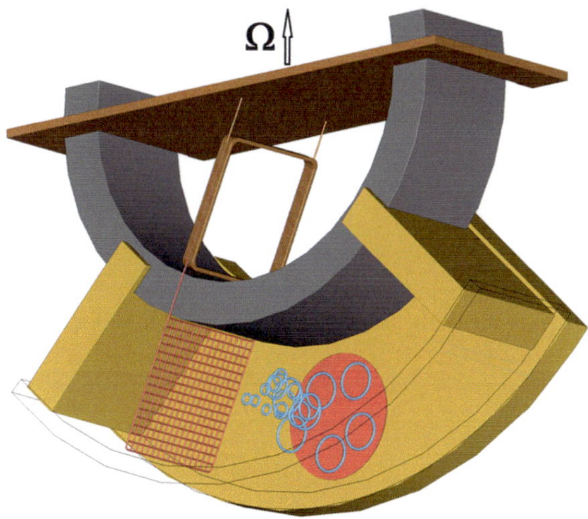

Figure 4.17 Experimental device combining the towed grid generation of quantum turbulence and its detection by ions, developed by Zmeev (2014). The front and bottom walls of the channel are not shown. Blue circles depict charged vortex rings (not to scale) used to probe quantum turbulence. The rings propagate from the injector (not shown) at the front wall to the collector inside a hole (shown by the red circle) in the back wall. The assembly could be rotated about the vertical axis. Reproduced from Zmeev *et al.* (2015), used under CC BY 3.0 (https://creativecommons.org/licenses/by/3.0).

been mounted to a rotating ^{3}He–^{4}He dilution refrigerator and is capable of working in He II down to almost the zero-temperature limit (Fig. 4.17).

4.8 Detection of Quantum Turbulence in Superfluid ^{3}He-B

Experimental techniques for the detection of quantum turbulence in the fermionic superfluid ^{3}He-B are generally very different from those used in bosonic He II. At relatively high temperatures (in terms of T/T_c, where the superfluid transition temperature T_c depends on pressure and is of the order of 1 mK), the flow of ^{3}He-B can be described by the two-fluid model. As we already discussed in Chapter 2, kinematic viscosity of the normal fluid in ^{3}He-B is high enough that in experimental containers of typical size 1 cm^3 the normal fluid hardly moves and a completely new class of quantum turbulence becomes possible when only the superfluid component is moving with respect to boundaries. One can use the fact that superfluid ^{3}He (in both A and B phases) is a magnetic liquid and thus obtain a lot of useful information by the nuclear magnetic resonance (NMR) technique. The relevant experiments have been carried out over many years by the Helsinki ROTA group using the rotating cryostat. This work has contributed a great deal to

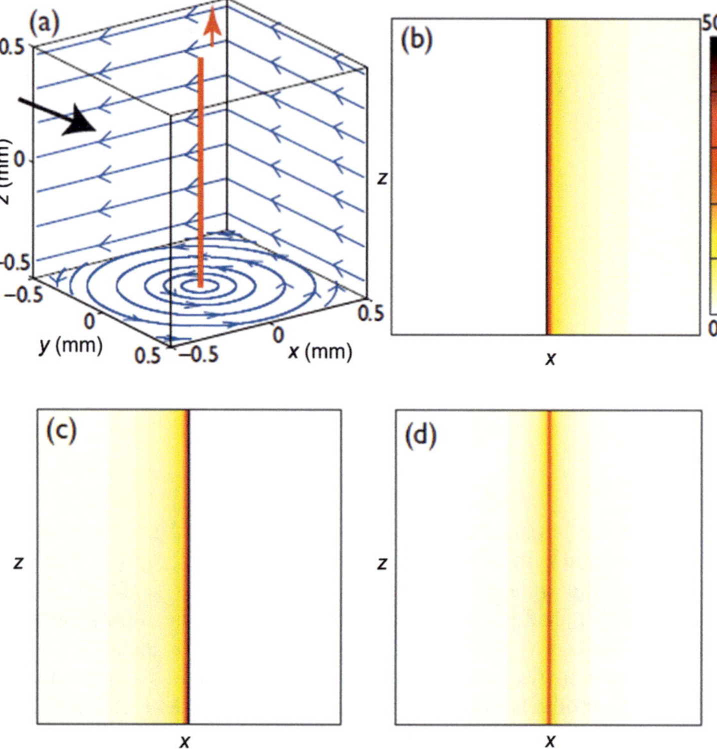

Figure 4.18 Schematic representation of Andreev scattering. (a) Isolated straight vortex at the center of a cube (the arrows show the direction of the superflow); quasiparticles and quasiholes are injected from "infinity," moving along the *y*-direction. (b) Quasiparticle reflection (darker/lighter regions correspond to larger/smaller reflection). (c) Quasihole reflection. (d) Combined reflection of quasiparticles and quasiholes. Reproduced from Tsepelin *et al.* (2017), used under CC BY 4.0 (https://creativecommons.org/licenses/by/4.0).

our knowledge of rotating superfluid ^{3}He phases. We shall describe these studies in Chapter 8.

For experimental reasons (which will not be discussed in this book), it is very difficult to apply the NMR detection technique for quantum turbulence in ^{3}He-B in the zero-temperature limit. But the Lancaster ^{3}He group invented an experimental technique based on the Andreev scattering of quasiparticles caused by the velocity field of quantized vortices. The underlying physics of the Andreev scattering detection technique is explained in the review by Fisher and Pickett (2009). Briefly, the equilibrium number of quasiparticles (and quasiholes, which are paired because ^{3}He-B is a fermionic superfluid) can be sensed by a vibrating wire resonator. The drag force exerted by these quasiparticles is reduced if the vibrating wire is surrounded by quantum turbulence (as some of incoming quasiparticles and quasiholes cannot reach it) being Andreev reflected by the energy barrier caused by the velocity field of the superflow near the vortex cores. The fractional decrease in damping can then be converted into the vortex line density in the tangle.

To understand the process of Andreev scattering, it is worth adding the following details about the motion of thermal excitations. A thermal excitation can be thought as an object of position $\boldsymbol{r}(t)$ and momentum $\boldsymbol{p}(t)$. Its dispersion (energy–momentum) curve (shown schematically in Fig. 2.11), $E = E(\boldsymbol{p})$, has a minimum at the Fermi momentum, p_F, corresponding to the Cooper pair binding energy Δ. Thermal excitations can be divided into quasiparticles, which have $|\boldsymbol{p}| > p_{\mathrm{F}}$, and quasiholes, which have $|\boldsymbol{p}| < p_{\mathrm{F}}$. In the reference frame of the superfluid moving with velocity u_{s}, the dispersion relation $E(\boldsymbol{p})$ becomes $E(\boldsymbol{p}) + \boldsymbol{p} \cdot u_{\mathrm{s}}$. The trajectory of an excitation is determined by the Hamilton equations

$$\frac{d\boldsymbol{r}}{dt} = \frac{\epsilon_{\mathrm{p}}}{\sqrt{\epsilon_{\mathrm{p}}^2 + \Delta^2}} \frac{\boldsymbol{p}}{m^*} + u_{\mathrm{s}}, \qquad \frac{d\boldsymbol{p}}{dt} = -\frac{\partial}{\partial \boldsymbol{r}}(\boldsymbol{p} \cdot u_{\mathrm{s}}), \tag{4.6}$$

where $\epsilon_{\mathrm{p}} = p^2/(2m^*) - \epsilon_{\mathrm{F}}$ is the kinetic energy relative to the Fermi energy ϵ_{F}, $m^* \approx 0.3 m_3$ is the effective mass of the excitation, m_3 the bare mass of a ^{3}He atom, Δ the superfluid energy gap, and u_{s} the superflow generated by the quantized vortices. Equation (4.6) is valid provided the distance to a vortex core is larger than the coherence length, $\xi \approx 50\,\mathrm{nm}$. Consider a beam of excitations that moves in the y-direction orthogonally to the (x, y) face of a cube containing a straight vortex, in the center, aligned in the z-direction, as shown in panel (a) of Fig. 4.18. The excitations move with constant energy. A quasiparticle arriving along the y-direction on the left of the vortex $(x < 0)$ has enough energy to go through but, to the right of the closely positioned vortex $(x > 0)$, it sees a potential barrier and will be reflected. Therefore, the reflection coefficient for quasiparticles is high to the right of the vortex, as shown in panel (b), where more intensely colored regions correspond to reflection. In contrast, a quasihole is reflected at the left of the vortex, as shown in panel (c). Panel (d) shows the total reflection of a beam consisting of both quasiparticles and quasiholes. Notice that the region of reflection is much larger than ξ_0 (the excitations are reflected by the velocity field, not by the vortex cores). Now assume that the cube contains a tangle of vortices rather than a single straight vortex. The reflection coefficient for quasiparticles incident on the (x, y) plane is shown in Fig. 4.19 together with the vortex lines, which are visible as thin intense lines. Panel (a) shows the reflection of quasiparticles: The regions of higher intensity correspond to large-scale flow into and out of the page. Panel (b) shows the reflection of quasiholes, with darker and lighter regions indicating large-scale flows in the direction opposite to those of (a). The combination of experiments

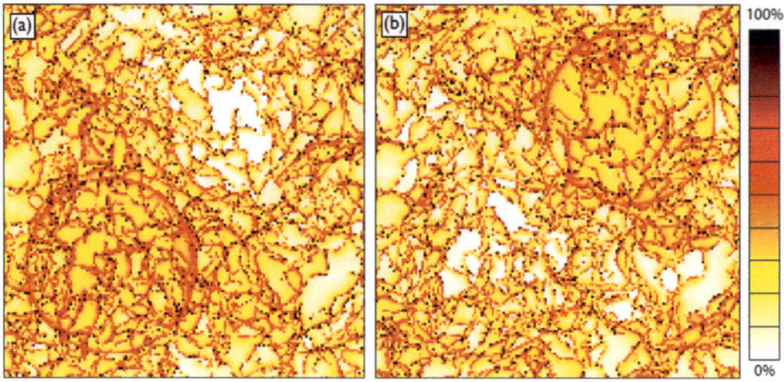

Figure 4.19 Andreev reflection coefficient on the (x, y) plane for thermal excitations incident from the y-direction on a vortex tangle. The quantized vortices are the thin dark lines. (a) Reflection of quasiparticles. Some vortex rings instantaneously injected in the turbulence to sustain it in statistically steady state are also visible as thin dark lines. (b) Reflection of quasiholes. Reproduced from Tsepelin *et al.* (2017), used under CC BY 4.0 (https://creativecommons.org/licenses/by/4.0).

and simulations of the reflection of excitations from vortex tangles shows that the Andreev reflection can be used to measure the vortex line density and its fluctuations (Baggaley *et al.*, 2015; Tsepelin *et al.*, 2017), although there are screening effects for dense vortex tangles.

In a more recent experiment, the Lancaster group succeeded in directly measuring the energy decay rate in quantum turbulence generated by the oscillating grid in the zero-temperature limit (Bradley *et al.*, 2011a) inside a box acting as a black-body radiator of quasiparticles. This experiment can be thought of as a complement to decaying classical turbulence: The quantity measured – the energy decay rate – is the same in both cases, although the detection technique is very different.

The Lancaster group is also developing a two-dimensional quasiparticle detector to visualize quantum turbulence in superfluid ^{3}He-B at ultra-low temperatures. The prototype detector (Ahlstrom *et al.*, 2014), shown schematically in Fig. 4.20, consists of a 5×5 matrix of pixels, each a 1 mm diameter hole in a copper block containing a miniature quartz tuning fork. The damping on each fork provides a measure of the local quasiparticle flux. The detector is illuminated by a beam of ballistic quasiparticles generated from a nearby black-body radiator. A comparison of the damping on the different forks provides a measure of the cross-sectional profile of the beam. If a tangle of vortices (quantum turbulence) is generated in the path of the beam, the vortices cast a shadow on the face of the detector due to the Andreev reflection of quasiparticles in the beam. This allows the imaging of the vortices and the investigation of their dynamics.

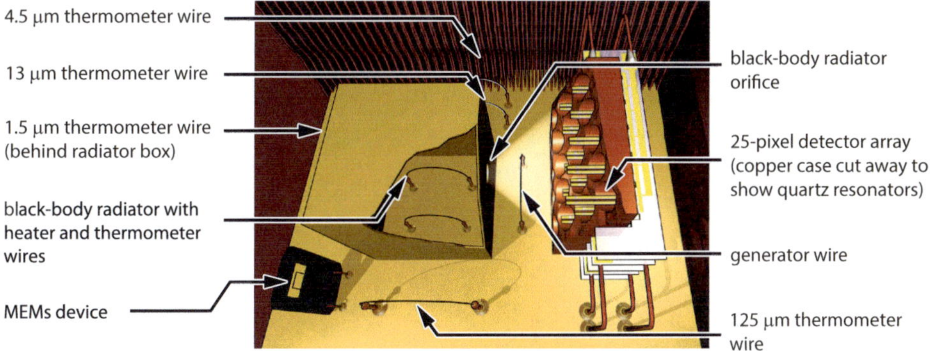

Figure 4.20 Schematic of the experimental volume for generation and detection of quantum turbulence in ^{3}He-B. The radiator is designed to illuminate the detector with a beam of quasiparticles centered on the central pixel. A vibrating wire in the path of the beam is used to generate vortices, which cast a quasiparticle shadow on the detector. The generator wire is approximately equidistant from the wall with the radiator orifice and the front face of the detector, which are approximately 2 mm apart. The cell contains additional thermometers and a MEMS device used for experiments in ^{3}He-B. From Ahlstrom *et al.* (2014).

4.9 Flow Visualization

Direct flow visualization is invaluable in classical turbulence research, where they have been developed to a high degree of precision and speed (Raffel *et al.*, 2007); methods such as particle image velocimetry (PIV) and particle tracking velocimetry (PTV) provide detailed quantitative data on turbulent flows. In brief, suitable particles (tracers) that are suspended in the fluid reflect the light of a laser sheet that illuminates the region of interest. The time-dependent positions of the particles are captured and analyzed by a digital imaging system. PIV estimates the fluid velocity (Eulerian approach) by assuming that the velocity field is varying sufficiently smoothly, whereas PTV allows the measurement of Lagrangian quantities – the fluid velocity and its derivatives.

4.9.1 PIV and PTV in Superfluid He II

Application of classical visualization methods to cryogenic flows is difficult for several reasons, and has been successful so far only in He I and He II. In He I, the difficulties are technical, such as the optical access to the experimental volume and the generation of suitable flows under cryogenic conditions. For its application in He I (and also liquid nitrogen), see White *et al.* (2002). In He II, these technical difficulties have limited the lowest working temperature to approximately 1.2 K. Indeed, in flows where the normal fluid, the superfluid, and the vortices have

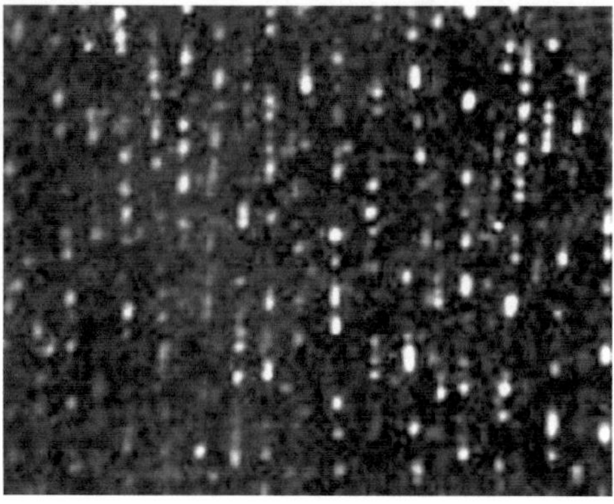

Figure 4.21 Direct visualization of vortex lattice in rotating He II (side view). The vortices are marked by hydrogen particles trapped in their cores. From Bewley *et al.* (2006).

different velocity fields, the behavior of particles is usually difficult to interpret. Besides tracing normal fluid flows, and also the superfluid due to the hydrodynamic added mass, the micron-sized particles currently used in PIV and PTV experiments can become trapped in vortex lines. In fact, a particle positioned on a vortex line displaces a volume of circulating superfluid component that has higher kinetic energy, because of its vicinity to the vortex axis, than the superfluid displaced by a farther particle; trapping is thus energetically favorable. Since the typical particle is about 10,000 times larger than the vortex core, it is better to think of particles trapped in vortex lines more as vortex lines pinned to particles. Pinning of one or several vortex lines to a particle may change the topology of the vortex tangle under study and mediate reconnections. Despite these difficulties, visualization experiments have already led to important results that will be discussed in later chapters. Here, we limit ourselves to describing the experimental principles of these visualization methods.

The usefulness of direct visualization is demonstrated by Fig. 4.21, a textbook example of the rotating vortex lattice in He II. This direct in situ visualization was achieved by Bewley *et al.* (2006) by trapping small particles of frozen hydrogen onto vortex cores and illuminating them by a laser sheet. The image is a side view of the vortex lattice, and complements top view images obtained three decades earlier using ions, already shown in Fig. 3.3. Bewley and Sreenivasan (2009) were also able to map and study the motion of a vortex ring decorated with particles.

It is very difficult to find suitable particles for PIV or PTV visualization of liquid helium flows (Bewley *et al.*, 2008a; Guo *et al.*, 2014). One can broadly classify particles into two categories: Solid particles (sometimes fluorescent), including plastic or (hollow) glass as used in classical fluid dynamics experiments at ambient tem-

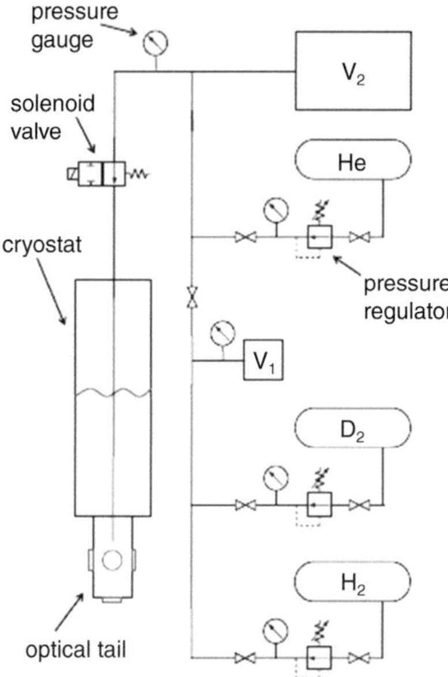

Figure 4.22 Schematic view of the specifically designed seeding system (not to scale). He, D_2, and H_2 indicate the high-pressure gas vessels containing helium, deuterium, and hydrogen, respectively. The chosen amount of gas (H_2, D_2) is collected into the volume V_1, about 50 times smaller than the volume V_2, where it is mixed with a chosen amount of helium gas. Reprinted from La Mantia *et al.* (2012) with the permission of AIP Publishing.

peratures, and solidified particles, produced by injecting gases (usually hydrogen, deuterium, and even nitrogen or air [Fonda *et al.*, 2016]) into liquid helium.

Attempts to use advanced PIV techniques to visualize He II flows have been made with micron-sized solid particles by Zhang *et al.* (2004, 2005); see Fig. 6.3 and also Fonda *et al.* (2016). These authors successfully observed average properties of turbulent states of superfluid helium and noticed peculiar flow patterns, such as two pairs of eddies, upstream and downstream of a cylinder placed in thermal counterflow (Zhang and Van Sciver, 2005b). However, such solid particles have proved to be too dense to explore the detailed structure of quantum turbulence. Experiments by Paoletti *et al.* (2008a) in Maryland, Chagovets and Van Sciver (2011) at Tallahassee, La Mantia *et al.* (2012) at Prague, and Kubo and Tsuji (2017) at Nagoya use solidified hydrogen or deuterium flakes of micron size; another tested possibility is to use solid fluorescent nanoparticles (Meichle and Lathrop, 2014).

To produce solid particles, a gaseous mixture of helium and hydrogen/deuterium prepared at room temperature in a volume ratio of approximately 100:1 is injected directly into liquid helium. As a result, a cloud of solid particles with diameters typically on the order of a micron is produced; these particles can be used for

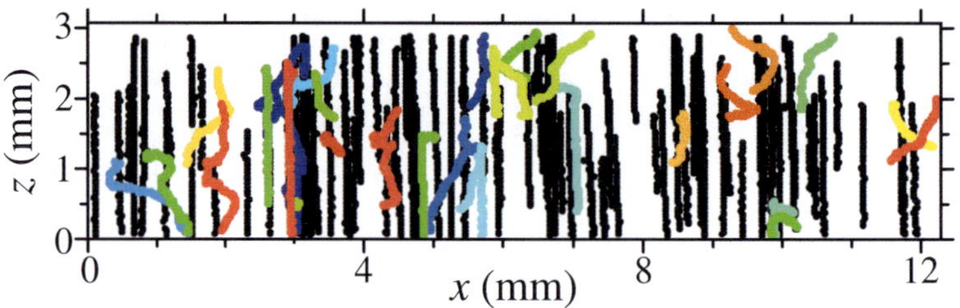

Figure 4.23 Example particle trajectories from a counterflow with $T = 1.95\,\mathrm{K}$ and heat input $36\,\mathrm{mW/cm^2}$. Trajectories that move upward with the normal fluid are shown in black while tracers trapped in the vortex tangle move downward and are shown in color. Reproduced with permission from Paoletti *et al.* (2008b).

both PIV and PTV techniques. The seeding system, designed by the Prague group to supply the helium bath with desired amounts of hydrogen and/or deuterium micrometer-sized solid tracers, is shown in Fig. 4.22. Solidified particles can be engineered so that their density is close to the temperature-dependent liquid helium density, about $145.6\,\mathrm{kg/m^3}$ for temperatures ranging between 1 and 2.5 K. Solid hydrogen has a smaller density, $88\,\mathrm{kg/m^3}$, while the density of solid deuterium is larger, about $200\,\mathrm{kg/m^3}$. Recently, the Prague group has started to use frozen deuterium hydride particles, the density of which closely matches that of He II; for earlier efforts, see Murakami *et al.* (1987). The tracer dimensions can be calculated using the recorded images beyond the point at which the dynamic effects of the injection are effectively erased; by assuming that the particles are spherical and that the buoyancy force is balanced by the Stokes drag, i.e., that the particle velocity is equal to their settling velocity, the radius r of a particle can be evaluated as

$$r = \sqrt{\frac{9\mu u_\mathrm{p}}{2g(\rho - \rho_\mathrm{p})}}, \tag{4.7}$$

where μ is the temperature-dependent dynamic viscosity of the normal component of He II, g is the acceleration due to gravity, ρ and ρ_p are, respectively, the densities of liquid helium and particles, and u_p denotes their vertical velocity. The diameters thus deduced of these deuterium hydride particles is typically on the order of a few μm.

Particle trajectories in a typical vertical counterflow are shown in Fig. 4.23, demonstrating that for low counterflow velocities some particles follow the upward moving normal fluid smoothly, while particles trapped in the vortex tangle move downward in a more erratic way. For higher counterflow velocities, however, all particles interact simultaneously with both the normal fluid and the superfluid, becoming intermittently trapped and untrapped: In this regime it is difficult to

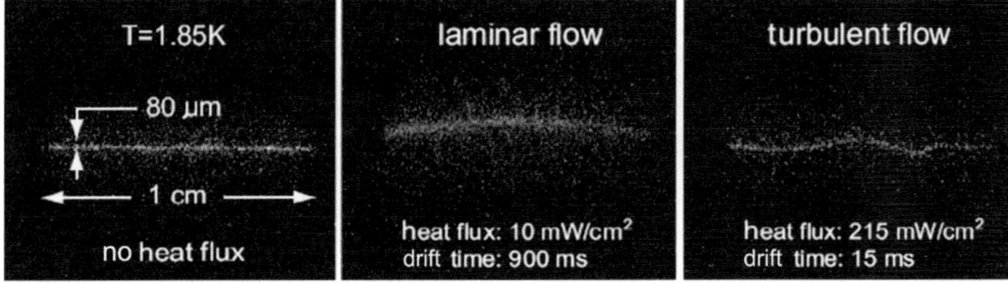

Figure 4.24 Fluorescence images showing the motion of a thin line of He_2^* tracers in thermal counterflow. The tracer line is created via laser-field ionization by focusing a femtosecond laser pulse into superfluid helium. The drift time denotes the time between the creation and the imaging of the tracer line. The second image was taken in steady-state flow, whereas the third image was taken when the heater was turned off. Reproduced with permission from Guo *et al.* (2014).

distinguish whether the superfluid velocity or the normal fluid velocity affects the particle motion more.

4.9.2 Neutral He_2^* Triplet Molecules

A more recent addition to visualization techniques is based on neutral He_2^* triplet molecules. These molecules can be produced in large numbers in liquid helium following the ionization or excitation of ground state helium atoms. The singlet state molecules radiatively decay in few nanoseconds, but the triplet state molecules are metastable with a radiative lifetime of about 13 s, allowing them to be useful in experiments. Similar to electrons, these triplet molecules form bubbles in liquid helium with a radius of about 6 Å and can be used as tracer particles. To image them, a sophisticated laser-induced fluorescence technique was developed by McKinsey *et al.* (2005). A laser pulse excites helium molecules from their triplet ground state to the excited electronic d-state; most of them quickly decay to an intermediate state emitting detectable fluorescent photons. A filter can be used to block unwanted laser light to achieve low background noise. From the intermediate state, molecules quench back to the triplet ground state, and the process can be repeated so that each molecule produces many fluorescence photons.

It is important to appreciate that, in the two-fluid temperature region above 1 K, larger particles (such as frozen flakes of hydrogen) interact with both the normal and superfluid velocity fields and with the vortex lines. Above 1 K, the smaller excited He_2^* triplet molecules are entrained solely by the normal fluid (they effectively are a part of the normal fluid), which makes them ideal for probing the patterns and properties of the normal component. It is not yet possible, however, to detect individual triplet molecules; clouds of typically a few hundred of them are detectable, which limits the length scale down to which the turbulent flows of He II can be probed to some tens of micrometers.

Due to the small energy needed to bind the particles to the vortices, thermal motion kicks them off the vortex at temperatures above 1 K. On the other hand, He_2^* triplet molecules can attach to quantized vortex lines below 0.2 K (Zmeev *et al.*, 2013), which has the potential for allowing direct vortex-line imaging in the zero-temperature limit.

An example of the application of this technique to thermal counterflow of He II is shown in Fig. 4.24. The Tallahassee group created a thin line of He_2^* molecules via laser-field ionization in helium. To achieve the required high electric field for ionizing ground state helium, a femtosecond laser pulse was applied. The molecular density was high enough to allow high-quality single-shot imaging of the tracer line. At low heat fluxes, a straight tracer line deformed into a parabolic shape, indicating the Poiseuille laminar velocity profile of the normal fluid. At large heat fluxes, the tracer line was distorted due to the "turbulence" in the normal fluid (Guo *et al.*, 2014).

Recently a new method of creation of He_2^* triplet molecules has been explored (Wen *et al.*, 2020) by ionizing radiation produced through neutron capture. Laser beams induce fluorescence of the excimers, which was recorded by a camera at a rate of 55.6 Hz. The location of the fluorescence was determined with an uncertainty of 5 μm. This promising new technique provides an opportunity to record the flow of He_2^* excimers and enables measurement of turbulence in three dimensions.

4.10 Summary

Our focus in this chapter has been the rich variety of methods of how quantum turbulence can be experimentally generated and detected, including flow visualization. Quantum turbulence in He II can be generated both mechanically and thermally, quite similar to classical turbulence. Thermal generation of quantum turbulence leads to a macroscopic counterflow of the normal and superfluid components of He II, a quantum effect with no direct classical analogue. However, owing to the two-fluid behavior of He II, thermal generation is possible only at nonzero temperatures because the normal fluid is absent in the important zero-temperature limit, and the superfluid component is incapable of carrying entropy.

In the superfluid ^{3}He-B phase, the high viscosity of the normal fluid, together with the experimental restrictions on the size of containers and the available cooling power, makes thermal generation of quantum turbulence practically impossible.

A rich variety of mechanical means can be used to generate turbulence in He II, often similar to those commonly used in classical turbulence research. Additionally, the quantum nature of He II allows rather extraordinary, nonclassical tools to both generate and detect quantum turbulence, such as second sound, negative ions, or neutral excimer molecules. A discussion of these techniques is presented in this chapter. In order to study quantum turbulence in the fermionic superfluid ^{3}He-B phase at submillikelvin temperatures, several remarkable nonclassical techniques have been developed based on nuclear magnetic resonance, black-body radiators, Andreev reflection, and so forth. As always in science, the fastest progress in understanding its particular research area is achieved when experimental investigations proceed hand in hand with theory and numerical simulations. It is therefore natural to discuss various theoretical approaches relevant to the problem of quantum turbulence. They will be discussed in Chapter 5.

5

Theoretical and Numerical Models

It is commonly believed that classical turbulence in viscous fluids is fully described by the Navier–Stokes equation with no-slip boundary conditions.[1] For an incompressible fluid of constant density and viscosity, these equations have the form

$$\frac{\partial \boldsymbol{u}}{\partial t} + (\boldsymbol{u} \cdot \nabla)\boldsymbol{u} = -\frac{1}{\rho}\nabla p + \nu \nabla^2 \boldsymbol{u}, \tag{5.1}$$

where $\boldsymbol{u} = \boldsymbol{u}(\boldsymbol{r}, t)$ is the velocity field, $p = p(\boldsymbol{r}, t)$ is the pressure field, $\boldsymbol{r}$ is the position, t is time, ρ is the density, and ν is the kinematic viscosity. The velocity field $\boldsymbol{u}$ must satisfy the solenoidal condition

$$\nabla \cdot \boldsymbol{u} = 0. \tag{5.2}$$

If the viscosity is zero, the Navier–Stokes equation reduces to the classical Euler equation for an inviscid fluid derived more than 250 years ago:

$$\frac{\partial \boldsymbol{u}}{\partial t} + (\boldsymbol{u} \cdot \nabla)\boldsymbol{u} = -\frac{1}{\rho}\nabla p. \tag{5.3}$$

Equations (5.1) and (5.3) are easily generalized to compressible fluids, in which case Eq. (5.2) must be replaced by the continuity equation for the density field ρ,

$$\frac{\partial \rho}{\partial t} + \nabla \cdot (\rho \boldsymbol{u}) = 0. \tag{5.4}$$

Unfortunately, when moving from classical fluids to superfluid helium, there is not a single model that adequately describes quantum turbulence. Instead, we have a hierarchy of physical models of increasing complexity that describe phenomena at different length scales. It is as if, unable to describe the trees and the forest in a unified way, we had one (microscopic) model for the details of an individual tree, a second (mesoscopic) model that does not resolve the bark of a tree, but still

[1] This view discounts the possibility that the equations might support singularities, and also that molecular dynamics is essential in describing very small scales of Navier–Stokes turbulence.

distinguishes individual trees as isolated sticks, and a third (macroscopic) model that does not resolve trees but recognizes where the forest is sparser or denser.

These microscopic, mesoscopic, and macroscopic levels of description are naturally defined by the following length scales: The vortex core radius a_0 (proportional to the healing length in bosonic or coherence length ξ in fermionic superfluids), the average distance between the vortices – the quantum length scale – ℓ, and the system's size D. In He II, these length scales are widely separated: $a_0 \approx 10^{-10}\,\mathrm{m} \ll \ell \approx 10^{-5}\,\mathrm{m} \ll D \approx 10^{-2}\,\mathrm{m}$, for typical experimental values of ℓ and D. In ^{3}He-B, the situation is similar, although a_0 is two orders of magnitude larger than in He II. Quantum turbulence in trapped atomic Bose–Einstein condensates (BECs) is different mainly because, as already stated in Chapter 3, atomic condensates at the present state of development lack the wide separation of length scales typical of turbulence in liquid helium: In BECs the size of the vortex core is smaller (but not orders of magnitude smaller) than the intervortex distance, which is smaller (but not much smaller) than the typical condensate's size. What makes current BECs relatively small is the (still significant) loss of atoms during the evaporation cooling stage of the trapped atoms.

5.1 Microscopic Level

The theoretical study of quantum turbulence requires models that are powerful enough to describe vortex lines in a superfluid, and also practical enough to describe vortex interactions. Most such models assume that the vorticity in a superfluid is localized along mathematical lines where the density vanishes and the superfluid velocity diverges (examples are models based on the Gross–Pitaevski equation and the Biot–Savart law, which we shall discuss later in this chapter). More physically refined models of vortex lines based on variational theory and quantum Monte Carlo methods give better physical information (Vitiello *et al.*, 1996; Galli *et al.*, 2014), but are not practical for addressing issues of vortex interaction and turbulence.

5.1.1 The Gross–Pitaevski Equation (GPE)

The most microscopic model that is practical enough to tackle turbulence in a quantum fluid is the *Gross–Pitaevski equation* (GPE) for a Bose–Einstein condensate, also called the cubic nonlinear Schrödinger equation (NLSE). The GPE is derived after suitable approximations are made from the Hamiltonian of a dilute Bose gas at zero temperature undergoing two-body collisions. It has the form (Barenghi and Parker, 2016)

$$i\hbar\frac{\partial\psi}{\partial t} = -\frac{\hbar^2}{2m}\nabla^2\psi + g|\psi|^2\psi, \tag{5.5}$$

where $\psi = \psi(\boldsymbol{r}, t)$, the complex wavefunction, $\hbar$ is Planck's constant divided by 2π, and g is the strength of the repulsive delta-function interaction between the bosons, often written as

$$g = \frac{4\pi\hbar^2 a_s}{m}, \tag{5.6}$$

where a_s is the scattering length and m is the bosonic mass. The GPE conserves energy and mass, as well as linear and angular momenta.

Without loss of generality, the wavefunction can be written with the prefactor $e^{-i\mu t/\hbar}$ where μ is the chemical potential (energy per boson); this choice reduces Eq. (5.5) to the form

$$i\hbar\frac{\partial\psi}{\partial t} = -\frac{\hbar^2}{2m}\nabla^2\psi + g|\psi|^2\psi - \mu\psi. \tag{5.7}$$

This is the form in which the GPE is often used.

5.1.2 Simple Solutions of the GPE

It is instructive to examine some simple solutions of the GPE that are relevant to quantum turbulence: the uniform solution, the wall solution, the vortex solution, and sound waves.

The uniform solution ψ_0 is the simplest solution of the GPE that is time-independent and space-independent: From Eq. (5.7) we find immediately $\psi_0 = \sqrt{\mu/g}$, which corresponds to bulk density $\rho_0 = m\mu/g$ and number density $n_0 = \mu/g$. This solution describes a superfluid at rest away from boundaries.

The wall solution describes a steady superfluid near a boundary. The simplest boundary is an infinite potential barrier that repels the atoms. From quantum mechanics, such a barrier is represented by the boundary condition $\psi = 0$. This means that the wavefunction ψ adjusts itself and drops from its bulk value ψ_0 (away from the boundary) to zero (at the boundary) over some characteristic distance ξ called the *healing length*. To estimate the magnitude of ξ we balance the first and second term at the right-hand side of the GPE (representing kinetic and interaction energies, respectively), and obtain

$$\xi \approx \frac{\hbar}{\sqrt{2m\mu}} = \frac{\hbar}{\sqrt{2mgn_0}}. \tag{5.8}$$

It must be stressed that, as in any estimates of this kind, the numerical factor $\sqrt{2}$ in the definition of ξ is arbitrary. To verify that our estimate is correct, we notice that the one-dimensional, time-independent GPE in the semi-infinite region $0 \leq x < \infty$ with boundary condition $\psi(0) = 0$ (corresponding to an infinite potential barrier at the origin) has the exact solution $\psi(x) = \psi_0 \tanh\left(x/(\xi\sqrt{2})\right)$, showing that indeed ξ is the characteristic length scale over which the wavefunction bends.

The vortex solution of the GPE is also readily found. We use cylindrical coordinates (r, θ, z) and set $\psi = \sqrt{n}\exp(i\phi) = F(r)\exp(iq\theta)$, where the integer q is the *winding number* or *charge* and represents the number of quanta of circulation carried by the vortex. Using Eq. (5.15) in cylindrical coordinates, we find that the corresponding velocity $\boldsymbol{u} = (0, q\kappa/(2\pi r), 0)$ is indeed a vortex line of q quanta of circulation κ each. Substitution into the GPE yields an equation that we use to determine the radial dependence of ψ. Since we require $\psi \to \psi_0$ for $r \to \infty$, it is convenient to define $F = \sqrt{g/\mu} f$ and measure the radial distance in units of the healing length. In most situations, certainly in turbulent He II flows, multiply quantized vortices are unstable (see Section 3.1), so we set $q = 1$ hereafter and find

$$\frac{d^2 f}{ds^2} + \frac{1}{s}\frac{df}{ds} + \left(1 - \frac{1}{s^2}\right) f - f^3 = 0, \tag{5.9}$$

where we have introduced the new radial variable $s = r/\xi$ because we know that $\psi = 0$ on the vortex axis and that it drops to zero, in the case of the one-dimensional wall solution, on a length scale of the order of the healing length, ξ. Equation (5.9) must be solved with boundary conditions $f \to 0$ for $s \to 0$ and $f \to 1$ for $s \to \infty$. The solution, first obtained numerically by Pitaevskii, confirms that a quantum vortex is a "hole" of thickness of the order of the healing length, around which the quantum mechanical phase changes by 2π. Berloff (2004) found good Padé approximations of Pitaevskii's numerical solution; the lowest order is $f(s) \approx \sqrt{s^2/(s^2+2)}$, shown in Fig. 5.1. The figure also shows that the vortex core radius a_0 (arbitrarily defined as the distance from the vortex axis where the density is about 90% of the bulk density) is approximately $a_0 \approx 4\xi$.

Small perturbations about the uniform solution ψ_0 of number density n_0 oscillate as waves of angular frequency ω and wavenumber $k = 2\pi/\lambda$ (where λ is the wavelength); their dispersion relation is

$$\hbar^2\omega^2 = \frac{\hbar^2 k^2}{2m}\left(\frac{\hbar^2 k^2}{2m} + 2n_0 g\right). \tag{5.10}$$

In the long-wavelength limit ($k \to 0$) the dispersion relation (5.10) reduces to

$$\omega \approx ck, \tag{5.11}$$

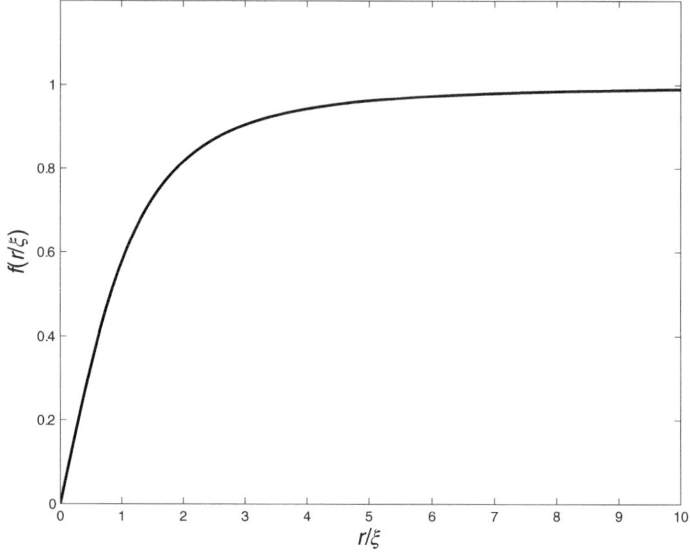

Figure 5.1 Radial dependence f vs. $s = r/\xi$ of the vortex wave function $\psi = \sqrt{\mu/g}f(r)e^{i\theta}$ (Padé approximation).

thus describing sound waves (phonons) having the speed

$$c = \sqrt{\frac{n_0 g}{m}}, \tag{5.12}$$

whereas, in the short-wavelength limit $k \to \infty$, the dispersion relation (5.10) becomes

$$\hbar\omega \approx \frac{\hbar^2 k^2}{2m}, \tag{5.13}$$

thus describing free particles of energy $E = p^2/(2m)$ and momentum p, where $E = \hbar\omega$ and $p = \hbar k$.

5.1.3 Fluid Dynamics Interpretation of the GPE

The fluid dynamics interpretation of the GPE is made apparent by the *Madelung transformation*

$$\psi = \sqrt{n}e^{i\phi}, \tag{5.14}$$

where $\phi = \phi(\boldsymbol{r}, t)$ is the phase of the wavefunction, $n = n(\boldsymbol{r}, t)$ is the number density (number of atoms per unit volume), and $\rho = \rho(\boldsymbol{r}, t) = m\, n(\boldsymbol{r}, t)$ is the mass density (mass per unit volume). Following standard quantum mechanical prescriptions, the velocity field $\boldsymbol{u} = \boldsymbol{u}(\boldsymbol{r}, t)$ is defined as

$$\boldsymbol{u} = \frac{\hbar}{m}\nabla\phi. \tag{5.15}$$

Substitution of Eqs. (5.14) and (5.15) into Eq. (5.5) yields the classical continuity equation (5.4) together with the "quasi-Euler" equation

$$\rho\left(\frac{\partial u_j}{\partial t} + u_k\frac{\partial u_j}{\partial x_k}\right) = -\frac{\partial p}{\partial x_j} + \frac{\partial P_{jk}}{\partial x_k} \tag{5.16}$$

(for $j, k = 1, 2, 3$), where u_j is the jth Cartesian component of $\boldsymbol{u}$ and we have used the summation convention. The pressure field p and the *quantum pressure* field P are defined as

$$p = \frac{g}{2m^2}\rho^2, \qquad P_{jk} = \left(\frac{\hbar}{2m}\right)^2 \rho\frac{\partial^2 \ln\rho}{\partial x_j \partial x_k}. \tag{5.17}$$

Equation (5.16) differs from the classical (compressible) Euler equation (5.3) because of the presence of the quantum pressure, which is responsible for phenomena outside the realm of Euler dynamics such as vortex nucleation and vortex reconnections, which are essential for quantum turbulence.

Let us examine the quantum pressure term in Eq. (5.16) more carefully. Its dependence on spatial derivatives of the density suggests that the term is not important away from the boundary, where the density is constant. More precisely, by balancing the first and second terms on the right-hand side of Eq. (5.16), we find that the quantum pressure term is negligible at distances d larger than the healing length ξ. In other words, in the limit $d \gg \xi$, the GPE reduces to the classical, compressible Euler equation (5.3), which must be accompanied by Eq. (5.4). Since the defining property of superfluidity is the absence of viscosity, recovering the classical inviscid limit was expected.

5.1.4 Energy Decomposition

The total energy of the condensate, defined as

$$E = \int \left(\frac{\hbar^2}{2m}|\nabla\psi|^2 + \frac{g}{2}|\psi|^4 + V|\psi|\right) d^3\boldsymbol{r}, \tag{5.18}$$

is conserved during the time evolution (for the sake of generality we have included the potential V, which confines an atomic condensate, discussed in Section 5.1.7). Following Nore *et al.* (1997), it is useful to identify the different contributions to E by writing

$$E = E_{\text{kin}} + E_{\text{q}} + E_{\text{int}} + E_{\text{trap}}, \tag{5.19}$$

where E_{kin}, E_{q}, E_{int}, and E_{trap} are, respectively, the kinetic energy, the quantum energy, the interaction energy, and the trap's potential energy (equal to 0 in a homogeneous system, e.g., a condensate in a periodic domain). The definitions are:

$$E_{\text{kin}} = \int \frac{1}{2}\rho \boldsymbol{u}^2, \qquad E_{\text{q}} = \int \frac{\hbar^2}{2m^2}(\nabla\sqrt{\rho})^2\, d^3\boldsymbol{r}, \tag{5.20}$$

$$E_{\text{int}} = \int \frac{g}{2}|\psi|^4\, d^3\boldsymbol{r}, \qquad E_{\text{trap}} = \int V|\psi|^2\, d^3\boldsymbol{r}, \tag{5.21}$$

where

$$E_{\text{kin}} + E_{\text{q}} = \int \frac{\hbar^2}{2m}|\nabla\psi|^2\, d^3\boldsymbol{r}. \tag{5.22}$$

Physically, E_{int} arises from the repulsion between the bosons and E_{q} from the presence of vortex cores (the tubular regions around the vortex axes where the density drops to zero). To help the comparison between GPE turbulence (which is compressible) and turbulence in helium II (which is considered incompressible), and incompressible Navier–Stokes turbulence (which we take as the reference), it is useful to separate the contributions of vortex lines and sound waves. We apply the orthogonal Helmholtz decomposition to the velocity field (Nore *et al.*, 1997), writing

$$\sqrt{\rho}\boldsymbol{u} = (\sqrt{\rho}\boldsymbol{u})^{\text{i}} + (\sqrt{\rho}\boldsymbol{u})^{\text{c}}, \tag{5.23}$$

where $\nabla \cdot (\sqrt{\rho}\boldsymbol{u})^{\text{i}} = 0$ and $\nabla \times (\sqrt{\rho}\boldsymbol{u})^{\text{c}} = \boldsymbol{0}$. In this way, the kinetic energy can be written as the sum of compressible (superscript c) and incompressible (superscript i) contributions,

$$E_{\text{kin}} = E^{\text{c}}_{\text{kin}} + E^{\text{i}}_{\text{kin}}, \tag{5.24}$$

where

$$E^{\text{c}}_{\text{kin}} = \int \frac{1}{2}\left[(\sqrt{\rho}\boldsymbol{u})^{\text{c}}\right]^2 d^3\boldsymbol{r}, \qquad E^{\text{i}}_{\text{kin}} = \int \frac{1}{2}\left[(\sqrt{\rho}\boldsymbol{u})^{\text{i}}\right]^2 d^3\boldsymbol{r}. \tag{5.25}$$

During the time evolution of the GPE, although the total energy E remains constant, the relative proportions of $E^{\text{c}}_{\text{kin}}$, $E^{\text{i}}_{\text{kin}}$, E_{int}, E_{q}, and E_{trap} change. For example, as vortices interact and decay they radiate sound waves, thus $E^{\text{c}}_{\text{kin}}$ increases while $E^{\text{i}}_{\text{kin}}$ decreases.

5.1.5 Limitations of GPE

One should not forget the limitations of GPE when applying it to quantum turbulence. The GPE is a very good quantitative model of the behavior of atomic Bose–Einstein condensates at sufficiently low temperatures ($T/T_{\text{c}} \ll 1$), when thermal excitations can be ignored. Indeed, the GPE has been used to study 2D (Neely *et al.*, 2013) and 3D (Kobayashi and Tsubota, 2007; White *et al.*, 2010) turbulence in atomic condensates. However, the validity of the GPE when applied to superfluid helium (in the limit $T/T_\lambda \ll 1$, of course) is only qualitative. Superfluid helium is a liquid, not a weakly interacting gas. If one put numbers in the parameters that appear in the GPE, it is impossible to match observed values of density, sound speed, and healing length simultaneously. Moreover, the dispersion relation of the GPE does not exhibit the roton minimum that is typical of He II; to include this effect, Berloff and Roberts (1999) proposed a modification of the GPE called the nonlinear nonlocal Schrödinger equation (NNLSE), in which the delta-function repulsion is replaced by a more realistic interaction potential (turning the GPE from a partial differential equation into an integro-partial differential equation).

5.1.6 Finite-Temperature Generalization of GPE

In Section 5.1.5, we stated that the GPE is a likely model only at low temperatures, $T \ll T_c$. The generalization of the GPE to finite temperatures is an open problem and we refer the reader to the review of Berloff *et al.* (2014). Here, it is worth recalling that the simplest approach consists in replacing $i\hbar$ on the left-hand side of Eq. (5.5) with $(i-\gamma)\hbar$, introducing the small, phenomenological damping parameter γ. A versatile approach is to turn the GPE into a stochastic equation (Cockburn and Proukakis, 2009). A physical approach worth mentioning is the Zaremba–Nikuni–Griffin formalism described by Proukakis and Jackson (2008), which couples the GPE to the Boltzmann equation for the thermal cloud of non-condensed atoms, allowing atomic collisions within the thermal cloud and between the thermal cloud and the condensate. This is similar to the two-fluid model, although we are still in a dilute ballistic regime rather than in a liquid regime. The crucial feature is that the condensate and the thermal cloud determine each other self-consistently, and from this interaction the dissipative effects of vortex motion emerge; as shown by Jackson *et al.* (2009), these effects follow the prediction of the vortex filament model. This approach provides an *ab initio* calculation of the mutual friction interaction between the thermal cloud of non-condensed atoms (i.e., the normal fluid component) and the condensate (i.e., the superfluid component, in the language of the two-fluid model), although the numerical values of the friction coefficients refer to atomic gases (which depend on the parameters chosen) rather than to superfluid helium.

5.1.7 The GPE for a Trapped Condensate

Turbulence in atomic Bose–Einstein condensate takes place in trapped atomic gases. At temperatures $T \ll T_c$, these systems are modeled accurately by the GPE, Eq. (5.5), provided that a realistic trapping potential $V_{\text{trap}} = V_{\text{trap}}(\boldsymbol{x})$ is included to confine the bosons (otherwise their collisions would make the gas to expand to infinity). The resulting equation is

$$i\hbar\frac{\partial\psi}{\partial t} = -\frac{\hbar^2}{2m}\nabla^2\psi + g|\psi|^2 + V_{\text{trap}}\psi, \tag{5.26}$$

where the trapping potential is usually harmonic and of the form

$$V_{\text{trap}} = \frac{m}{2}\left(\omega_x x^2 + \omega_y y^2 + \omega_z z^2\right). \tag{5.27}$$

The harmonic oscillator frequencies ω_x, ω_y, and ω_z are parameters that can be chosen in order to confine the trapped condensate in the desired shape (spherical, cigar-like, pill-like), or to reduce the effective dimensionality (from three to two to one).

The main feature of solutions of Eq. (5.26) is that the density $n = |\psi|^2$ is not uniform. For a spherically symmetric trapping potential

$$V_{\text{trap}} = \frac{m\omega^2 r^2}{2}, \tag{5.28}$$

where $r^2 = x^2 + y^2 + z^2$ and $\omega^2 = \omega_x^2 + \omega_y^2 + \omega_z^2$, the density of the ground state has a maximum at the origin, the center of the trap, and goes to zero at distance of the order of the Thomas–Fermi radius

$$r_{\text{TF}}(r) = \left(\frac{15N\hbar^2 a_s}{m^2\omega^2}\right)^{1/5}, \tag{5.29}$$

where a_s is the s-wave scattering length and $N = \int |\psi|^2 dV$ is the number of atoms in the trap. Figure 5.2 shows the typical profile of a harmonically trapped atomic condensate that contains a single off-center vortex: Note that the density vanishes at the edge of the condensate and at the vortex core.

The nonuniform ground state density makes harmonically trapped atomic condensates quite different from the situation in liquid helium. Fortunately, Gaunt *et al.* (2013) succeeded in trapping atomic condensates in *box traps*: To model them, we replace Eq. (5.28) (for example) with $V_{\text{trap}}(\boldsymbol{x}) = 0$ for $r < R_0$ and $V_{\text{trap}} = V_0 \gg \mu$ for $r > R_0$; the resulting density $n(\boldsymbol{x})$ is uniform within the sphere $r < R_0$ (apart from a small correction near the edge at $r = R_0$), and is hence a better model of superfluid helium. To model a box trap with an infinite potential barrier ($V_0 \to \infty$), the trapping potential is omitted from Eq. (5.26) and is replaced by the boundary conditions $\psi \to 0$ for $r \to R_0$. Another popular way to model superfluid helium is to solve Eq. (5.5) without the trapping potential in a computational domain with periodic boundary conditions, again yielding a ground state with constant density: This is the case of the homogeneous condensate described in previous sections.

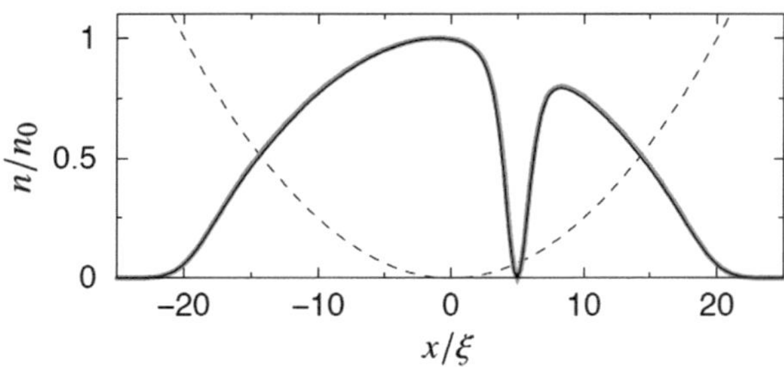

Figure 5.2 Typical number density profile n/n_0 vs. x/ξ in the x-direction across a condensate that contains a single off-center vortex, where $n = |\psi|^2$ and n_0 is the number density at the center of the trap. The dashed line is the trapping potential V_{trap}. Reproduced with permission from Parker (2004).

5.2 Mesoscale Level

5.2.1 The Vortex Filament Model (VFM)

A mesoscale approach that bypasses effects at the scale of the vortex core is the *vortex filament model* (VFM) developed by Schwarz (1988). The model represents vortex lines as space curves of infinitesimal thickness and circulation κ. This limit is recovered from the GPE level of description under the assumptions that

(i) the length scales are much larger than the healing length ξ (so that the quantum stress term can be neglected); and
(ii) the velocity is much smaller than the sound velocity c; the governing superfluid equation is then the Euler equation (5.3) under the solenoidal condition $\nabla \cdot \boldsymbol{u} = 0$.

Let the vector $\boldsymbol{s} = \boldsymbol{s}(\zeta, t)$ be the position along a vortex line, where ζ is the arc length and t is time. The equation of motion of the vortex line has been derived by Barenghi *et al.* (1983), who analysed the forces acting on the vortex line: The Magnus force (which arises when a body with circulation around it moves in a flow) and the drag force (which arises from the scattering or absorption of thermal excitations by the vortex line). The sum of these forces is negligible because the hydrodynamical mass of the vortex is negligible. One finds that the velocity of the vortex line at the point $\boldsymbol{s}$ is given by the *Schwarz equation*

$$\frac{d\boldsymbol{s}}{dt} = \boldsymbol{u}_{\text{s}}^{\text{tot}} + \alpha \boldsymbol{s}' \times \left(\boldsymbol{u}_{\text{n}} - \boldsymbol{u}_{\text{s}}^{\text{tot}}\right) - \alpha' \boldsymbol{s}' \times \left[\boldsymbol{s}' \times \left(\boldsymbol{u}_{\text{n}} - \boldsymbol{u}_{\text{s}}^{\text{tot}}\right)\right], \tag{5.30}$$

where $\boldsymbol{s}' = d\boldsymbol{s}/d\zeta$ is the unit tangent vector to the vortex line at the point $\boldsymbol{s}$. Here, the total superfluid velocity, $\boldsymbol{u}_{\text{s}}^{\text{tot}}$, has been decomposed into the sum of any externally imposed superflow (e.g., by bellows or a heater), $\boldsymbol{u}_{\text{s}}^{\text{ext}}$, and the self-induced velocity,

$\boldsymbol{u}_{\rm s}^{\rm self}$, arising from the curvature of the line itself and the presence of other vortex lines in the vicinity, which we write as

$$\boldsymbol{u}_{\rm s}^{\rm tot}(\boldsymbol{s}) = \boldsymbol{u}_{\rm s}^{\rm ext} + \boldsymbol{u}_{\rm s}^{\rm self}. \tag{5.31}$$

The quantity $\boldsymbol{u}_{\rm s}^{\rm self}$ at the point $\boldsymbol{s}$ is obtained from the classical Biot–Savart law as

$$\boldsymbol{u}_{\rm s}^{\rm self}(\boldsymbol{s}) = -\frac{\kappa}{4\pi}\oint_{\mathcal{L}} \frac{(\boldsymbol{s}-\boldsymbol{r})}{|\boldsymbol{s}-\boldsymbol{r}|^3} \times d\boldsymbol{r}, \tag{5.32}$$

where the line integral extends over the entire vortex configuration $\mathcal{L}$. The quantities α and α' in Eq. (5.30) are the known temperature-dependent mutual friction coefficients tabulated by Donnelly and Barenghi (1998) related to the mutual friction coefficients B and B' originally introduced by Vinen (see Chapter 2) via the relations $2\alpha = \rho_{\rm n}B/\rho$ and $2\alpha' = \rho_{\rm n}B'/\rho$. The quantity $\boldsymbol{u}_{\rm n}$ is the normal fluid velocity field evaluated at the point $\boldsymbol{s}$. In the original approach of Schwarz, for simplicity $\boldsymbol{u}_{\rm n}$ is taken as a constant, representing the normal fluid externally applied by a heater (or by mechanical means) rather than a velocity field obtained by solving its own governing equation of motion. Finally, the quantity $\boldsymbol{s}' = d\boldsymbol{s}/d\zeta$ is the unit vector tangent to the vortex line at the point $\boldsymbol{s}$.

Note that if $T = 0$, Schwarz's equation (5.30) reduces to

$$\frac{d\boldsymbol{s}}{dt} = \boldsymbol{u}_{\rm s}^{\rm ext} + \boldsymbol{u}_{\rm s}^{\rm self}, \tag{5.33}$$

in agreement with the classical Helmholtz theorem that a vortex line moves with the (super)flow.

When implementing the VFM, the configuration of vortex lines is discretized over a (large) number of points $\boldsymbol{s}_j$ ($j = 1, \ldots, N_p$) that move according to Eq. (5.30). Usually, this Lagrangian discretization depends on the local curvature: Regions along a line where the local curvature is large (or the local radius of curvature is small) evolve rapidly and require more discretization points than in regions where the local curvature is small (or the local radius of curvature is large) and the evolution is slow. Typically, the numerical scheme uses variable meshing along the lines (discretization points $\boldsymbol{s}_j$ are added or removed), and a minimum allowed distance $\Delta\zeta_{\rm min}$ between the discretization points is set at the beginning of the calculation; this spatial cutoff, which sets the shortest (or most rapid) Kelvin wave allowed in the calculation, also determines the size of the time step (or the minimum time step in the case of adaptive time-stepping).

When writing a VFM code, three difficulties become immediately apparent: the singularity of the Biot–Savart law, what to do when vortex lines collide, and the computational cost.

(i) Desingularization of the Biot–Savart integral. The Biot–Savart integral Eq. (5.32) diverges as $\boldsymbol{r} \to \boldsymbol{s}_j$. The difficulty is well known in the classical literature of thin-cored vortices, and is taken care of by introducing a cut-off parameter based on the vortex core thickness a_0. More precisely, Eq. (5.32) at the point $\boldsymbol{s}_j$ is replaced (Schwarz, 1988) by

$$\boldsymbol{u}_\mathrm{s}^{\mathrm{self}}(\boldsymbol{s}_j) = \boldsymbol{u}_\mathrm{s}^{\mathrm{loc}}(\boldsymbol{s}_j) + \boldsymbol{u}_\mathrm{s}^{\mathrm{non}}(\boldsymbol{s}_j), \tag{5.34}$$

where the local and nonlocal contributions to $\boldsymbol{u}_\mathrm{s}^{\mathrm{self}}$ are

$$\boldsymbol{u}_\mathrm{s}^{\mathrm{loc}} = \frac{\kappa}{4\pi} \ln\left(\frac{\sqrt{\Delta\zeta_+ \Delta\zeta_-}}{a_0}\right) \boldsymbol{s}_j' \times \boldsymbol{s}_j'', \qquad \boldsymbol{u}_\mathrm{s}^{\mathrm{non}} = \frac{\kappa}{4\pi} \oint_{\mathcal{L}'} \frac{(\boldsymbol{s}_j - \boldsymbol{r})}{|\boldsymbol{s}_j - \boldsymbol{r}|^3} \times d\boldsymbol{r}. \tag{5.35}$$

Here, $\Delta\zeta_+$ and $\Delta\zeta_-$ are the arc lengths between the point $\boldsymbol{s}_j$ and the adjacent points $\boldsymbol{s}_{j-1}$ and $\boldsymbol{s}_{j+1}$ along the vortex line. The term $\mathcal{L}'$ is the original vortex configuration $\mathcal{L}$, but now without the section between $\boldsymbol{s}_{j-1}$ and $\boldsymbol{s}_{j+1}$. Note that the local contribution to the self-induced velocity is inversely proportional to the local radius of curvature, $R_j = 1/|\boldsymbol{s}_j''|$ $\left(\text{where } \boldsymbol{s}'' = d^2\boldsymbol{s}/d\zeta^2\right)$, and points in the binormal direction $\boldsymbol{s}' \times \boldsymbol{s}''$.

(ii) Vortex reconnections. The second difficulty concerns the collisions of two vortex lines. In the absence of friction (zero-temperature limit), two isolated lines will try to become locally antiparallel to conserve energy, moving rapidly away from each other. If the interaction with the other lines prevents this alignment, the two lines (locally parallel) will rotate rapidly around each other; in either case, smaller (faster) length scales will be produced, requiring smaller time steps for the numerical method of integration to fully resolve the motion of the lines. This effect is expected because the Euler equation lacks the mechanisms to dissipate the short length scales that are created, and the Biot–Savart law simply reformulates the Euler equation for thin-cored vortices in integral form.

Following an original insight of Feynman (1955), Schwarz (1988) understood that vortex reconnections are essential features of superfluid vortex dynamics, and assumed that when two vortex lines become sufficiently close they reconnect, changing the topology of the flow, as shown in Fig. 3.7. The reconnection procedure, which goes beyond Euler dynamics, must be implemented by a suitable algorithm. The various reconnection algorithms described in the literature (Schwarz, 1988; Tsubota *et al.*, 2000; Kondaurova and Nemirovskii, 2005; Tsubota and Adachi, 2011) have been reviewed by Baggaley (2012). There is no significant physical difference arising from using any of these numerical procedures; in other words, the existing experimental information is not detailed enough to decide which algorithm is more realistic.

(iii) Computational cost. The third difficulty of the VFM is the computational cost of evaluating Biot–Savart integrals, which increases rapidly with the square of the number of discretization points along the vortex lines, N_p^2. To speed up his calculations, Schwarz replaced the Biot–Savart law with its local induction approximation, essentially neglecting the second term in Eq. (5.34). Physically, this approximation neglects any vortex–vortex interaction with the exception of vortex reconnections, which are taken care of by the vortex reconnection algorithm. Unfortunately, as clearly demonstrated by Adachi *et al.* (2010), Schwarz's approximation requires an arbitrary mixing step to achieve a statistically steady state of turbulence – something that is regarded as being unsatisfactory by today's standards. The problem of the computational cost of the VFM was addressed and vastly improved by Baggaley and Barenghi (2011a, 2012) by adapting to vortex dynamics the $N_p \log N_p$ tree algorithm created for N-body gravitational interactions in computational astrophysics (Barnes and Hut, 1986).

5.2.2 Limitations of VFM

The VFM is perhaps the most flexible numerical tool for quantum turbulence in He II and ^{3}He-B, and is thus used quite often. It is therefore important to appreciate the limitations of this model. We discuss a few essentials here and refer the reader to the state-of-the-art review by Baggaley and Hänninen (2014).

The first limitation of VFM is that, unlike the GPE, it does not describe acoustic losses of energy by rapidly rotating Kelvin waves, which are thought to be important at very low temperatures. It must be said that dissipation is present in the VFM even at $T = 0$ because of the finite Lagrangian discretization of the vortex lines. Kelvin waves of wavelength shorter than the minimum discretization are effectively dissipated away; the vortex reconnection algorithm also introduces dissipation of vortex length. But these effects are unlike the physical reality of phonon emission.

The second limitation is that, in the original formulation of Schwarz and in most recent implementations, the VFM determines the evolution of vortex lines in the presence of an *imposed* normal fluid velocity field, without accounting for the back reaction of the vortex lines on the normal fluid velocity via mutual friction. Examples include the uniform velocity field of Schwarz (1988), the Poiseulle profile to model channel flows (Baggaley and Laurie, 2015), the single vortex tube (Samuels, 1993) used to study the behavior of vortex lines in the presence of concentrated normal fluid vorticity, the ABC flow proposed by Barenghi *et al.* (1997) to model multiple coherent structures, and the time-dependent synthetic turbulence introduced by Baggaley and Barenghi (2011a) to investigate the effect on normal fluid turbulence with a prescribed energy spectrum.

In fluid dynamics, the *dynamo problem* (Dormy and Soward, 2007) provides a useful analogy. In this problem, the issue is the mechanism that generates the

(a) $u_n \longrightarrow u_s$ (b) $u_n \rightleftarrows u_s$

Figure 5.3 (a) Kinematic approach in which vortex lines evolve in the presence of a given normal fluid $\boldsymbol{u}_n$; (b) self-consistent approach in which vortex lines and normal fluid affect each other.

observed magnetic fields in planetary and stellar interiors (including the Earth and the Sun) and in the interstellar medium. In their simplest form, the governing dynamo equations are:

(i) the magnetic induction equation for the magnetic field $\boldsymbol{B}$,

$$\frac{\partial \boldsymbol{B}}{\partial t} = \eta_m \nabla^2 \boldsymbol{B} + \nabla \times (\boldsymbol{u} \times \boldsymbol{B}), \tag{5.36}$$

where η_m is the magnetic diffusivity; and

(ii) the Navier–Stokes equation for the velocity field $\boldsymbol{u}$,

$$\frac{\partial \boldsymbol{u}}{\partial t} + (\boldsymbol{u} \cdot \nabla)\boldsymbol{u} = -\frac{1}{\rho}\nabla p + \nu \nabla^2 \boldsymbol{u} + \boldsymbol{j} \times \boldsymbol{B}, \tag{5.37}$$

where $\boldsymbol{j} = (1/\mu_m)\nabla \times \boldsymbol{B}$ and μ_m is the magnetic permeability, alongside the solenoidal conditions $\nabla \cdot \boldsymbol{u} = \nabla \cdot \boldsymbol{B} = 0$; in the presence of convective forces, when ρ is not constant, an energy equation linking the temperature and density fields is also required.

Note that Eq. (5.36) determines $\boldsymbol{B}$ given $\boldsymbol{u}$, whereas Eq. (5.37) determines $\boldsymbol{u}$ given $\boldsymbol{B}$; in both equations, the terms that couple $\boldsymbol{B}$ and $\boldsymbol{u}$ (the advection term $\nabla \times (\boldsymbol{u} \times \boldsymbol{B})$ and the Lorentz force $\boldsymbol{j} \times \boldsymbol{B}$) are nonlinear.

Until the 1990s, when computers became sufficiently powerful, it was common to distinguish between the *kinematic approach* and the *fully self-consistent approach*. In the kinematic approach, Eq. (5.36) was solved to find $\boldsymbol{B}$ given a prescribed $\boldsymbol{u}$, or Eq. (5.37) was solved to find $\boldsymbol{u}$ given a prescribed $\boldsymbol{B}$. The drawback of this approach is that, given the nonlinear nature of the coupling, the interpretation of the results is difficult. After the 1990s, when more computational power became available, it became more common to pursue the more physically realistic, fully self-consistent approach in which Eqs. (5.36) and (5.37) were solved simultaneously.

In the same spirit, it is useful to distinguish between *kinematic* and *fully self-consistent* models of quantum turbulence. In the former case, the motion of vortex lines is determined by a prescribed normal fluid as in the original approach of Schwarz (1988), see Fig. 5.3(a), or the stability of the normal fluid is investigated given a vortex tangle of prescribed intensity (Melotte and Barenghi, 1998), as indicated by Fig. 5.3(b). In the latter case, the vortex lines and the normal fluid affect

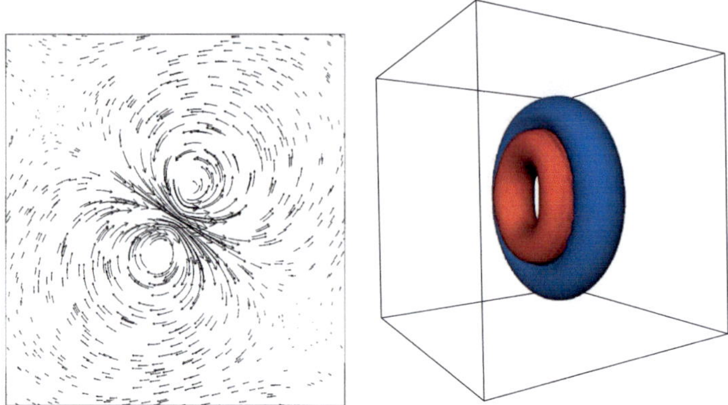

Figure 5.4 (Left) Two-dimensional dipolar structure in the normal fluid velocity field (represented by arrows) induced by the presence of a moving vortex line (unmarked, at the center of the figure) via the friction force. Reproduced with permission from Barenghi *et al.* (2000). (Right) Three-dimensional double ring structure in the normal fluid (represented by isosurfaces of normal fluid vorticity) surrounding a (hidden) vortex ring in the middle (Kivotides *et al.*, 2000) that moves to the right; the positive (blue) outer vortex ring leads the motion, the negative (red) smaller vortex ring trails behind. From Kivotides *et al.* (2000). Reprinted with permission from AAAS.

each other simultaneously via the friction force; this requires one to solve Schwarz's equation (5.30) for the vortex lines given the normal fluid velocity field $\boldsymbol{u}_{\mathrm{n}}(\boldsymbol{x}, t)$ as well as the Navier–Stokes equation for $\boldsymbol{u}_{\mathrm{n}}(\boldsymbol{x}, t)$, suitably modified by the addition of the friction force. The form of the friction is well established for Schwarz's equation, but is not in the Navier–Stokes equation. More precisely, it depends on the interpretation of the normal fluid velocity field $\boldsymbol{u}_{\mathrm{n}}$. If $\boldsymbol{u}_{\mathrm{n}}(\boldsymbol{x}, t)$ is meant to be the microscopic velocity at length scales less than the typical intervortex distance, the motion of a vortex line with respect to the normal fluid will induce a microscopic dipole-like normal fluid perturbation around the vortex line, as in the 2D numerical simulation showed in Fig. 5.4 (left). In 3D, a vortex ring will be surrounded by two normal fluid vortex rings, one slightly ahead and one slightly behind, as seen in Fig. 5.4 (right), thus creating a triple ring structure (Kivotides *et al.*, 2000).

Recently, Mastracci *et al.* (2019) argued that alongside this double normal fluid structure there should be a jet; the jet has been identified by Galantucci *et al.* (2020), as shown in Fig. 5.5. This last work is based on a new approach for the localized (therefore numerically difficult) friction coupling between a vortex line and the normal fluid based on a recent result in classical fluid dynamics about a similar coupling between active particles and a viscous fluid. Recent advances in particle tracking velocimetry are motivating studies of the perturbations that quantized vortices induce in the normal fluid (Mastracci *et al.*, 2019), as such perturbations

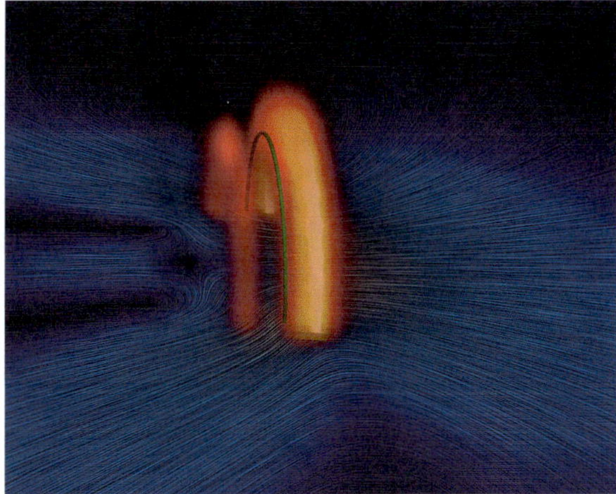

Figure 5.5 A superfluid vortex ring (thin green line) moving (from left to right) in a normal fluid initially at rest. The two larger (orange) tubes are normal fluid enstrophy structures (one slightly ahead, one slightly behind the superfluid ring) generated by the friction with the superfluid ring (corresponding to the red/blue tubes of Fig. 5.4). Only one half of each of the three rings is shown. The magnitude of the normal fluid velocity is represented on the plane to show the jet structure. From Galantucci *et al.* (2020).

may affect the motion of tracer particles when present. Galantucci *et al.* (2020) followed the fully self-consistent evolution of a superfluid vortex ring in a turbulent normal fluid. The ring, initially circular, is stretched, twisted, distorted, and turned into a small tangle by the intense turbulence; the effect is captured in Fig. 5.6, which shows the superfluid vortices (in green) and the coherent enstrophy structures of the normal fluid, or “worms” (in orange).

Alternatively, if $\boldsymbol{u}_{\rm n}(\boldsymbol{x}, t)$ represents the normal fluid velocity field coarse-grained over a region containing many vortex lines, each Cartesian component of the friction on the normal fluid will arise from the net number of vortex lines aligned in that particular direction like in the macroscopic HVBK model (see the next section). This second approach, implemented in 2D, determined the normal fluid velocity profiles in thermal counterflow (Galantucci *et al.*, 2015) and in pure superflows (Galantucci *et al.*, 2017).

5.3 Macroscopic Level

5.3.1 The HVBK Equations

The problem of self-consistency of the vortex filament model of Schwarz is solved, at a more *macroscopic* level, by the Hall–Vinen–Bekharevich–Khalatnikhov (HVBK)

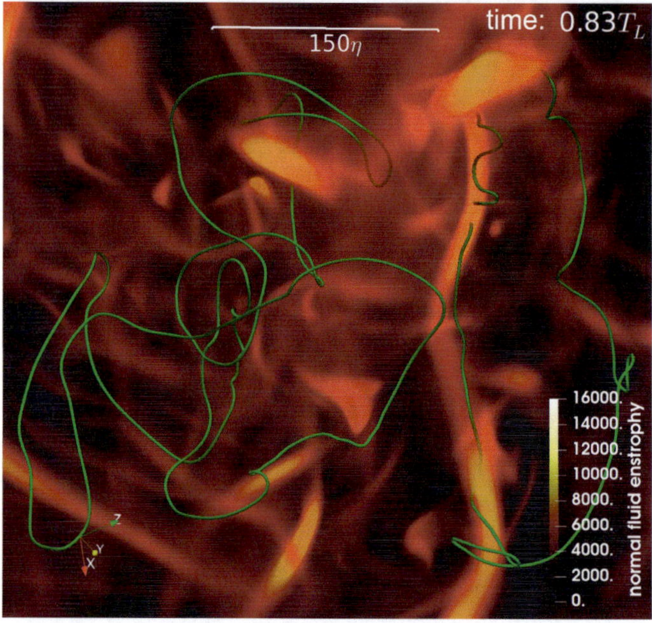

Figure 5.6 Quantized vortex lines (green) and intense coherent regions of normal fluid enstrophy (orange) computed fully self-consistently. From Galantucci *et al.* (2020).

equations (Hills and Roberts, 1977). The idea underlying these equations is the following. In the original two-fluid model of Landau and Tisza, the superfluid is irrotational ($\nabla \times \boldsymbol{u}_s = \mathbf{0}$), consistent with the more microscopic GPE model. The HVBK equations generalize the two-fluid equations to account for the presence of vortex lines by introducing the concept of a *coarse-grained superfluid vorticity field* $\boldsymbol{\omega}_s = \nabla \times \boldsymbol{u}_s$. This idea is similar to the familiar continuum approximation for a gas of molecules. The HVBK equations thus describe the motion of parcels of fluid containing a sufficiently large number of vortex lines. Unlike the molecules, the vortex lines are oriented, so we also require that in each fluid parcel there is a well-defined orientation of the parcels of vortex lines; the density of vortex lines and their orientation may, however, change from one fluid parcel to the next. Therefore, the HVBK equations have no meaning at length scales smaller than the average intervortex spacing ℓ, and require a high degree of polarization of the vortex lines. More precisely, consider a small (cubic) fluid parcel of size $\Delta > \ell$ centered around the point $\boldsymbol{r}$; the parcel is threaded by a number of vortex lines that are locally parallel; see Fig. 5.7. The x-component of $\boldsymbol{\omega}_s$ can be defined by counting the number of lines that cross the y-z cross section of the parcel in the x-direction,

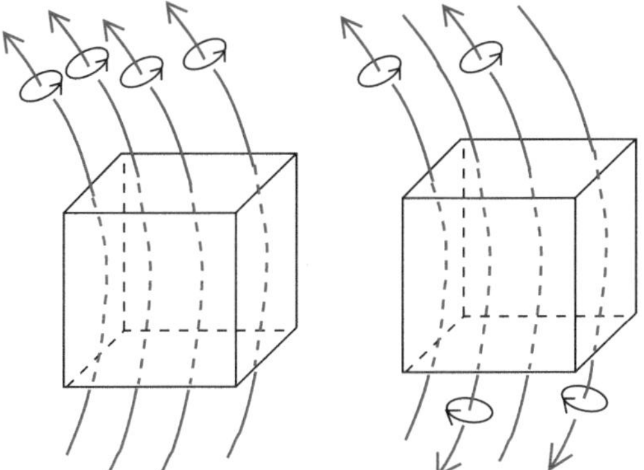

Figure 5.7 Schematic fluid parcel containing vortex lines. (Left) The lines are polarized (their vorticity points up) and one can define a coarse-grained vorticity field for the HVBK equations with a positive component in the vertical direction. (Right) The same procedure yields zero coarse-grained vorticity (equal numbers of lines point up and down), although the vortex length is not zero.

and multiplying this number by κ/Δ^2. The y- and z-components are defined in a similar way.

The vortex lattice in the helium bucket, which rotates at angular velocity Ω shown in Fig. 3.3, is a good example for explaining the equations further: Here, the number of vortex lines per unit area is $n = 2\Omega/\kappa$, and the intervortex distance is $\ell = \sqrt{\kappa/(2\Omega)}$. It is clear that the vortex lines are totally polarized. At length scales $d \ll \ell$ the superflow is irrotational, contains singularities (the vortex lines), and is quite complicated spatially; at length scales $d \gg \ell$ the coarse-graining described above yields a simpler large-scale azimuthal superflow $u_\theta = \Omega r$, which increases with the radial distance and is rotational. The coarse-grained vorticity is uniform, with the value $\boldsymbol{\omega}_\mathrm{s} = 2\Omega\hat{\mathbf{z}}$.

The above coarse-grained procedure has been applied to tangles of vortex lines calculated by the vortex filament model. The left panel of Fig. 5.8 shows vortex lines (approximately) driven into an Arnold–Beltrami–Childress (ABC) flow configuration consisting of three bundles aligned along the Cartesian directions within a periodic computational domain (Barenghi *et al.*, 1997). The coarse-graining procedure described above yields the expected 3D superfluid vorticity field $\boldsymbol{\omega}_\mathrm{s}(\boldsymbol{x}, t)$ shown in the right panel; the vortex bundles are clearly visible.

A second coarse-graining procedure (Baggaley *et al.*, 2012b) is based on Gaussian smoothing. Given a tangle of vortex lines modeled as space-curves of position $\boldsymbol{s}(\zeta, t)$ where ζ is arc length, we define the coarse-grained superfluid vorticity as

$$\boldsymbol{\omega}_\mathrm{s}(\boldsymbol{r}, t) = \kappa \sum_{i=1}^{N_p} \frac{\boldsymbol{s}_i'}{(2\pi\sigma^2)^{3/2}} e^{(-|\boldsymbol{s}_i - \boldsymbol{r}|^2/2\sigma^2)} \Delta\zeta, \tag{5.38}$$

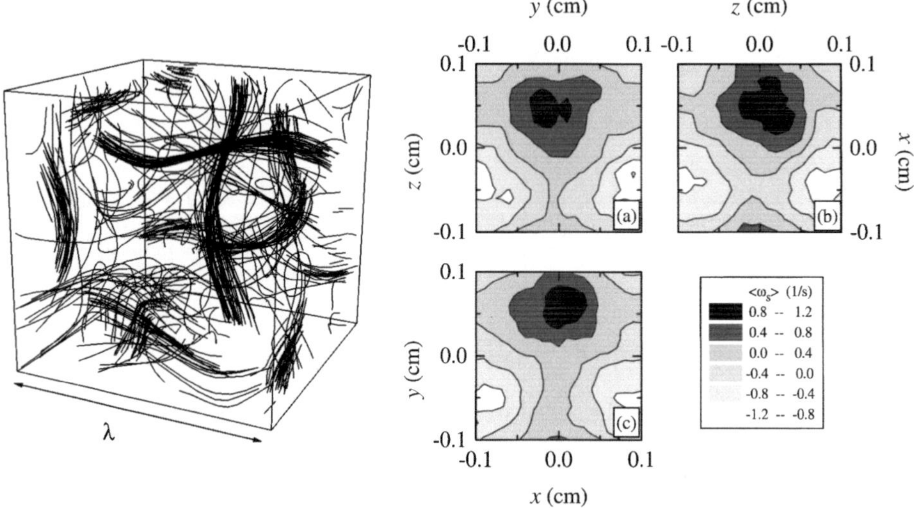

Figure 5.8 Example of the coarse-graining procedure illustrated in Fig. 5.7. (Left) Bundles of vortex lines computed by the vortex filament model. (Right) Contour plots of the resulting coarse-grained superfluid vorticity $\boldsymbol{\omega}_s$ in the x-, y- and z-direction, respectively. Reprinted from Barenghi *et al.* (1997) with the permission of AIP Publishing.

where N_p is the number of discretization points along a line, $\boldsymbol{s}_i' = d\boldsymbol{s}_i/d\zeta$ is the unit tangent vector along the vortex at the point $\boldsymbol{s}_i(\zeta, t)$, and the smoothing length σ is of the order of the inter-vortex distance ℓ. Under this smoothing procedure, a collection of randomly oriented vortex lines whose separation is of the order of ℓ yields vorticity $\boldsymbol{\omega}_s \approx \mathbf{0}$; in contrast, a bundle of parallel vortex lines yields a nonzero smooth vorticity distribution. Figure 5.9 illustrates (Baggaley *et al.*, 2012b) Gaussian smoothing of two wiggly antiparallel vortex strands (top) and two wiggly parallel vortex strands (bottom).

The HVBK equations read (Khalatnikov, 1965; Hills and Roberts, 1977):

$$\frac{\partial \boldsymbol{u}_s}{\partial t} + (\boldsymbol{u}_s \cdot \nabla)\boldsymbol{u}_s = -\frac{1}{\rho_s}\nabla p_s + \boldsymbol{G} - \frac{\rho_n}{\rho}\boldsymbol{F}, \tag{5.39}$$

$$\frac{\partial \boldsymbol{u}_n}{\partial t} + (\boldsymbol{u}_n \cdot \nabla)\boldsymbol{u}_n = -\frac{1}{\rho_n}\nabla p_n + \nu_n \nabla^2 \boldsymbol{u}_n + \frac{\rho_s}{\rho}\boldsymbol{F}, \tag{5.40}$$

where $p_s = (\rho_s/\rho)p - \rho_s ST$ and $p_n = (\rho_n/\rho)p + \rho_s ST$ are efficient pressures, and $\hat{\boldsymbol{\omega}}_s = \boldsymbol{\omega}_s/|\boldsymbol{\omega}_s|$; further,

$$\boldsymbol{G} = \nu_s \boldsymbol{\omega}_s \times (\nabla \times \hat{\boldsymbol{\omega}}_s) \tag{5.41}$$

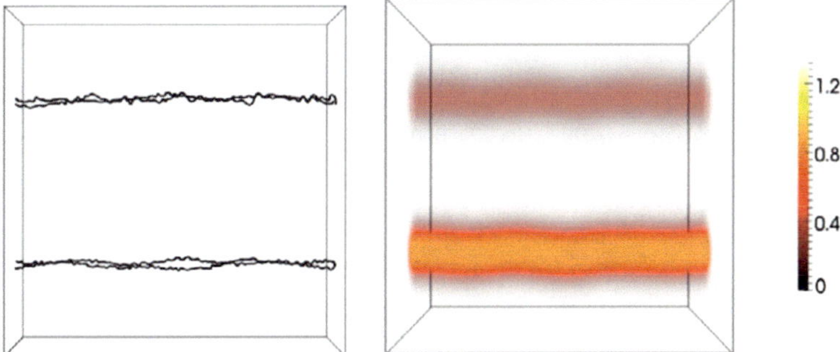

Figure 5.9 Examples of Gaussian smoothing. (Left) Two antiparallel vortex lines (top) and two parallel vortex lines (bottom). Noise has been superimposed to the centerlines of vortices to make them slightly wiggly. (Right) The magnitude of the coarse-grained superfluid vorticity. The two antiparallel vortices correspond to vanishing coarse-grained vorticity; in contrast, the coarse-grained vorticity resulting from the two parallel vortex lines is large. Reproduced with permission from Baggaley *et al.* (2012b).

is the tension force (the energy per unit length of a vortex line),

$$\nu_s = \frac{\kappa}{(4\pi)} \ln(\ell/a_0) \tag{5.42}$$

is the vortex tension parameter, and

$$\boldsymbol{F} = \frac{B}{2}\hat{\boldsymbol{\omega}}_s \times (\boldsymbol{\omega}_s \times (\boldsymbol{u}_n - \boldsymbol{u}_s - \nu_s \nabla \times \hat{\boldsymbol{\omega}}_s)) + \frac{B'}{2}\boldsymbol{\omega}_s \times (\boldsymbol{u}_n - \boldsymbol{u}_s - \nu_s \nabla \times \hat{\boldsymbol{\omega}}_s) \tag{5.43}$$

is the mutual friction force.

The HVBK equations successfully predict the oscillation of a rotating vortex lattice (Henderson & Barenghi, 2004), the Donnelly–Glaberson instability (Tsubota *et al.*, 2003a), the instability of helium Couette flow (Barenghi, 1992), and the transition to Taylor vortices (Henderson *et al.*, 1995) up to the weakly nonlinear regime: In all these flows, the assumption is valid that vortex lines within each fluid parcel are locally aligned.

5.3.2 Application and Limitations of HVBK Equation

Strictly speaking, the direct application of the HVBK equations to turbulence is not justified. Numerical simulations performed using the vortex filament model suggest a large amount of randomness of the vortex lines. Consider a fluid parcel threaded by vortex lines oriented in random direction, as shown to the right in Fig. 5.7: The net superfluid vorticity (obtained by the coarse-grained procedure) may be zero, but the vortex length (which must contribute to the friction and the dissipation) is not.

It would seem that to apply the HVBK equations to turbulence some modifications are necessary to account for the (local) polarization of the vortex lines. Waiting for further theoretical progress, the existing HVBK equations have been applied to homogeneous isotropic turbulence only in simplified form: The vortex tension has been neglected, and an approximate form of the mutual friction has been assumed (Roche *et al.*, 2009; Salort *et al.*, 2011b), resulting in

$$\frac{\partial \boldsymbol{u}_{\mathrm{s}}}{\partial t} + (\boldsymbol{u}_{\mathrm{s}} \cdot \nabla)\boldsymbol{u}_{\mathrm{s}} = -\frac{1}{\rho_{\mathrm{s}}}\nabla p_{\mathrm{s}} - \frac{\rho_{\mathrm{n}}}{\rho}\boldsymbol{F}_{\mathrm{ns}} + \boldsymbol{f}_{\mathrm{s}}^{\mathrm{ext}}, \tag{5.44}$$

$$\frac{\partial \boldsymbol{u}_{\mathrm{n}}}{\partial t} + (\boldsymbol{u}_{\mathrm{n}} \cdot \nabla)\boldsymbol{u}_{\mathrm{n}} = -\frac{1}{\rho_{\mathrm{n}}}\nabla p_{\mathrm{n}} + \nu_{\mathrm{n}}\nabla^2\boldsymbol{u}_{\mathrm{n}} + \frac{\rho_{\mathrm{s}}}{\rho}\boldsymbol{F}_{\mathrm{ns}} + \boldsymbol{f}_{\mathrm{n}}^{\mathrm{ext}}, \tag{5.45}$$

and

$$\boldsymbol{F}_{\mathrm{ns}} = -\frac{B}{2}|\boldsymbol{\omega}_{\mathrm{s}}|(\boldsymbol{u}_{\mathrm{n}} - \boldsymbol{u}_{\mathrm{s}}), \tag{5.46}$$

under the solenoidal conditions $\nabla \cdot \boldsymbol{v}_{\mathrm{s}} = \nabla \cdot \boldsymbol{u}_{\mathrm{n}} = 0$. The external forcing terms $\boldsymbol{f}_{\mathrm{s}}^{\mathrm{ext}}$ and $\boldsymbol{f}_{\mathrm{n}}^{\mathrm{ext}}$ are necessary to sustain turbulence and achieve a statistically steady state in the presence of viscous losses. Connection with the experiment is achieved by identifying the vortex line density L as the root mean square superfluid vorticity $\omega_{\mathrm{rms}} = \kappa L$, for which the friction force becomes

$$\boldsymbol{F}_{\mathrm{ns}} = -\frac{\kappa L B}{2}(\boldsymbol{u}_{\mathrm{n}} - \boldsymbol{u}_{\mathrm{s}}). \tag{5.47}$$

The strength of this model is that the superfluid and normal fluid affect each other, but results are meaningful only at length scales larger than ℓ.

5.3.3 Shell Models

Numerical calculations of quantum turbulence based on Eqs. (5.44) and (5.45) suffer the same practical limitations that constrain direct numerical solutions of the Navier–Stokes equation: the limited range of length scales achieved on existing computers. The typical computational domain is a periodic cubic box that is discretized over N^3 mesh points in the three Cartesian directions. For practical reasons, the resolution ranges from about hundred to about a few thousand points in each direction. Therefore, even the most powerful computers provide only about three orders of magnitude in k-space, which is not sufficient if one is looking for scaling laws, especially because parts of the k-space are used to include forcing and viscous dissipation.

In classical turbulence, shell models (Frisch, 1995; Biferale, 2003) provide very large k-space by giving up the spatial information altogether. They are idealized truncated models of the Navier–Stokes equation that neglect the geometry of the flow but describe the Richardson cascade and its properties (for example, the Kolmogorov energy spectrum) in a relatively simple way over a wide range of length and time scales. A popular shell model is the GOY model, named after its inventors, Gledzer (1973) and Yamada and Ohkitani (1987). It has the form

$$\left(\frac{d}{dt} + \nu k_m^2\right) u_m = G_m + f\delta_{m,m'}, \tag{5.48}$$

where u_m $(m = 1, \dots, M)$ represents the idealized (complex) Fourier component of the velocity corresponding to wavenumber k_m, $\delta_{m,m'}$ is the Kronecker delta, and f is the amplitude of the external forcing applied to some particular shell $m = m'$ in order to prevent the turbulence from decaying. The inertial term G_m is quadratically nonlinear and local in k-space, coupling u_m with its nearest neighboring shells via

$$G_m = i\left(c_m^{(1)} + \bar{u}_{m+1}\bar{u}_{m+2} + c_m^{(2)} + \bar{u}_{m-1}\bar{u}_{m+1} + c_m^{(3)} + \bar{u}_{m-1}\bar{u}_{m-2}\right), \tag{5.49}$$

where $\bar{u}_m$ is the complex conjugate of u_m, and

$$c_m^{(1)} = ak_m, \qquad c_m^{(2)} = bk_{m-1}, \qquad c_m^{(3)} = ck_{m-2}. \tag{5.50}$$

The boundary conditions are $u_m = 0$ for $m \leq 0$, $m > M$, and $c_1^{(2)} = c_1^{(3)} = c_2^{(3)} = c_{M-1}^{(1)} = c_M^{(2)} = 0$. The coefficients $a = 1$, $b = -1/2$, $c = -1/2$, and $\lambda = 2$ are chosen so that in the steady $(d/dt = 0)$, inviscid $(\nu = 0)$, and unforced $(f = 0)$ case the nonlinear interaction conserves the two quadratic invariants of the three-dimensional Euler equation: energy and helicity.

Wacks and Barenghi (2011) generalized the classical GOY model to He II and obtained

$$\left(\frac{d}{dt} + \nu_{\rm n} k_m^2\right) u_m^{(n)} = G_m^{(n)} + \frac{\rho_{\rm s}}{\rho_{\rm n}} F_m + f_{\rm n}\delta_{m,m'}^{(n)}, \tag{5.51}$$

$$\frac{d}{dt} u_m^{(s)} = G_m^{(s)} - F_m + f_{\rm s}\delta_{m,m'}^{(s)}, \tag{5.52}$$

where the inertial terms $G_m^{(n)}$ and $G_m^{(s)}$ are as in Eq. (5.49), the friction is

$$F_m = \alpha\kappa L\left(u_m^{(s)} - u_m^{(n)}\right), \tag{5.53}$$

and the vortex line density L is identified as $L = \sqrt{Q}/\kappa$. Here, Q, given by

$$Q = \sum_{m=1}^{M} \frac{1}{2}\left|u_m^{(s)}\right|^2, \tag{5.54}$$

is the enstrophy. The original *GOY model* suffers a small period-three oscillation of the energy spectrum that arises from an artificial correlation of triads of shells; this effect is eliminated by a slight modification of the nonlinear term, Eq. (5.49), resulting in the so-called *Sabra model* (L'vov *et al.*, 1998). Sabra models were implemented by L'vov and collaborators first for ^{3}He-B (Boue *et al.*, 2012), and later for He II (Boue *et al.*, 2013). Note the lack of a viscous term in the superfluid equation (5.52): At relatively high temperatures, the superfluid energy that cascades to large k does not pile up at the largest wavenumbers as friction transfers it to the normal fluid where it is dissipated. At relatively low temperature and in a pure superfluid, however, some artificial damping of the form $-\nu_s k_m^2 u_m^{(s)}$ must be introduced in Eq. (5.52), where the small "kinematic viscosity of turbulent superfluid" $\nu_s \ll \nu_n$ models acoustic emission (Boue *et al.*, 2013). Another strategy (Boue *et al.*, 2012) is the introduction of a superfluid hyperviscosity (Boue *et al.*, 2013) of the form $\nu_s k_m^4 u_m^{(s)}$ to suppress energy strongly at large k. In any case, as long as $\nu_s \ll \nu_n$, the artificial superfluid damping does not affect the energy spectrum at length scales larger than ℓ, where the hydrodynamical approach is valid.

5.3.4 Other Models

Other macroscopic models have been proposed to describe the properties of a turbulent superfluid. One approach is based on an alternative to the two-fluid model: the *one-fluid extended model* of extended thermodynamics (Mongiovì *et al.*, 2018). In this model, which emphasizes nonequilibrium thermodynamics and heat transport, the fundamental variables are directly measurable: the total mass density ρ, the barycentric velocity $\boldsymbol{u} = (\rho_n/\rho)\boldsymbol{u}_n + (\rho_s/\rho)\boldsymbol{u}_s$, and the heat flux vector $\boldsymbol{q}$. The model can be adapted (Sciacca *et al.*, 2014) to include a turbulent tangle of vortex lines described in terms of a vortex line density L obeying a modified Vinen equation – defined in Eq. (6.5). The resulting solution describes the crossover between two regimes: laminar heat flux proportional to the temperature gradient (as predicted by the two-fluid model) in which the thermal resistance is limited by the viscosity of the normal component, and turbulent heat flux proportional to the cube root of the temperature gradient in which a tangle of vortices increases the thermal resistance.

Lipniacki (2006) developed another macroscopic model of turbulent helium II based on the two-fluid description. In this model, the polarized part of the vortex lines contributes to the curl of the superfluid velocity $\boldsymbol{u}_s$, and the random part is described by a modified Vinen equation.

5.4 The Topology of Quantum Turbulence

When studying a physical system, it is instructive to distinguish three aspects: the geometry, the dynamics, and the topology. For many systems we understand how geometry determines the dynamics and vice versa; in general, the aspect that is less understood is the topology, but we also know that in some cases the topology places constraints on the dynamics: As a somewhat trivial example, a single straight vortex between two infinite plates in a superfluid otherwise at rest cannot decay. The apparent similarity between the turbulent flows shown in Figs. 3.15 and 3.16 suggests that vorticity in the shape of tubes or filaments is perhaps the "skeleton" of turbulence, thus bearing its structure. In the case of pure quantum turbulence (He II or ^{3}He-B at very low temperature) the skeleton counts for a lot: The filaments are all the same (in the sense that the circulation κ and the vortex core radius a_0 are the same), and what is between the filaments is very simple (inviscid incompressible irrotational flow, solution of the Laplace equation). In Chapters 10 and 11 we shall see that, at least for Kolmogorov-type turbulence, the apparent similarity between quantum turbulence and classical turbulence is deep, in the sense that, at length scales larger than the vortex separation, the presence of enough quanta of circulation guarantees the same distribution of kinetic energy and velocity statistics. It is interesting, therefore, to inquire into the topology of the vortex tangle, what it tells us about the physical properties of quantum turbulence, and whether this topology is similar to that of classical turbulence. Images of tangles of quantized vortices computed either via the VFM or the Gross–Pitaevskii equation suggest that turbulence may contain knots and links. Since this area is still rather unexplored and the best tools to pursue it are unclear, we describe in this section the techniques that have been proposed and the few results that have been achieved.

The possibility that slender vortex structures may be knotted or linked has been considered since the time of Lord Kelvin, who proposed the vortex theory of atoms (Thomson, 1867). Topologically, the simplest vortex structure is the vortex ring, also called the *unknot*. The first nontrivial knot is the trefoil. Indeed, there have been various studies of the motion and stability of trefoil vortex knots (Kida, 1981; Ricca *et al.*, 1999; Proment *et al.*, 2012), but no success was attained for a long time in creating vortex knots in the laboratory in a controlled way. Moreover, despite the fact that direct numerical simulations of the Navier–Stokes equation shows highly complex, possibly knotted vorticity and streamlines, there had been no attempt to quantify this topology.

This situation changed in 2013 when Kleckner and Irvine (2013) generated trefoil vortex knots in water in a controlled way using a specially designed airfoil; see Fig. 5.10. This experiment stimulated further work in the topological aspects of fluid mechanics, such as the definition and the meaning of helicity in superfluids (Hänninen *et al.*, 2016; diLeoni *et al.*, 2017; Salman, 2018; Zuccher and Ricca, 2018; Galantucci *et al.*, 2021). In the classical context, the helicity is defined as (Moffatt, 1969)

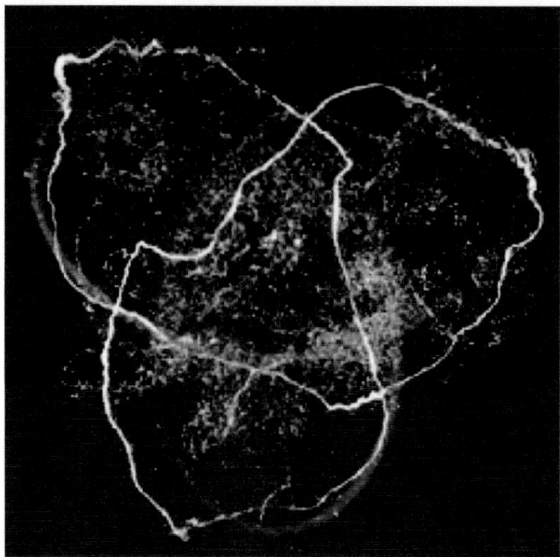

Figure 5.10 Trefoil vortex knot generated in water. Visualization via buoyant gas bubbles that concentrated in regions of high vorticity. The three-dimensional nature of the trefoil vortex knot was verified by high-speed scanning tomography. Reproduced with permission from Kleckner and Irvine (2013).

$$H = \int_V \boldsymbol{\omega} \cdot \boldsymbol{u} \, dV, \tag{5.55}$$

which, for slender vortex tubes, can be decomposed in writhe, link, and twist (Moffatt and Ricca, 1992). At the mesoscopic level of description of the turbulence provided by the VFM method, Galantucci *et al.* (2021) showed that helicity can be interpreted as a measure of the nonlocal contributions to the velocity field: $H = \kappa L \langle \boldsymbol{s}' \cdot \boldsymbol{u}_{\mathrm{s}}^{\mathrm{non}} \rangle$ where $\boldsymbol{u}_{\mathrm{s}}^{\mathrm{non}}$ is defined in Eq. (5.35). The vortex lines displayed in Fig. 3.15 are color coded using the magnitude of H.

Kleckner *et al.* (2016) have wondered if the decay of knotted turbulent structures occurs following preferred topological pathways, and also provided some evidence that this is the case, at least for the moderately complex vortex knots that they explored. A trivial example is the decay of a trefoil knot into a Hopf link and then into two free rings.

We emphasize that a quantitative characterization of the topology of turbulent flows is much needed. In classical turbulence, the identification of instantaneous vortex lines from the numerical solution of the incompressible Navier–Stokes equation has a potential difficulty: Starting from the initial position, one can numerically trace a vortex line, but numerical noise across regions of large and small vorticity may prevent the traced vortex line from joining with itself to form a closed loop required by $\nabla \cdot \boldsymbol{\omega} = 0$. Also, the fact that the classical vorticity is a continuous field means that there are infinite vortex lines to consider and a statistical approach is required.

Because of the discrete nature of quantized vorticity, quantum turbulence is the ideal context for topological fluid dynamics. However, to avoid ambiguities in the definition of knots and links, all vortex lines must be closed loops; thus the usual periodic boundary conditions used, for example, in Fig. 3.15, are not suitable for topological studies. Moreover, to interpret the results and relate the observed topology to the vortex line density, or other average properties of the turbulence, it is convenient to maintain it in a statistically steady state. Thus motivated, Cooper *et al.* (2019) performed numerical simulations of vortex tangles in an infinite domain (to make sure that all vortex lines are closed). In order to achieve a statistically steady state, they seeded a few random vortex lines at nonzero temperature using a prescribed, time-dependent, solenoidal normal fluid velocity field $\boldsymbol{u}_{\mathrm{n}}(\boldsymbol{r}, t)$, which is concentrated in a spherical region of radius D (a flow combining toroidal and poloidal motions at the scale D (Dudley and James, 1989)). After an initial transient, they found that the vortex tangle settles down to a steady regime in which vortex length and energy fluctuate around average values independent of the initial condition as the driving and dissipation balance out. A snapshot of the vortex tangle in this regime is shown in Fig. 5.11. Small vortex loops that move away from this spherical region shrink and vanish due to friction in the quiescent normal fluid, unlike that seen in Section 6.6, where such small vortex loops fly away ballistically without shrinking at $T = 0$. Essentially, Cooper *et al.* (2019) modeled situations such as the experiment of Schwarz and Smith (1981) in which quantum turbulence was generated by focusing ultrasound at the center of a large cell, far away from its boundaries.

Cooper *et al.* (2019) verified that the energy spectrum in this regime of tangles obeys the classical Kolmogorov $k^{-5/3}$ scaling (where k is the wavenumber) in the inertial range $2\pi/D < k < 2\pi/\ell$ (for details on the spectra, see Chapter 10). To quantify the complexity of the vortex tangle, they computed the Alexander polynomial $\Delta(\tau) = c_0 + c_1\tau + \cdots + c_\nu\tau^n$ of each vortex loop. For example, a ring has Alexander polynomial of degree $n = 0$, a trefoil has degree $n = 2$, etc; the higher n, the more complex the knot type. It appears that, although vortex reconnections keep breaking vortex loops, loops are also continually created by other reconnections. Surprisingly, at each time, the vortex tangle always contains a few long, highly knotted vortex loops with high degree of Alexander polynomial (from $n \approx 10$ to about 100). Indeed, the probability of a vortex knot with degree n appears to scale as $P(n) \sim n^{-1.5}$, independent of the tangle's total vortex length. Moreover, the probability that a vortex loop is knotted increases with the length, as for macromolecules. In a similar calculation, the same authors showed that a

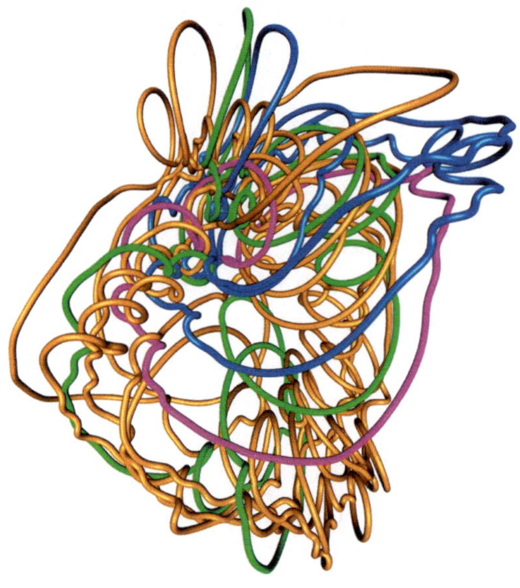

Figure 5.11 Snapshot of numerically computed turbulent tangle of vortex lines. For clarity, a tubular region of arbitrary radius has been drawn around the axis of each vortex line; the colors distinguish the different vortex loops.

vortex tangle contains a large number of links (Mesgarnezhad *et al.*, 2018). Since vortex reconnections (which create knots and links) have energy costs (Leadbeater *et al.*, 2001), these results suggest that a typical superfluid vortex tangle packs some energy into its rich topology. A natural question is whether these findings apply to ordinary Navier–Stokes turbulence.

5.5 Summary

Contrary to classical turbulence, which is studied theoretically on the well-established grounds of the Navier–Stokes equation with no-slip boundary conditions, no single theoretical model adequately describes all aspects of quantum turbulence. What is available is a toolkit of microscopic, mesoscopic, and macroscopic models describing phenomena at different length scales: the Gross–Pitaevski equation (which captures phenomena at the microscopic scale of the vortex core radius such as vortex nucleation and vortex reconnections), the mesoscopic vortex filament model of Schwarz (which performs reconnections algorithmically), and the macroscopic Hall–Vinen–Bekharevic–Khalatnikhov equations (which assume that the local polarization of the vortex lines is total). Additionally, as in classical turbulence, shell models have been developed that capture a wide range of k-space but give up on the spatial resolution.

The choice of which model to use depends therefore on the nature of the problem. In the following chapters, we will therefore employ various theoretical models that describe the problem at hand in the most adequate convenient way.

Finally, we have drawn attention to important but elementary aspects of the topology of vortex lines in quantum turbulence, where it appears to be more directly applicable than in Navier–Stokes turbulence.

6

Transitions and Steady-State Turbulence

In classical fluids, the transition to turbulence generally occurs upon exceeding some critical parameter, such as the Reynolds number. This process includes the growth of perturbations of an underlying basic laminar flow (for example, mechanical or thermal noise) and often represents a complicated process; in some cases the perturbations grow exponentially from infinitesimal values, in others they destabilize the basic flow only by exceeding some threshold amplitude. The ultimate connection of the stability of the initial laminar state to turbulence still remains elusive. Transition to quantum turbulence, because of the two-fluid nature of quantum fluids, is potentially even more complex.

Does quantum turbulence always contain vortex lines? Strictly speaking, if we define quantum turbulence as the turbulence occurring in quantum fluids displaying two-fluid behavior, vortex lines are, in principle, not a necessary ingredient of turbulence in quantum fluids: We can imagine a two-fluid flow consisting of turbulent normal flow and potential superflow. Indeed, consider the hypothetical case of a macroscopic sample of He II free of vortex lines (i.e., without mutual friction). In isothermal conditions and in the absence of friction the normal and superfluid components move independently, and the transition criteria ought to be applied to them separately. We know from Chapter 2 that in this hypothetical case vortex lines must be nucleated intrinsically by a process requiring critical velocities of the order of 10 m/s or larger. In He II, at typical experimental temperatures not far below T_λ, the kinematic viscosity of the normal fluid is approximately $\nu_n \approx \kappa/6$ (L'vov *et al.*, 2014; Donnelly and Barenghi, 1998). For typical laboratory conditions (pipe diameter $\approx$ 1 cm, normal fluid velocity of the order of 1 cm/s), the Reynolds number of the normal fluid, approximately 6000, is large enough to expect turbulence: We should then have a flow in which the normal fluid is turbulent but the superfluid is not. In practice, however, this situation is unlikely to occur.[1] When preparing

[1] Investigation of this type of quantum flows is still in its infancy, but later in this chapter we shall show that experimental evidence for such flows does exist (Schmoranzer *et al.*, 2019).

any macroscopic sample of He II, it is difficult to avoid the presence of remanent vortex lines that are left over by the process of cooling the helium sample through the phase transition. By favoring the extrinsic nucleation of further vortex lines, remanent vortex lines tend to induce the transition to turbulence at small velocities of the order of a few cm/s.

In this chapter we discuss various cases of transition to quantum turbulence, and describe the properties of turbulent flows held in a statistically steady state by mechanical, thermal, and other means outlined in Chapter 4. We start with conceptually simple cases of steady flows past bluff bodies of various shapes, before moving to pipe and channel flows driven mechanically and thermally, then to inhomogeneous turbulence, and finally to turbulence generated by injecting vortex rings.

6.1 Steady Flow Past an Obstacle

6.1.1 Classical Fluid Flows

The character of the incompressible steady flow of a classical fluid past a bluff obstacle (such as a sphere or a cylinder) is determined by a single dimensionless parameter, the Reynolds number, defined as $Re = UR/\nu$, where U is the characteristic speed of flow past the obstacle, R is the characteristic dimension of the obstacle, and ν is the kinematic viscosity of the fluid. The nature of the flow at increasing Re is shown in Fig. 6.1, and can be understood physically as follows. There are two forces acting on the fluid immediately behind the cylinder in the laminar regime: a viscous force, in the direction towards the stagnation point at the rear of the cylinder, and a Bernoulli pressure force acting in the opposite direction. The viscous force per unit volume has a magnitude of order $\eta U/R^2$; the Bernoulli force per unit volume has a magnitude of order $\rho U^2/R$ (η being the dynamic viscosity of the fluid and ρ its density). As long as the Bernoulli force is much less than the viscous force, its effect is negligible, but as soon as it exceeds the viscous force, the flow near the rear surface of the cylinder tends to be reversed with the consequent formation of an eddy. This change in regime occurs when the two forces become roughly equal, corresponding to $Re \simeq 2$, which agrees fairly well with experiments.

At high values of Re the flow separates into two distinct regions: an inner region at the rear of the cylinder where the turbulence is fully developed, and an outer region, where (except for a thin layer close to the cylinder surface) the flow behaves approximately as an ideal fluid (in spite of viscosity). The drag force is

$$F = \frac{1}{2} C_{\mathrm{D}} \rho A U^2, \tag{6.1}$$

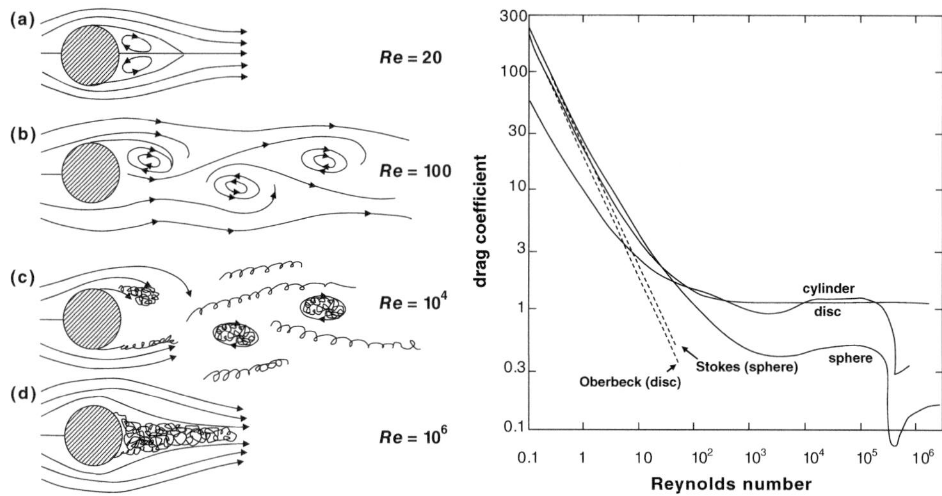

Figure 6.1 (Left) Schematic diagrams depicting the evolution of the flow past a circular cylinder as the Reynolds number is increased. For $Re \ll 1$, the flow is laminar; at $Re \sim 10$ a pair of eddies appears behind the cylinder; at $Re \sim 100$ these eddies separate from the cylinder and a so-called *Kármán vortex street* is formed consisting of eddies of alternating directions of rotation that are shed away. Further increase in *Re* leads to a region of fully developed turbulence in the wake behind the cylinder. (Right) Drag coefficient C_D vs. *Re* for steady flow past obstacles of various shapes. Stokes' law for a sphere and the Oberbeck law for a disc $\left(C_D \sim Re^{-1}\right)$ are represented by the broken lines. From Feynman (2011), copyright © 2011. Reprinted by permission of Basic Books, an imprint of Hachette Book Group, Inc.

where A is the projected area of the obstacle on a plane normal to the incoming flow. For laminar viscous flow for small *Re*, the dimensionless drag coefficient C_D is approximately proportional to the inverse of the velocity U, as given by the well-known formula for Stokes' law around a sphere of radius R, $F = 6\pi\eta RU$ in the right panel of Fig. 6.1.

For high values of *Re*, the character of the turbulent drag is such that C_D is approximately constant, its exact value depending on the position of the surface that separates the inner and outer flow regions. In the case of a flat disc placed normal to the flow, the drag coefficient is approximately $C_D \approx 1$. In the case of a cylinder or a sphere, the behavior is more complex. Experiments reveal the existence first of the so-called *Kármán vortex street*, a time-dependent flow pattern consisting of eddies of alternating directions of rotation that are shed away. Additionally, at relatively high *Re*, the so-called *drag crisis* occurs: the sudden fall of the value of C_D at $Re \approx 3 \times 10^5$ caused by the sudden shift of the surface of separation towards the rear of the obstacle; see Fig. 6.1.

6.1.2 General Remarks on He II Flows

The classical result known as *d'Alembert's paradox* states that, in incompressible inviscid potential flow, the drag force is zero on a body moving with constant velocity relative to the fluid.[2] Thus, D'Alembert's paradox ought to be applicable to steady flows past bluff bodies in both He II and ^{3}He-B in the zero-temperature limit (characterized by the absence of normal fluid) under the assumption that no vortex lines are present. Such experimental conditions are, however, difficult to achieve for objects of macroscopic size for reasons already discussed, and there is no unequivocal experimental evidence of this important result, except for helium ions of atomic size, which move without measurable drag up to a certain critical velocity (Allum *et al.*, 1976). Experimentally, in the zero-temperature limit, it is much easier to study time-dependent flow created by mechanical oscillators than steady flows.

For this reason, quantum steady flows of He II past bluff bodies have been studied at finite temperature, where He II displays two-fluid behavior, and in experimental arrangements in which a bluff body is at rest in the laboratory reference frame, placed inside an essentially steady (usually turbulent) pipe flow of He II. Note that there are no comparable studies in ^{3}He-B.

We start with the important experiment of Smith *et al.* (1999) designed to detect the drag crisis. The pressure distribution on the surface of a sphere was measured in flowing He I and He II as a function of Reynolds number. The drag coefficient was extracted by integrating the pressure distribution using some assumptions about the symmetry of the flow field. The deduced drag coefficients are plotted against Reynolds number for both He I and He II in Fig. 6.2 and compared to classical data for both smooth and non-smooth spheres. The results in He II suggest that the drag crisis occurs at Reynolds number $Re \approx 2 \times 10^5$, in fair agreement with classical data for rough surface of the sphere. This result suggests that, at high Reynolds numbers, the macroscopic, forced He II coflow behaves as a single component fluid possessing an effective viscosity.

On the other hand, the behavior of He II counterflow past a bluff body seems different from classical flows. Indeed, Zhang and Van Sciver (2005b) visualized thermal counterflow past a cylinder using a particle image velocimetry (PIV) technique, and observed the existence of large turbulent eddies both behind (downstream) and in front (upstream) of the cylinder (in the direction of heat flow), as shown in Fig. 6.3. It is tempting to interpret this surprising flow pattern as the superfluid and the normal fluid undergoing some type of flow separation as they pass over the cylinder. The actual situation, however, seems more complex. When

[2] This result is valid only in the case of unbounded fluid with no free surface, as the propagating surface waves generated by motion of the submerged body carry its energy away.

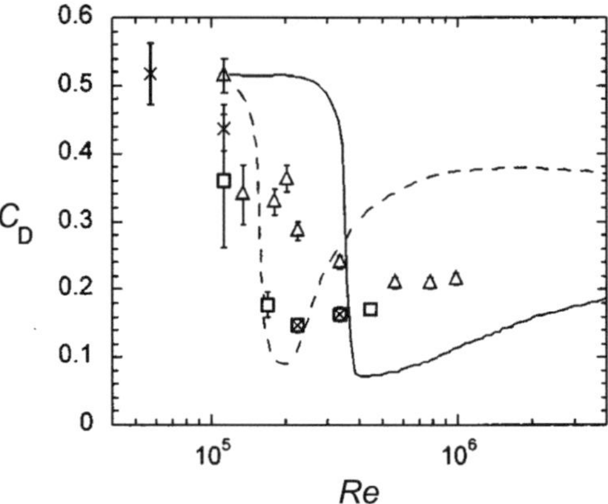

Figure 6.2 Drag coefficient C_{D} vs. Reynolds number *Re*. Open squares and crosses correspond to temperatures 2.54 K and 4.2 K, respectively. The triangles were recorded in He II at 1.8 K. The solid line represents commonly accepted drag crisis data, and the dashed line shows the effect of surface roughness (Achenbach, 1974). Reprinted from Smith *et al.* (1999) with the permission of AIP Publishing.

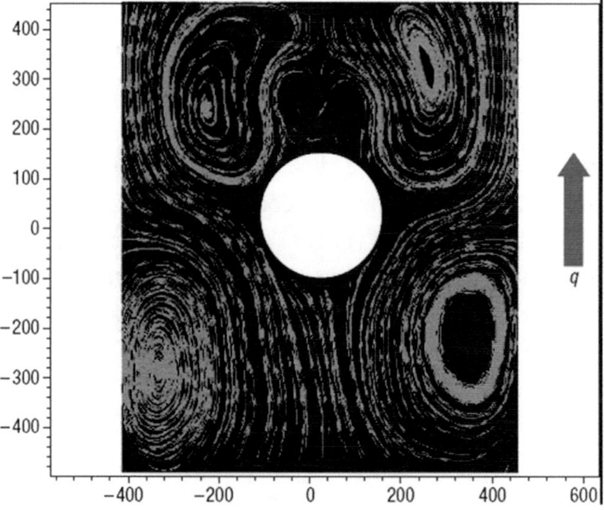

Figure 6.3 Computed streamlines of particle motion at heat flux $\dot{Q} = 1.12\ \mathrm{W/cm^2}$ and $T = 2.03$ K corresponding to $Re = 21{,}000$. Reproduced with permission from Zhang and Van Sciver (2005b).

the experiment was repeated by Chagovets and Van Sciver (2011) using a modified PTV technique based on solid hydrogen tracer particles, the authors found that, for a relatively small velocity of the normal fluid, the flow was similar to classical laminar flow, as in Fig. 6.1. Increasing the flow velocity led to the appearance of a pair of eddies downstream of the cylinder; the surprising eddies at the front of the cylinder appeared only at the highest attainable velocities. Additionally, Chagovets and Van Sciver (2011) claimed the existence of a state in which more than one pair of eddies coexisted downstream of the cylinder.

Sergeev and Barenghi (2009a) have remarked that, in the case of 2D potential flow past a cylinder, there exist four positions in front and behind the cylinder (corresponding to the four eddies shown in Fig. 6.3) where vortex–antivortex pairs would be stable for long enough time; these positions may therefore act as attractors of vortex lines. Clearly, this interesting experiment deserves further study as the flow is not yet fully understood.

6.2 Pipe and Channel Flows

6.2.1 Recapitulation of Some Results for Classical Flows

The flows of viscous fluids through pipes and channels of various cross sections have been studied since the seminal work of Reynolds (1883). Despite the effort of many investigators and the knowledge accumulated from a plethora of studies (experimental, theoretical, and numerical), the classical pipe flow problem cannot be considered solved: There are a number of outstanding open questions on the transition to turbulence, the boundary layer structure, the behavior of fluctuations, and the role of surface roughness on these properties.

We illustrate the complexity of transition by briefly considering only the pipe flow of the classical Navier–Stokes fluid. We first note that the parabolic laminar velocity distribution is stable to all infinitesimal axisymmetric perturbations, so the transition to turbulence must arise by some other means. Clearly, finite amplitude instability is a likely scenario, with the transition Reynolds number decreasing as the perturbation increases in amplitude. However, more recent work has suggested that nonnormality of the linear operator could be a likely explanation because nonnormality implies the potential for transient growth. The linear transient amplification of disturbances could trigger nonlinearities that would prevent the eventual viscous decay of those disturbances. This is primarily linked to the three-dimensional character of the transition problem.

In the alternative scenario, an inquiry into the transition to turbulence by ignoring the entry length is likely to be an error. The parabolic pipe flow develops as a result of the radial growth of the boundary layer until there is a merger towards the pipe axis. Unlike the parabolic flow, the boundary layer region is not stable to all infinitesimal fluctuations, and it is possible for perturbations in the boundary to grow to produce transition. What is clear is that all perturbations introduced into the boundary layer region will die out eventually if the Reynolds number *Re*, defined above, falls below about 1500.

Finally, if we treat the surface of the pipe to be rough, as one must in the case of helium flows, linear stability analysis leads to stable vortex formation at low enough Reynolds numbers, but these vortices can become dislodged as the Reynolds numbers exceeds a certain threshold value, thus triggering the sequence of events needed for transition.

In practice, experiments dedicated to the study of turbulent pipe flows shorten the boundary layer development length by artificially thickening the boundary layer by means of deliberately planted roughnesses. In any case, no matter how the turbulent pipe flow is generated, there is a finite development length (of the order of several tens of pipe diameters) that one has to allow for all kinds of transients to die out, so the flow can be considered a standard, fully developed pipe flow. There are many rules of thumb that a serious experimentalist follows in generating such turbulent flows, but comparable care and concern has not always been evinced in helium flows because the emphasis has been different. Nevertheless, there is much to be gained by studying the available literature on the subject, as we shall do now.

6.2.2 Varieties of He II Flows

In liquid helium, the pioneering pipe flow experiment of Allen and Misener (1938) belongs to the early history of superfluidity of ^{4}He. Since then, superfluid pipe/channel flows have continued to be extensively studied, addressing issues of stability, transition to turbulence, boundary conditions, and properties of fully developed turbulence. Using mechanical and/or thermal drives, a variety of two-fluid channel flows can be generated, as illustrated in Fig. 6.4.

Mechanical forcing by the action of compressing bellows almost always results in *coflows*, the closest analogues to classical viscous channel flows, in which normal fluid and superfluid move (on average) with the same mean velocity in the same direction. Quite generally, the two components of He II can also be made to flow (on average) with different velocities, which is the situation of *counterflow*. A special case of counterflow is *thermal counterflow*, in which one end of the channel is heated, the opposite end is open to the helium bath, and there is no net mass flow: In this case, the ratio of normal fluid and superfluid velocities is not arbitrary but is set by the temperature.[3]

Another special case is *pure superflow*, when a net flow of the superfluid component occurs in a channel, while the normal component remains (on average) stationary. Pure superflows can be generated both mechanically (e.g., by compressing

[3] Generally, the temperature and flow velocities are not homogeneous and change along the channel for reasons that will be discussed later.

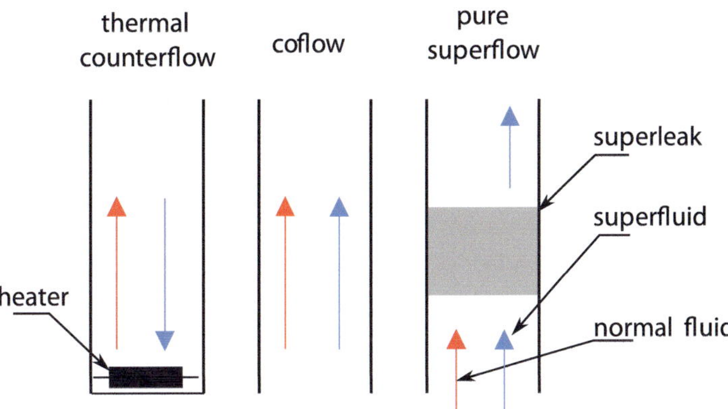

Figure 6.4 Illustration of the three basic types of quantum two fluid flows of He II. Coflow can be either a pressure driven (e.g., forced by a compressible bellows) or result from mechanical stirring (e.g., a grid towed through stationary helium). Pure superflow could also be pressure driven by bellows (not shown), with the flow of the normal component filtered out using a superleak, in a form of a porous plug with sub-μm channels made by sintering silver powder to about 50% of bulk silver density that allows the through-flow of only the superfluid component.

bellows, as for coflow) and thermally (as for counterflow), using suitable *superleaks*. A superleak is a filter located at one end of the channel, with sub-micrometer-sized holes that are permeable only to the inviscid superfluid component, as sketched in the rightmost panel of Fig. 6.4. Note that if an additional heater is placed above the superleak, converting a part of the superfluid through-flow into normal fluid, any flow ratio of the two components can be achieved (Baehr *et al.*, 1984).

In summary, in flow channels with or without superleaks, by varying the temperature and by mechanical and thermal forcing, either singly or in some combination, a rich variety of two-fluid flows can be arranged and experimentally investigated. It should be noted that, due to technical difficulties, the conceptually simpler single-component pipe/channel flow of He II in the zero-temperature limit is yet to be experimentally investigated.

We now describe, keeping a historical perspective, the most important observations that have led to the current understanding of the underlying physical processes relevant to two-fluid channel flows of He II.

6.3 Thermal Counterflow of He II

We shall describe counterflow turbulence in some detail in this section and relegate discussion of the coflow case to Section 6.5. As already explained in Section 2.3.2, thermal counterflow can easily be set up by applying a voltage to a resistor (heater) located at the closed end of a channel that is open to a helium bath at the other end, as illustrated in Fig. 2.9. We also remarked that thermal counterflow is specific to two-fluid superfluid hydrodynamics (Tilley and Tilley, 1986; Landau and Lifshitz, 1987), and its steady state has no obvious analogue in flows of ordinary viscous fluid. For this reason it has often been stated in the literature that counterflow turbulence and any associated convective heat transfer in He II have nothing to do with classical turbulence in viscous fluids. Some experiments, however, suggest a different possibility, discussed briefly in Section 6.3.1.

6.3.1 Basic Principles

As noted by Skrbek *et al.* (2003), if one attempts to find a possible connection between thermal counterflow in He II and classical turbulence, the likely candidate would be turbulent thermal convection. Visualization of thermal counterflow reveals the existence of coherent structures similar to those known to exist in classical convective turbulent flows (La Mantia and Skrbek, 2014b; Rusaouen *et al.*, 2017b). Until recently, it was believed that the mean temperature distributions in turbulent convection (He I) and in turbulent counterflow (He II) were very different. In turbulent thermal convection at high Rayleigh numbers, it is known that the temperature in the bulk region between the hot plate and the cold plate is approximately constant – the temperature gradients being concentrated almost entirely near the horizontal plates (Chillà and Schumacher, 2012); in thermal counterflow, it was thought that the temperature gradient is approximately constant in the direction of the main flow (Henberger and Tough, 1982). This picture has changed due to more recent visualization experiments of counterflow (Hrubcová *et al.*, 2018; Švančara *et al.*, 2018), which have revealed a region near the heat source where the vortex line density is strongly enhanced. This boundary layer has been directly probed by a miniature temperature sensor measuring the temperature profile in the proximity of a resistive heater by Varga and Skrbek (2019). Using three different heater geometries, these authors directly observed the existence of a thermal boundary layer, further confirmed by complementary numerical simulations. However, they also found that the observed temperature gradient enhancement seems to be partly caused by the inhomogeneity of the areal distribution of input power. Complete solution of the boundary layer problem in thermal counterflow must await further dedicated experiments, but the observations just cited are already very intriguing.

Counterflow turbulence in He II has a long history since the discovery of superfluidity, beginning with heat transport experiments through capillaries and channels of various sizes. Already in 1949, it was observed by Gorter and Mellink (1949) that, in wide channels with flow velocities that are not very low, there is a dissipative process corresponding to a force (per unit volume) between normal fluid and superfluid of the form (see Eq. (3.15))

$$F_{\mathrm{ns}} = A_{\mathrm{GM}}\rho_{\mathrm{s}}\rho_{\mathrm{n}}(u_{\mathrm{s}} - u_{\mathrm{n}})^3, \tag{6.2}$$

where the Gorter–Mellink constant A_{GM} is temperature dependent. These early experiments revealed that, above a certain critical heat flux, the heat conductivity of He II is strongly nonlinear, and the temperature gradient along the counterflow channel depends on the supplied heat flux as

$$\nabla T \propto \dot{Q}^m, \tag{6.3}$$

where the exponent m varies between approximately 3 and 3.5, depending on the temperature (Van Sciver, 1986). The more recent study of Sato *et al.* (2005) quotes $m = 3.4$ for a wide range of pressure up to 1.5 MPa. The origin of the dissipative process causing this temperature gradient was not known during the early studies but, from the ideas of Onsager (1949) and Feynman (1955), it soon became clear that a mutual friction between the superfluid component and the normal fluid component arises from the interaction between the phonons and rotons (constituents of the normal fluid) and vortex lines in the superfluid.

Counterflow turbulence in a wide channel was first investigated in detail in an outstanding series of papers by Vinen (1957). Vinen also introduced a phenomenological model of counterflow turbulence based on the concept of a random vortex tangle characterized by a single variable, the (approximately homogeneous) vortex line density L, defined as the total length of vortex lines per unit volume: see Eqs. (3.22) and (3.23).

In a steady-state thermal counterflow, L was found to scale with the counterflow velocity u_{ns} as

$$L = \gamma^2(u_{\mathrm{ns}} - u_{\mathrm{c}})^2, \tag{6.4}$$

where u_{c} is the critical velocity typically of the order of 1 mm/s and γ is a temperature-dependent parameter. The temperature gradient in thermal counterflow stems from the balance between the friction that the vortices exert on the normal fluid (and vice versa) through the effect of mutual friction and the mechanocaloric force (Landau and Lifshitz, 1987), which is linear in ∇T and responsible for the fountain effect. In light of this result, the Gorter–Mellink expression can be rewritten as $\nabla T \propto L u_{\mathrm{ns}}$ for a steady-state homogeneous counterflow. This implies that any thermal boundary layer (a region of enhanced temperature gradient) must be accompanied locally by enhanced vortex line density.

Vinen argued that L obeys the following equation (now called the *Vinen equation*):

$$\frac{\partial L}{\partial t} = \frac{\rho_{\mathrm{n}} B}{2\rho}\chi_1 u_{\mathrm{ns}} L^{3/2} - \frac{\kappa}{2\pi}\chi_2 L^2 + g(u_{\mathrm{ns}}). \tag{6.5}$$

Here, χ_1 and χ_2 are undetermined dimensionless constants and B is the (dimensionless) mutual friction parameter, tabulated by Donnelly and Barenghi (1998). The arguments leading to Eq. (6.5) are dimensional. The two terms on the right-hand side describe the *production* and the *decay* of turbulence, respectively. The unspecified function $g = g(u_{\rm ns})$ (which, for dimensional reasons, must depend also on κ) was included to account for the observation that no vortex lines are observed if $u_{\rm ns} = |u_{\rm n} - u_{\rm s}| < v_{\rm c1}$ where $v_{\rm c1}$ is a small critical velocity u_{c1} (corresponding to the critical heat flux $\dot{q}_{\rm c1}$).

Vinen's approach accounts fairly well for most of the phenomena observed in steady state counterflow turbulence in relatively wide channels (of order 100 μm–1 cm) as reviewed by Tough (1982), whose group contributed many experimental investigations. In particular, assuming that $g(u_{\rm ns})$ is negligible, the steady-state solution of the Vinen equation is

$$L^{1/2} = \frac{\pi \rho_{\rm n} B \chi_1}{\rho \kappa \chi_2} u_{\rm ns}, \tag{6.6}$$

which means that L is proportional to the square of the counterflow velocity, in agreement with experimental observations.[4]

Equation (6.5) was later deduced by Schwarz (1988) under a number of assumptions involving the local induction approximation (LIA) to the laws of vortex dynamics, introduced in Chapter 5. Under the assumption that the vortex tangle is homogeneous and that the vortex line density L is adequate to characterize the turbulence, the only relevant length scale of the turbulence is $\ell \approx L^{-1/2}$, which we refer to as the *quantum length scale*, whose importance will be explained in later chapters. For tangles of relatively low density and small deviations from the steady state, Schwarz obtained

$$\frac{\partial L}{\partial t} = \alpha I_l v_{\rm ns} L^{3/2} - \beta \alpha c_2^2 L^2, \tag{6.7}$$

where

$$\beta = \frac{\kappa}{4\pi} \log\left(\frac{1}{c_1 L^{1/2} a_0}\right). \tag{6.8}$$

Here, a_0 denotes the vortex core parameter, I_l one of the anisotropy parameters introduced by Schwarz (1988), and c_1 and c_2 are defined through $\bar{S} = c_1 L^{1/2}$, $\tilde{S} = c_2 L^{1/2}$ where $\bar{S}$ and $\tilde{S}$ are the mean and root mean square (RMS) curvatures, respectively, of vortex lines in the turbulent tangle.

Recently, this form of the dynamical equation as well as its very appropriateness was challenged on theoretical and numerical grounds by Nemirovskii (2016) and

[4] Note that the measured (or numerically computed) steady-state value of L determines only the ratio of the parameters χ_1 and χ_2; a different approach (e.g., the study of fluctuations as in Barenghi *et al.* (1982)) is needed to determine χ_1 and χ_2 separately.

Khomenko *et al.* (2015, 2016) with several new proposals. Restricting attention to cases of homogeneous turbulence (i.e., no vortex flux term), and without the new terms that are experimentally inaccessible until now, the proposals can be summarized in the form

$$\frac{\partial L}{\partial t} = \mathcal{A}_n v_{\mathrm{ns}}^n L^{2-n/2} - \mathcal{B}_n L^2, \tag{6.9}$$

where $\mathcal{A}_n$, $\mathcal{B}_n$ are adjustable parameters and n is 1, 2, or 3. Note that all three cases satisfy the crucial condition that L is proportional to the square of the counterflow velocity.

A more recent second sound experiment Varga and Skrbek (2018) found that for small overall tangle densities all three proposed forms provide, within experimental uncertainty, adequate description. For higher vortex line densities and across all temperatures, these forms of the dynamical equation cannot account for the significant slow down in the tangle growth rate as the steady state is approached. At the same time, it was found that the region of validity of Eq. (6.7) is perhaps greater than expected. Moreover, the agreement with experiments (including those on early decay; see Chapter 9) can be significantly improved if one varies the geometrical factor c_2, connecting the RMS curvature and inter-vortex distance of the tangle. This is remarkable, as the local induction approximation introduced by Schwarz neglects nonlocal interactions between vortex lines. This is equivalent to neglecting the motion of a vortex line induced by the presence of another line (e.g., the rotation of vortex lines in a bucket of rotating helium), and the stretching of vortex bundles; both these effects are essential in classical turbulence. An unavoidable question therefore emerges: Why does this approximation describe the *steady* state of thermal counterflow quantitatively?

As for the numerical approach, Adachi *et al.* (2010) performed vortex filament simulations on steady-state counterflow turbulence using the full Biot–Savart law and compared the results with the simulations of Schwarz (1988) and their own, based on local induction approximation. They claimed that local induction approximation is not suitable for simulations of turbulence, as such calculations construct a layered structure of vortices that does not proceed to a turbulent state.[5] While it is clear that the full Biot-Savart approach is certainly better, there are still other aspects such as the approach to vortex reconnections and influence of possible normal fluid turbulence that limit the predictive power of these simulations, which are crucially based on an imposed normal fluid flow.

In Chapter 2, we discussed the use of negative ions in thermal counterflow. Using this technique, Awschalom *et al.* (1984) found that within the experimental

[5] Schwarz actually recognized this problem in his original calculations but solved it using an artificial randomizing procedure.

uncertainty of up to few percent, the (rather small) vortex line density L in a wide channel is constant over the cross section of the channel, except for a small region of about 1 mm close to the walls where the technique fails.

On the other hand, more recent numerical studies, such as those of Baggaley and Laurie (2015), found that L peaks close to solid boundaries where reconnections occur more often. The position of the largest local vortex line density L moves closer to the walls with increasing temperature. This behavior is a consequence of the normal fluid profile, which, due to no slip conditions for the normal fluid velocity, cannot be flat across the channel. We shall return to this issue when considering recent visualization studies of the normal fluid profile in various channel flows together with relevant numerical simulations.

6.3.2 T I, T II, and T III States in Counterflow Turbulence

We now continue with the description of some important experimental details of steady-state thermal counterflow turbulence in He II. When a heater situated at the end of the counterflow channel is switched on, it does not take long before a steady-state counterflow u_{ns} becomes established. If the applied power density $\dot{q}$ is less than a certain critical value $\dot{q}_{c1}$, there are no vortex lines in the flow (except for the remanent ones). Without vortex lines, no mutual friction acts on the two fluid components, which means, in particular, that the normal fluid flows as an ordinary fluid of viscosity η. Taking into account the usual no-slip boundary conditions, the normal fluid acquires the usual parabolic profile, and the pressure drop along the channel agrees with the classical formula

$$\Delta p_{\mathrm{lam}} = \frac{G\eta L_{\mathrm{ch}} u_{\mathrm{n}}}{d^2}, \tag{6.10}$$

where G is a dimensionless factor that depends on the channel's geometry, L_{ch} is the length of the channel, and d its smallest dimension. The temperature difference along the channel, ΔT, is naturally determined by Δp_{lam} via London's Eq. (3.16),

$$\Delta T_{\mathrm{lam}} = \frac{\Delta p}{\rho S} = \frac{G\eta L_{\mathrm{ch}} u_{\mathrm{n}}}{\rho S d^2}. \tag{6.11}$$

If the heat flux exceeds a critical value, $\dot{q} > \dot{q}_{c1}$ (hence $u_{ns} > u_{c1}$), the heat transport is affected by the appearance of a turbulent tangle of vortex lines. The intensity of the turbulence is characterized by the vortex line density L, as argued by Vinen (1957). The pressure drop Δp_{turb} along the channel increases with the applied heat flux $\dot{q}$, but only slightly, an effect called the *Allen–Reekie rule* (Allen and Reekie, 1939). On the other hand, chemical potential and temperature differences along the channel increase dramatically in the turbulent state, exceeding their laminar values in rough proportion to $\dot{q}^3$. The increased dissipation in the turbulent state is described as a stronger mutual friction force between normal fluid and superfluid components, which also affects sound propagation; in particular, it gives rise to an excess second sound attenuation, increasing approximately as $\dot{q}^2$ (Vinen, 1957).

Historically, careful and extensive measurements of temperature differences ΔT along counterflow channels performed by Tough and coworkers (Tough, 1982) clearly indicate that, depending on the geometry, there are several steady counterflow turbulent states. In rather large channels (from the perspective of helium experiments) of circular or square cross section, upon increasing the applied power per unit area of the channel, $\dot{q}$, the laminar flow undergoes the first transition at $\dot{q}_{c1}$, above which the measured ΔT exceeds the extrapolated laminar value only slightly, until at $\dot{q}_{c2}$ a second transition occurs, above which the measured ΔT is dramatically enhanced. These turbulent states, illustrated in Fig. 6.5, are, respectively, referred to as T I and T II.

The existence of the state T I for $\dot{q}_{c1} < \dot{q} < \dot{q}_{c2}$ was recognized by early investigators by its signature in the second sound attenuation data of Vinen (1957), in the pressure difference data of Brewer and Edwards (1961), and in the temperature difference data by Chase (1963). But it was Tough's group that experimentally investigated it in considerable detail. Much less is known about T I than about the state T II (especially about the scaling of observed variables with the applied heat flux), as it exists only in the rather narrow, temperature-dependent region $\dot{q}_{c1} < \dot{q} < \dot{q}_{c2}$ (see Fig. 6.5).

Apparently, the geometry of the flow apparatus has a profound effect on thermal counterflow turbulence. In rectangular channels of high aspect ratio (1:10) with the small dimension less than 100 μm, only one transition has been observed at a critical heat flux $\dot{q}_{c3}$. This steady-state turbulence was denoted by Tough as T III. No hysteresis associated with the T I to T II transition has been observed, but metastable laminar states can be observed for heat fluxes considerably larger than $\dot{q}_{c1}$ or $\dot{q}_{c3}$.

Although the functional form of Eq. (6.4) holds in both states T I and T II, the observed (as well as numerically calculated) values of $\gamma = \gamma(T)$, denoted, respectively, as γ_1 and γ_2, are distinctly different, as illustrated in Fig. 6.6. Indeed, in the T I state, the experimental values of γ_1 measured by Martin and Tough (1983) (solid magenta diamonds) and Childers and Tough (1973) (open circles), as well as the numerical data by Adachi *et al.* (2010) (dashed line, calculated by assuming a laminar normal fluid flow) are for the same temperature appreciably lower than γ_2, observed by several groups in the T II state.

The existence of the two distinctly different steady turbulent states in He II counterflow turbulence, T I and T II, has long been a puzzle. One possibility was considered by Melotte and Barenghi (1998), who addressed the issue of the velocity profile of the normal fluid and its stability. Their calculation suggests that in the

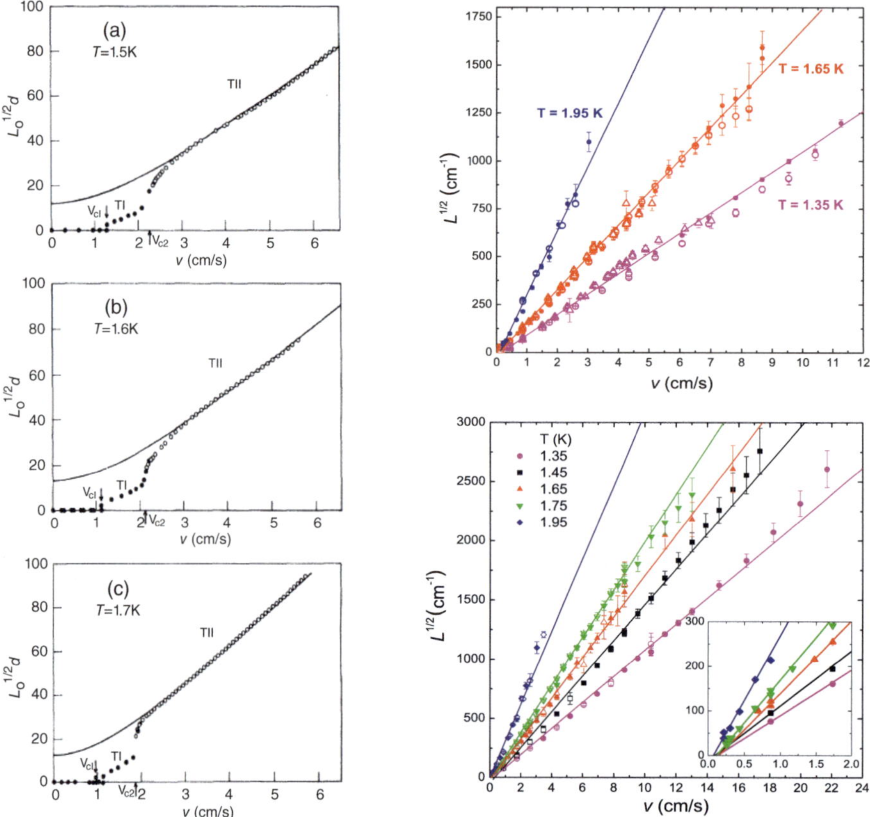

Figure 6.5 Comparison of thermal counterflow with pure superflow. (Left) Vortex-line density, shown as the dimensionless number $L_0^{1/2}d$ as a function of counterflow velocity at 1.5, 1.6, and 1.7 K. Open circles are determined from the heat pulse measurements and represent state T II. Solid circles are taken from the second sound measurements of state T I. Two critical velocities, marked v_{c1} and v_{c2}, are indicated. The solid line is a best fit of the following equation to data in state T II: $(L_0)^{1/2}d = \left(\gamma_2^2(T)u_{ns}^2d^2 + C^2\right)^{1/2} - 1.48\alpha_2$, where d denotes the channel diameter, C and α_2 are fitting constants, and γ_2 has the same meaning as γ in Equation (6.4) for the T II state. Reprinted figure with permission from Martin and Tough (1983). Copyright 1983 by the American Physical Society. (Top right) Comparison of steady-state vortex line density in pure superflow obtained in three different channel configurations: (i) solid circles, pure superflow in 7 mm channel with superleaks both below and above the channel; (ii) open circles, with the downstream superleak removed; (iii) open triangles, 10 mm channel with both superleaks. Straight lines guide the eye. (Bottom right) The square root of the steady-state vortex line density L, as a function of mean superflow velocity, for the 7 mm wide channel at different temperatures. Open symbols are for L obtained from the second sound amplitude as extracted from a Lorentzian fit for the full resonant curve, while solid symbols relate to a direct measurement at standing second sound resonance. Straight lines are fits to the data weighted by the uncertainty in L. The inset highlights the existence of a single critical velocity in pure superflow. The right panels are reproduced with permission from Babuin *et al.* (2012). Copyright 2012 by the American Physical Society.

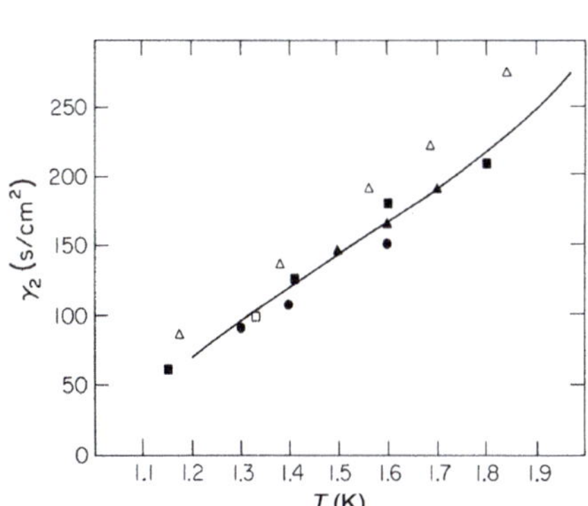

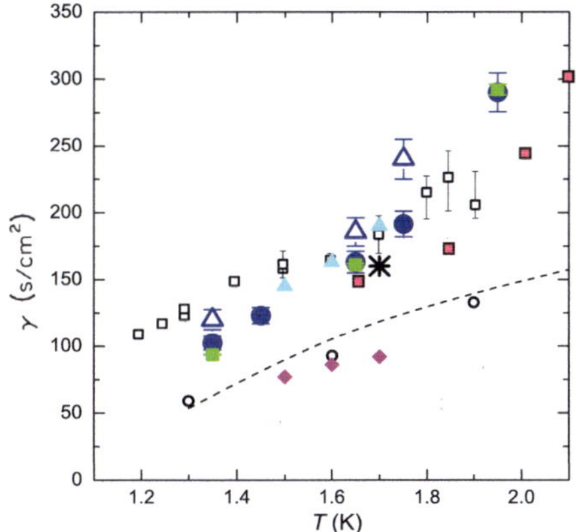

Figure 6.6 Comparison of the temperature-dependent parameter γ in thermal counterflow and pure superflow. (Left) When state T II of thermal counterflow is fully developed, $\gamma_2 = \gamma$ of Eq. (6.4). The values shown are extracted from experiments of Martin and Tough (1983) (full triangles), Dimotakis and Broadwell (1973) (full circles), Peshkov and Tkachenko (1962) (empty squares), Brewer and Edwards (1961, 1981) (empty triangles), and Chase (1963) (full squares). The solid line is the calculation (local induction approximation, vortex filament method) of Schwarz (1982). (Right) Comparison of γ obtained in available experiments in states T I and T II of thermal counterflow and pure superflow. Experimental values of γ in the T I state, Martin and Tough (1983) (solid magenta diamonds) and Childers and Tough (1973) (open circles), as well as numerical data by Adachi *et al.* (2010) (dashed line), are appreciably lower than those in the T II state, represented here by experiments of Martin and Tough (1983) (open blue up-triangles) and of Gao *et al.* (2017) (red-filled squares). Pure superflow data of Babuin *et al.* (2012) (solid blue circles, 7 mm channel; solid green squares, 7 mm channel with the downstream superleak removed; open blue up-triangle, 10 mm channel) as well as of Ashton *et al.* (1981), (open squares, 0.13 mm diameter glass channel) are in fair agreement with the T II state thermal counterflow data. (Left) Reprinted figure with permission from Martin and Tough (1983). Copyright 1983 by the American Physical Society. (Right) Reprinted figure with permission from Babuin *et al.* (2012). Copyright 2012 by the American Physical Society.

T I state the superfluid is turbulent but the normal fluid is still laminar, and the T I-to-T II transition corresponds to the onset of turbulence in the normal fluid, which would enhance the vortex line density. This idea is supported by recent visualization experiments of Guo *et al.* (2010) who used excimer molecules that trace the normal fluid and showed it to be turbulent at sufficiently high value of heat flux. We shall discuss this visualization experiment in detail in Chapter 10.

Along with studies of steady-state thermal counterflow, it is also interesting to study transient effects such as spatiotemporal development of temperature difference ΔT, and the dynamics of generation of vortex line density, L, as the latter builds up in different places along the channel. Depending on experimental conditions such as the applied heat load, geometry of the channel, wall material, and/or surface roughness, various scenarios of vortex tangle development were observed or theoretically predicted (for review of early works, see Nemirovskii and Fiszdon (1995) and references therein), from stationary turbulent plugs to propagating fronts, as claimed by Mendelssohn and Steele (1959). Many investigations have followed. We mention detailed investigations of propagating and stationary turbulent fronts in narrow counterflow channels, both uniform and nonuniform, investigated by Tough's group. These studies of counterflow turbulence in narrow channels were based primarily on measurements of the temperature difference ΔT over the counterflow channel and therefore cannot provide direct information on the spatiotemporal development of L in the channel. Recently, the Prague group designed and manufactured an approximately 20 cm long, 7 mm wide multipurpose counterflow channel of square cross section, equipped along its length with three pairs of second sound sensors and seven additional ports for thermometers. The data show that L builds up at different locations almost simultaneously, at variance with earlier results that the turbulent front moves from one side of the channel with either superfluid or normal mean velocities, as was indirectly observed and claimed in long but very thin channels. The complementary numerical simulations, performed in collaboration with L'vov's Rehovot group, show that the experimentally observed complex patterns of the early buildup of vortex line density are consistent with the tangle growth from multiple localized sources of remanent vortices (Varga *et al.*, 2017).

6.4 Pure Superflow

The simplest approach to understanding pure superflow of He II is to ignore boundaries and treat it as thermal counterflow in the frame of reference of the normal fluid, and then apply Vinen's (or Schwarz's) equations. The problem, however, is more complicated than this Galilean transformation suggests for two reasons: (i) the profile of the normal fluid induced by the no-slip boundary conditions cannot be ignored; (ii) the superfluid exerts a friction on the normal fluid, perhaps creating spatial structures. In either case, no frame of reference exists in which the normal fluid is at rest.

The experimental observations are, in fact, rather complex. Let us start with the critical velocity for the onset of quantum turbulence in pure superflows and thermal counterflows, as measured in a variety of experimental geometries and illustrated in Fig. 6.7. According to the review of Tough (1982), the early attempts to account theoretically for its scaling with channel geometry did not lead to a satisfactory understanding of the problem. In view of results of recent experiments with oscillating objects discussed above, however, we can offer at least some hints for a qualitative understanding of this problem.

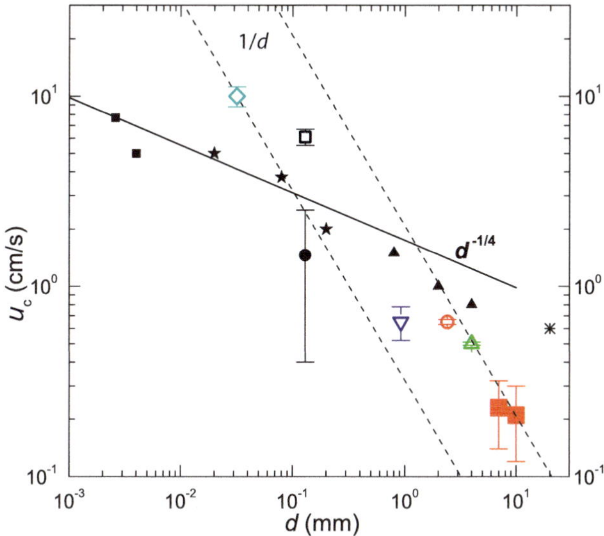

Figure 6.7 Critical velocities for the onset of turbulence as a function of the smallest dimension of channel cross section, d, for different systems. Pure superflow experiments (solid symbols); Hammel and Keller (1965) (black squares, isothermal flow); Vermeer *et al.* (1965) (stars, adiabatic flow rate); Baehr *et al.* (1983) (circle, temperature average of all data); Chase (1963) (up-triangles, heat conduction); Craig and Pellam (1957) (asterisk, superfluid wind tunnel); Babuin *et al.* (2012) (red squares, bellows-driven pure superflow). Counterflow experiments (open symbols): Ladner and Tough (1979) (open blue diamond); Childers and Tough (1973) (black open square); Yarmchuk and Glaberson (1979) (blue down-triangle); Vinen (1957) (open red circle, open green up-triangle). Solid line represents the Kruglov (2011) theory, and dashed lines illustrate lower and upper bounds for the experimental data for $1/d$ scaling. Reprinted figure with permission from Babuin *et al.* (2012). Copyright 2012 by the American Physical Society.

We have already mentioned that, in rectangular channels of high aspect ratio with the smaller dimension d less than 100 μm, only one transition, at $\dot{q}_{c3}$ (denoted by Tough as T III), has been observed in thermal counterflows. Assuming the existence of two independent velocity fields before the critical velocity is reached (i.e., no quantized vortices in the flow), the Reynolds number, Re_n, based on the normal fluid velocity u_n, channel size d, and kinematic viscosity of the normal fluid ν_n, typically does not exceed about 100. In analogy with the analysis of oscillatory He II flows in the limit of high Stokes number, as discussed in Chapter 7, we may regard this dimensionless number as the Donnelly number rather than Re_n. The condition $\mathrm{Dn} \lesssim 100$ suggests that, upon reaching the observed critical counterflow velocity, the normal fluid flow is laminar and the instability (most likely of the Donnelly–Glaberson type) first occurs in the superflow. In contrast, in large enough channels (corresponding to larger Dn, say, of order 1000) the observed critical velocity is temperature dependent (due to strong temperature dependence of the kinematic viscosity of the normal fluid), in accordance with the idea that instability first develops in the normal flow.

In pure superflows the situation is similar in that only one transition must occur in the superfluid, as the normal flow remains quiescent until the critical velocity is reached (i.e., Dn = 0). A large compilation of results obtained by various groups, in channel sizes ranging from a few Å up to a few cm, was already provided half a century ago by Van Alphen *et al.* (1966). These pure superflow results are temperature independent, supporting the idea that there is a critical superfluid velocity, rather than a critical Donnelly number, which would be strongly temperature dependent. Further, the collected data suggest that the superfluid critical velocity varies perhaps with the fourth root of the channel size, i.e., very weakly, as illustrated in Fig. 6.7. In contrast, critical velocity in thermal counterflow displays much steeper dependence, varying roughly as the inverse of the channel size, in qualitative agreement with the expected scaling for the critical Donnelly number Dn.

In spite of this plausible explanation, a complete quantitative understanding of critical velocities and underlying physical processes associated with the transition to turbulence has yet to be achieved for both the thermal counterflow and the pure superflow. Experimentally, careful design is needed in order to take into account various aspects known from classical studies, such as surface roughness and entry length. A theoretical study of Bertolaccini *et al.* (2017) predicts the need for disproportionately long entry lengths for both normal and superfluid components in thermal counterflow, though for rather artificial treatment of the channel ends. A very difficult problem is to assure that the incoming flow is laminar.

The right panels of Fig. 6.5 show that pure superflow displays only one turbulent state, in which vortex line density scales according to Eq. (6.4), with temperature-independent critical velocity. Babuin *et al.* (2012) verified that this scaling holds even if one removes the downstream channel superleak. Moreover, the observed $\gamma(T)$ in steady states of pure superflow agrees fairly well with $\gamma(T)$ in thermal counterflow. We shall return to this aspect in Chapter 11 when discussing the temporal decays of these two quantum flows, which are distinctly different.

6.5 Coflow, Counterflow, and Superflow Compared

The best way to experimentally compare various types of flows of He II is to generate and study them in the same flow channel. An example of such a complex and controllable experimental configuration designed to study problems of superfluid hydrodynamics is shown in Fig. 4.4. This experimental set-up allows a comparative experimental study of quantum turbulence in coflows, thermal counterflows, and pure superflows as schematically shown in Fig. 6.4 for a channel of square cross section. The counterflow is prepared by closing the bottom end of a channel with a

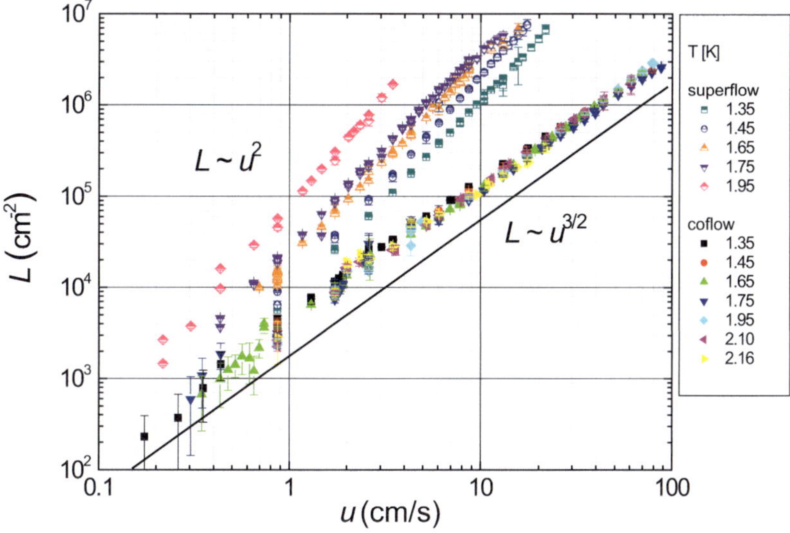

Figure 6.8 Vortex line density vs. mean flow velocity for superflow and coflow experiments by Babuin *et al.* (2014a) and Varga *et al.* (2015). When the two helium components are forced to undergo relative motion (superflow and counterflow), the scaling is quadratic and temperature dependent. The coflow (of the grid turbulence character, produced by the flow conditioner at the channel entrance) displays classical 3/2-power scaling that is temperature independent.

seal fitted with a resistive wire heater and leaving the other end open to a helium bath. For superflow, superleak filters block the viscous normal component. For coflow, the superleaks are removed and the lower one replaced by a flow conditioner made as a dense pack of 10 mm long capillaries of 1 mm diameter, intended to disintegrate the oncoming larger scale turbulent eddies. Moreover, in this case the turbulence can be intensified (Varga *et al.*, 2015) by a grid, as shown in Fig. 6.9, placed roughly 25 grid meshes upstream from the center of the detection region. The coflow with and without the grid is referred, respectively, as "near wake" and "far wake" limits of grid turbulence since, even in the absence of the grid, the upstream flow conditioner effectively produces grid turbulence. The turbulent coflow along the channel can be understood loosely as a mixture of grid and channel turbulence. In all cases, the main observable is the density of vortex lines in the channel, L, as a function of mean flow velocity, geometry, temperature, and time.

Figure 6.8 shows that, in coflows, the dependence $L \propto (u - u_c)^{3/2}$ (where u is the mean flow velocity) is observed in both the near and far wake limits, with a near collapse in temperature of many datasets. These datasets span the temperature range of $1.17\,\mathrm{K} \leq T \leq 2.16\,\mathrm{K}$, with a ten-fold variation of the superfluid fractional density, in the range $0.98 \geq \rho_s/\rho \geq 0.09$. Considering that the velocity varies

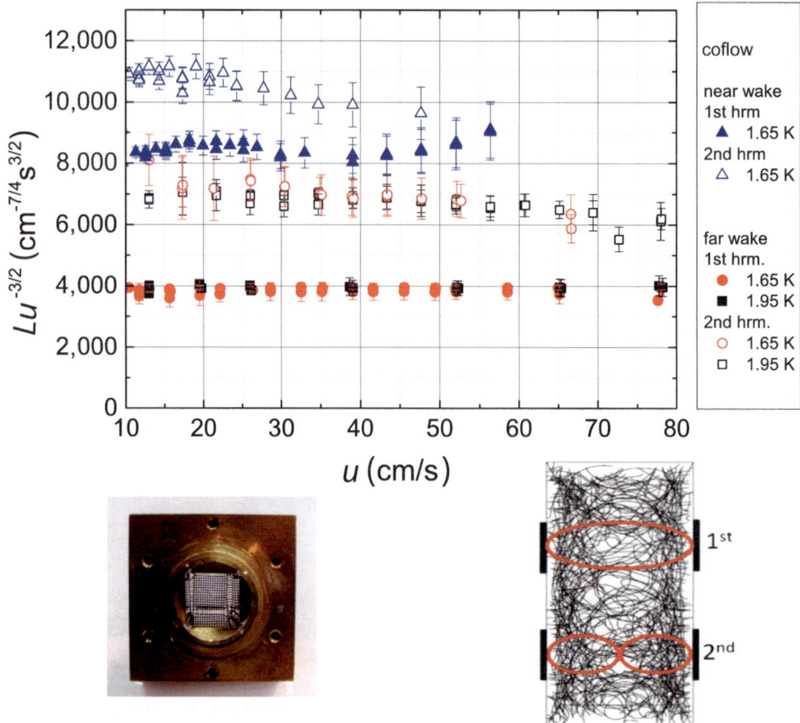

Figure 6.9 The use of different second sound harmonics to sample vortex tangle spatial distribution in coflow. To emphasize the effect, the vertical axis shows vortex line density compensated by the steady-state scaling of the mean velocity, as discussed in the text. The second harmonic detects considerably higher L than does the first, in both cases with and without an added grid, shown in the bottom left. The two transverse second sound standing modes have maximum amplitudes and therefore maximum sensitivity at different locations of the channel width. The first harmonic has a maximum at $x = 0.5\,D$, the second at $x = 0.25\,D$ and $x = 0.75\,D$ (where D is the distance between second sound resonators). Their sensitivity is qualitatively illustrated by red ellipses in the tangle image at the bottom right, where the tangle image was adapted from numerical simulation (Baggaley and Laurie, 2015). Because of this feature of probing second sound harmonics, the vortex tangle in coflow is found to be inhomogeneous. In this particular example, the second harmonic provides a stronger signal than the first.

over three orders of magnitude, the figure provides robust evidence of temperature independence of the observed quantum turbulence. Detecting turbulence near or far from the grid does not change the scaling or the temperature dependence to first approximation (except for some minor details on scaling exponents and tangle inhomogeneity, to be discussed in Section 6.9). The main difference is that the vortex line density is about twice as high at 24 grid meshes compared to that at 55 meshes downstream of the grid.

Can one understand the observed $L \propto (u - u_c)^{3/2}$ dependence of Fig. 6.8? Although the rigorous microscopic theory is yet to be developed, a qualitative phenomenological understanding is at hand (Babuin *et al.*, 2014a). This understanding is based on the analogy between coflow and classical homogeneous isotropic turbulence (HIT), where one usually connects energy dissipation to the velocity and lengths of the large-scale flow – known as the dissipative anomaly (Sreenivasan, 1984, 1998). On the other hand, in analogy with classical turbulence where $\varepsilon = \nu\omega^2$, following Vinen (2000) we postulate (and discuss in Section 11.6) that $\varepsilon = \nu_{\mathrm{eff}}(\kappa L)^2$. Here, ν and ν_{eff} stand for kinematic and effective viscosities, respectively, and ω and κL for vorticity and effective vorticity of turbulent superfluid. Remembering that the quantum length scale $\ell = 1/\sqrt{L}$, we arrive at the quasi-classical scaling of the form

$$\frac{\ell}{H} = \left(\frac{\nu_{\mathrm{eff}}}{\kappa}\right)^{1/4} Re_{\mathrm{s}}^{-3/4}. \tag{6.12}$$

With suitable choices for H and Re_{s}, one achieves a collapse of the fourteen “near wake” and “far wake” datasets shown in Fig. 6.8 (see Fig. 3 in Babuin *et al.* (2014a)). This collapse strongly supports the quasi-classical character of the observed $L \propto (u - u_c)^{3/2}$ scaling. One might conclude that coflowing He II behaves, at least approximately, as a single-component viscous fluid possessing an effective kinematic viscosity ν_{eff}.

In both pure superflow and counterflow (the latter data sets are not shown in Fig. 6.8 to avoid clutter, but the datasets nearly collapse), the quadratic scaling of vortex line density with counterflow velocity is confirmed, $L = \gamma^2(u_{\mathrm{ns}} - u_c)^2$, with a pronounced temperature dependence. The quantity u_c is a small critical velocity for the onset of turbulence in the superfluid component, which can be determined either from fitting the data by a power law and requiring it to pass the origin, or by direct measurement of the minimum velocity at which the second sound attenuation departs from noise level in stationary helium. Both the direct measurement and the fits give comparable values or the order of 0.1 cm/s for pure superflow, thermal counterflow, and coflow, which, to a first approximation, does not depend on temperature or on whether the flow velocity is ramped up or down.

Limited but useful information on the inhomogeneity of the vortex tangle along the direction of the channel width can be deduced by varying the mode number of second sound standing wave resonance, because inhomogeneous tangles attenuate different harmonics differently. For superflow at $T = 1.65$ K, no appreciable change was observed in the deduced L, while using second sound resonant mode number

1, 2, or 10 for its detection, suggesting strongly that there is no or very little inhomogeneity in this flow. As shown in Fig. 6.9, this is not so in turbulent coflow, where the second harmonic detects considerably higher L than the first. Similar measurements for thermal counterflow have not yet been made.

6.6 Inhomogeneous Turbulence

In most experiments described so far, the quantum turbulence is fairly homogeneous. In practice, however, as in classical flows, the presence of boundaries always results in inhomogeneous turbulence. In this section, we discuss examples of inhomogeneous quantum turbulence displaying interesting special features.

6.6.1 Turbulent Fronts

Observation of turbulent fronts filling a long thin counterflow pipe/channel, propagating into it from both the cold and hot ends, was first made by Mendelssohn and Steele (1959). Many investigations followed; we mention here detailed investigations of propagating and stationary turbulent fronts in rather narrow counterflow channels, uniform as well as nonuniform, performed by Tough's group. Kafkalidis and Tough (1991) and Kafkalidis *et al.* (1994) used a sensitive Au–Fe thermocouple/SQUID to measure the temperature difference between various positions along the diverging channel and observed both laminar and turbulent dissipation regimes. Murphy *et al.* (1993) found that in a nonuniform channel of circular cross section both T I and T II turbulent states exist and the transition (substantially broadened) from the T I to the T II state occurs at stationary turbulent fronts in the channel; this study was followed by a more detailed one by Tough *et al.* (1994), who found that the front occurs at the position in the channel where the local velocity reaches the critical value.

Rather surprisingly, Castiglione *et al.* (1995) have observed that the critical heat current for the transition of thermally generated counterflow to turbulence in weakly nonuniform circular channels depends on the flow direction. This observation is surprising since no other property of the counterflow turbulence appears to have such a dependence. The authors claimed that critical heat current is associated with a stationary front between the laminar and turbulent flow and proposed a model for superfluid turbulent fronts.

Experimental studies performed in capillaries and narrow channels were based primarily on measurements of the temperature difference (in some cases the chemical potential difference [Murphy *et al.*, 1993]) over the counterflow channel and therefore cannot provide direct information about the spatiotemporal development of vortex line density in the channel. In order to study the development of the turbulent front, its propagation, and the streamwise structure of the vortex tangle in counterflow channels, several localized probes are required. Such a setup is very difficult to realize experimentally in narrow channels of characteristic cross section below 1 mm^2, where two turbulent states T I and T II have been observed. On the other hand, multiple turbulent states are generally not observed in channels of larger cross section of order 1 cm^2.

Varga *et al.* (2017) manufactured an approximately 20 cm long, 7 mm wide multipurpose channel of square cross section, equipped with three pairs of second sound sensors at various positions along its length. Experimental data, obtained at 1.65 K, clearly show that upon abrupt switching of the heater the vortex line density builds up at different locations almost simultaneously, most likely with the speed of the second sound shock wave (Nemirovskii and Fiszdon, 1995). The complementary numerical simulations performed by L'vov and Pomyalov in Rehovot, utilizing the vortex filament method, show that the experimentally observed buildup of complex patterns of the early vortex line density is consistent with the tangle growth from multiple localized sources of remanent vortices (extrinsic vortex nucleation). Initially, the vortex loops grow preferentially across the channel. At later stages, the tangle anisotropy becomes typical for the counterflow conditions and homogenous along the channel, despite inhomogeneous streamwise distribution of vortex line density. These results are at variance with the earlier observations that the turbulent front moves from any side of the channel with either superfluid or normal mean velocities, as indirectly observed and claimed in channels that are long and thin.

6.6.2 Spherical and Cylindrical Counterflow

A spherical counterflow was examined by Varga (2019) using the vortex filament method. In this configuration, a small spherical heater drives a spherically radial counterflow of the superfluid and normal fluid components, which (in the laminar regime) decay as r^{-2}. The surprising finding is the lack of a steady-state regime of turbulence independent of initial conditions, unlike in all other counterflow configurations previously examined. This preliminary study also suggests the existence of a peculiar phase diagram for the flow, with the temperature-dependent region of quantum turbulence bounded from both below and above by critical velocities.

Sergeev and Barenghi (2019) used the Hall–Vinen–Bekharevich–Khalatnikov (HVBK) equations to model inhomogeneous turbulent flows of He II and showed that a steady spherical isothermal counterflow is not possible, in agreement with Varga (2019). They also examined another simple configuration – a heated cylinder, which drives a cylindrically radial counterflow of the two fluid components; in the laminar regime, such counterflow decays as r^{-1}. This flow configuration is also relevant to hot-wire anemometry (Duri *et al.*, 2015). Sergeev and Barenghi (2019)

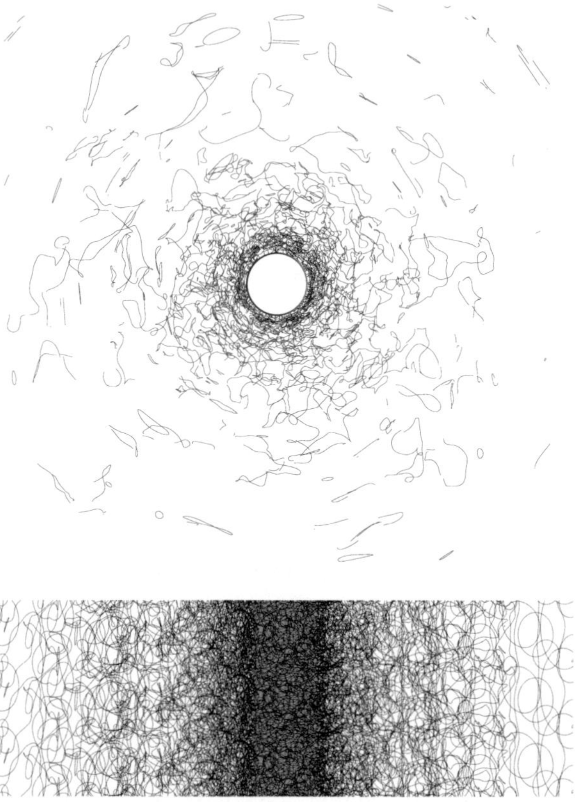

Figure 6.10 Cylindrical counterflow. Vortex tangle in a statistically steady state around a heated cylinder of radius $a = 0.1$ cm computed using the vortex filament method by Rickinson *et al.* (2020). The normal fluid velocity at the cylinder surface has the value $u_{\mathrm{n}} = 0.6$ cm/s. The temperature of the helium in the bulk is $T = 1.3$ K. Following Sergeev and Barenghi (2019), in the thermal layer region $r \lesssim 0.13$ cm, the friction coefficients α and α' are raised from the bulk value to those corresponding to the surface of the cylinder, $T = 2.15$ K. (Top) Top view of the cylinder. (Bottom) Side view. Reprinted figure with permission from Rickinson *et al.* (2020). Copyright 2020 by the American Physical Society.

predicted that a time-independent solution exists only if one accounts for the spatial nonuniformity of temperature and the thermodynamic properties. Essentially, a thermal layer forms near the cylindrical boundary, whose thickness grows with the temperature of the cylinder surface. The problem has been tackled numerically by Rickinson *et al.* (2020) using the vortex filament model. These authors argued that, from the point of view of the dynamics of vortices, the thermal layer is sufficiently well accounted for by an imposed radial dependence of the friction coefficients α and α' in the near vicinity of the cylinder, as calculated by Sergeev and Barenghi (2019). Figure 6.10 shows an example of the resulting statistically steady vortex

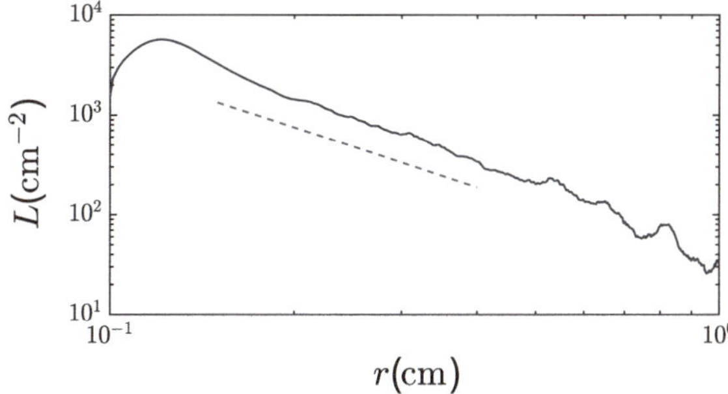

Figure 6.11 Cylindrical counterflow, as in Fig. 6.10. Vortex line density $L(r)$ vs. radial distance r; the dashed straight line shows the expected r^{-2} scaling. Reprinted figure with permission from Rickinson *et al.* (2020). Copyright 2020 by the American Physical Society.

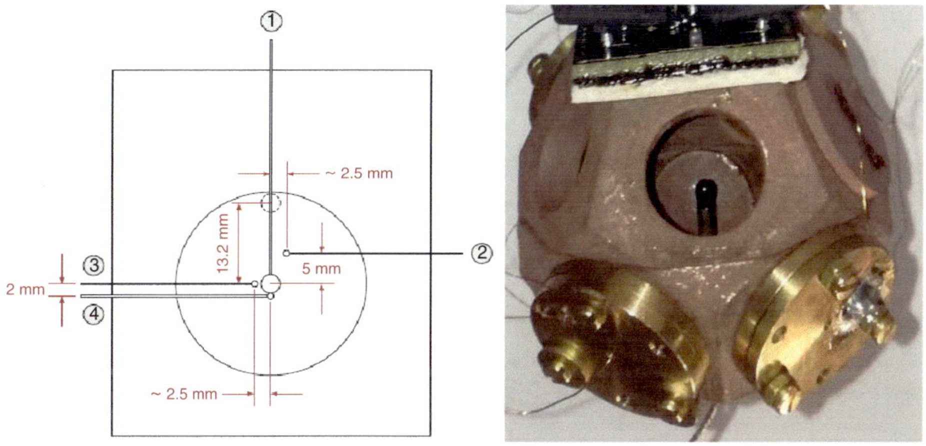

Figure 6.12 (Left) Schematic diagram of the temperature profile measurement setup. Labels denote the connection to the linear motor on the cryostat flange at room temperature (1); and to the temperature sensors, thermometers 2, 3, and 4. (Right) Photograph of the experimental cell of the form of a regular dodecahedron outside, containing a nearly spherical sample of liquid helium inside, with second sound transducers attached and the spherical heater in its center.

tangle. Note the large concentration of vortex lines near the cylinder. Away from the cylinder, large irregular vortex loops are held in balance by the counterflow. Further away, small vortex loops shrink to zero. The radial distribution of the vortex line density follows the expected r^{-2} dependence fairly well, as shown in Fig. 6.11, a result that should facilitate further theoretical progress on the problem.

Experiments on spherical counterflow in a spherical second sound resonator shown in Fig. 6.12 are under progress in Prague (Xie *et al.*, 2022). The experimental cell is a 3D-printed regular dodecahedron surrounding a spherical cavity

16 mm in diameter that contains He II. A small spherical resistive heater 1.8 mm in diameter is placed in the center, suspended by a cylindrical holder with a diameter of 1 mm. There is experimental evidence that the heater generates a vortex tangle in its vicinity, as expected. Although similar experiments have been performed and analyzed in rectangular channels, complementary numerical data analysis for the spherical geometry is still under development, and detailed quantitative information about $L(r)$ and structure of the vortex tangle surrounding the heater is not yet available. On the other hand, it is confirmed that steady heating produces a dynamically steady tangle. As we just discussed, the numerical prediction strongly suggests that this would not have been possible unless a temperature gradient existed in the radial direction from the heater. In a complementary experiment performed in an open He II bath, Xie *et al.* (2022) measured this temperature gradient directly (see Fig. 6.12), and claimed that, for not very high heat inputs $\dot{Q} = 4\pi r^2 \rho\sigma T u_{\rm n}$, it agrees with the analytical prediction (derived on the basis of HVKB equations) of the form

$$T(r) = \frac{\alpha\kappa\gamma^2}{5\sigma^4\rho_{\rm s}^3}\left(\frac{\dot{Q}}{4\pi T}\right)^3 \frac{1}{r^5} + T_0, \tag{6.13}$$

where σ stands for the entropy of He II and T_0 is the bath temperature at infinity. This approximation is sufficiently accurate for typical experimental conditions, i.e., when $\dot{Q}$ is of the order of 100 mW and the distance r varies from millimeters to centimeters. For high values of $\dot{Q}$, the steep temperature profile $T(r)$ in the vicinity of the heater must be calculated numerically.

6.6.3 Turbulence Generated by Vortex Rings

The generation of quantum turbulence via the injection of vortex rings from one side of the experimental cell, shown in Fig. 4.1, was analyzed by Baggaley *et al.* (2012a) in the context of the He II experiments of Walmsley *et al.* (2007) and Walmsley and Golov (2008). The key question here is why the injection of small vortex rings generates a large-scale flow. Indeed, in a confined cell, a strong and long enough injection may drive a circulating flow on the scale of the cell, as illustrated schematically in Fig. 11.7. In Chapter 11, we shall discuss the consequences of such a large-scale flow on the temporal decay of vortex line density; here we simply focus on its origin.

Baggaley *et al.* (2012a) claimed that with periodic boundary conditions (which in some sense corresponds to a very large container), the confinement effect of the boundary is not possible. In their simulations, these authors noticed that if two vortex rings of approximately equal size travel in opposite directions and collide head on, the reconnection will create two vortex loops of approximately the same size, as shown schematically in Fig. 6.13 (top). If the two rings of slightly different

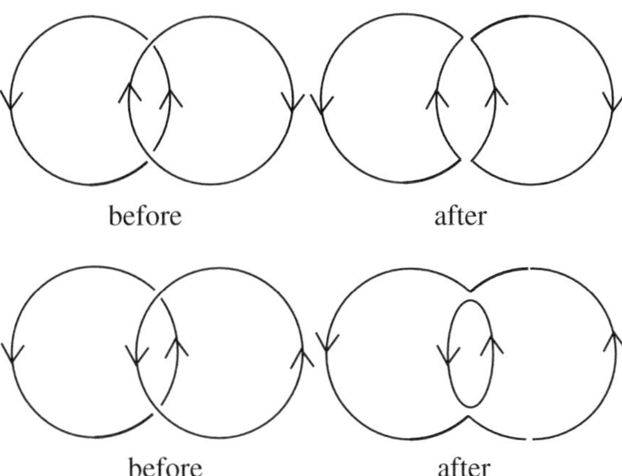

Figure 6.13 Schematic collisions of vortex rings and the outcome of the reconnection. (Top) Head-on collision of two vortex ring traveling in opposite directions; the two vortex loops that are generated have approximately the same size. (Bottom) Collision of two vortex rings traveling in the same direction, with one ring slightly smaller and moving faster than the other. The two vortex loops generated have very different sizes. Figure constructed on the basis of Baggaley *et al.* (2014).

size with a relative velocity travel in the same direction and collide, the reconnection will create two vortex loops, one small and one large; see Fig. 6.13 (bottom). The small vortex loop will quickly move away faster than the average velocity of the other injected vortex rings, whereas the large loop, moving more slowly, is likely to be hit by another ring from the back, resulting in the creation of further large loops. This runaway process creates a tangle of vortices containing large-scale velocity fields due to the large radii of curvature of these loops. To test this idea, Baggaley *et al.* (2012a) repeated their numerical calculation replacing the Biot–Savart law by LIA. Under LIA, the vortices interact only when they collide and the reconnection algorithm switches on. They noticed that the creation of the large-scale flow (hence the build-up of the energy spectrum at small wavenumbers) depends on the reconnections, not on using Biot–Savart or LIA. The key ingredient of this effect is the asymmetry of the ring injection, which boosts the number of collisions that induce larger radii of curvature.

It seems useful to point out an interesting link to classical turbulence. It is known that the dynamics of Navier–Stokes turbulence is determined by the nonlinear triadic interaction of modes that transfers energy to larger k. Biferale *et al.* (2012) wondered why in some situations (quasi-2D geometry, stratification, rotation, etc) a reverse energy transfer to smaller k is observed. To answer this question they numerically solved a decimated version of the Navier–Stokes equation such that only modes of a given sign of helicity are retained. They demonstrated a build-up of energy to low k by an inverse transfer process, shown in Fig. 6.14. They argued

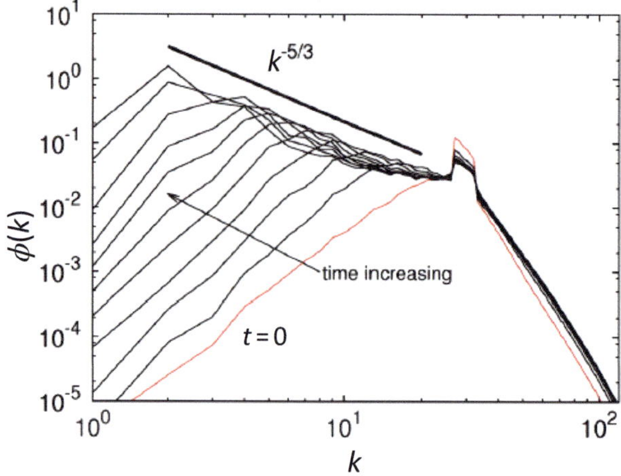

Figure 6.14 Time evolution (right to left) of the energy spectrum displaying an inverse energy cascade caused by the decimation of helical modes. Reprinted figure with permission from Biferale *et al.* (2012). Copyright 2012 by the American Physical Society.

that all 3D flows in nature possess nonlinear interactions that may lead to an inverse energy cascade if particular conditions of the flow can occur. In particular, the extreme concentration of vorticity in thin filaments typical of He II, together with the asymmetry of the experiments in which vortex rings are injected from one side (rather than isotropically), might provide a physical example of the effect discussed by Biferale *et al.* (2012).

6.7 Summary

The rich variety of cases that we have discussed in this chapter do not, by any means, cover all possibilities. Yet, they clearly demonstrate that the underlying physical processes that lead to statistically steady quantum turbulence are more complex than those leading to its counterpart in classical fluids. These processes are only partially understood, which highlights the fact that there is enormous room for innovative research in this area.

The conceptually simplest but experimentally the most difficult case to study is the transition to quantum turbulence from steady potential superflow in the zero-temperature limit. The difficulty is that the experimental realization of such steady potential flow is nearly impossible to achieve.

At nonzero temperatures, above about 1 K in He II, the two-fluid properties make the problem of transition to quantum turbulence in coflows, counterflows, or pure superflows quite diverse and complex. In general, it seems that turbulent motion of either the normal fluid or the superfluid serves as a trigger to generate turbulence in the other. On increasing the turbulence intensity, the normal and superfluid turbulent velocity fields become more and more tightly coupled by the action of the mutual friction force. In the particular case of isothermal coflow, this coupling results in quasi-classical behavior of turbulent He II, which can be described as one-fluid flow possessing an effective kinematic viscosity. This will be detailed in Chapter 11.

7
Oscillatory Turbulence

For practical cryogenic reasons, it is relatively difficult to generate quantum turbulence involving steady flows; this experimental difficulty has led to the widespread use of oscillating structures of various shapes for generating flows in spite of the difficulty in the interpretation of the results. Indeed, small mechanical oscillators immersed in liquid helium have been used to study quantum hydrodynamics since the discovery of superfluidity (for recent reviews, see Skrbek and Vinen (2009), Vinen and Skrbek (2014), and references therein). These devices can be used at any temperature, including the zero-temperature limit in both He II and ^{3}He-B (see Chapter 8) and thus play a valuable role in the investigation of quantum turbulence. Moreover, they work in classical fluids such as gaseous ^{4}He or normal liquid He I, which enables comparison of quantum oscillatory turbulence with its classical counterpart. In this section, after reminding the reader of the basic facts about oscillatory flows of viscous fluid, we discuss results obtained in He II with oscillating wires, grids, spheres, cantilevers, torsional pendulums, and tuning forks.

7.1 Brief Consideration of Oscillatory Classical Fluids

Oscillatory flows are generally more complicated than steady flows because a second characteristic length scale, L_2, emerges in addition to the linear size of the obstacle, L_1. To describe an oscillatory flow, we follow Schmoranzer *et al.* (2019) and write the governing Navier–Stokes equation in terms of dimensionless velocity $\boldsymbol{u}' = \boldsymbol{u}/U$, time $t' = t/\Upsilon$, and positions $\boldsymbol{r}' = \boldsymbol{r}/L_i$ as

$$\omega U \frac{\partial \boldsymbol{u}'}{\partial t'} + \frac{U^2}{L_1}(\boldsymbol{u}' \cdot \nabla' \boldsymbol{u}' + \nabla' p') = \frac{\nu U}{{L_2}^2} \Delta' \boldsymbol{u}', \tag{7.1}$$

where the characteristic length scales $L_{1,2}$ are used together with the characteristic velocity U to estimate the maximum magnitude of the respective velocity derivatives. An independent time scale, Υ, is introduced, given (in the continuum limit) by the angular frequency of oscillation, ω. Generally, the choice of L_1 and L_2 depends on body shape and flow parameters. Candidates may include the typical body size D, the surface roughness R_q, or the Stokes boundary layer thickness (viscous penetration depth), defined as $\delta = \sqrt{2\eta/(\rho\omega)}$, where η denotes the dynamic viscosity of the working fluid. For a given body $\delta \ll D$, we say that the body oscillates in the high-frequency regime if the Stokes number $\mathrm{St} = D^2/\left(\pi\delta^2\right) \gg 1$ is large. As we discuss below, this regime allows us to understand many features of oscillatory quantum flows.

In the high-frequency limit, depending on body geometry (especially surface roughness and sharp corners), δ or D may take the part of L_1 (related to a typical tangential velocity derivative) in the Navier–Stokes equation, but it is always δ that takes the part of L_2 (related to the largest velocity derivative in any direction); see Fig. 7.1. When $R_q \gg \delta$ (Fig. 7.1(d)) or when sharp corners are present (Fig. 7.1(a)), we can safely put $L_1 = L_2 = \delta$, and the Navier–Stokes equation contains only one dimensionless parameter. Conversely, for a hydrodynamically smooth body without any sharp corners, such as a cylinder (Fig. 7.1(b)), we obtain the Navier–Stokes equation with the Keulegan–Carpenter number $\mathrm{Kc} = U\Upsilon/D$ as the only relevant dimensionless parameter (Keulegan and Carpenter, 1958).

7.2 High-Stokes-Number Oscillatory Flows of He II

We apply the above considerations to He II in the case of small oscillatory velocities, assuming that the flow of the normal component is laminar and the flow of the superfluid component is potential. We replace ρ by ρ_n, decompose the pressure into partial pressures of the normal and superfluid components, and replace δ by $\delta_n = \sqrt{2\eta/(\rho_n\omega)}$, where η denotes the dynamic viscosity of He II. Again, if $\delta_n \ll D$ and $R_q \gg \delta_n$, one can set $L_1 = L_2 = \delta_n$, and the Navier–Stokes equation depends on a single dimensionless parameter, the *Donnelly number* $\mathrm{Dn} \equiv (\delta_n \rho_n U)/\eta$,[1] with the resulting equation,

$$2\frac{\partial \boldsymbol{u}'}{\partial t'} + \mathrm{Dn}\,(\boldsymbol{u}' \cdot \nabla' \boldsymbol{u}' + \nabla' p_n') = \Delta' \boldsymbol{u}'. \tag{7.2}$$

Note that Dn becomes equivalent to Re_δ at the superfluid transition temperature T_λ, allowing direct comparison between He II and classical oscillatory flows.

[1] The Donnelly number was first introduced by Schmoranzer *et al.* (2019) in memory and honor of Russell J. Donnelly, who, in a joint work with A. C. Hollis-Hallett (Donnelly and Hollis-Hallet, 1958), first introduced this dimensionless parameter as a “Reynolds number” based on the viscous penetration depth of the normal fluid; this idea did not appear in the previous work of Hollis-Hallet (1952) on the subject. Unfortunately, the importance of the Donnelly number was neglected for years, as it failed to describe the onset of turbulence for a torsionally oscillating sphere in He II. We now know that this was because the turbulent transition in the experiment occurred first in the superfluid component.

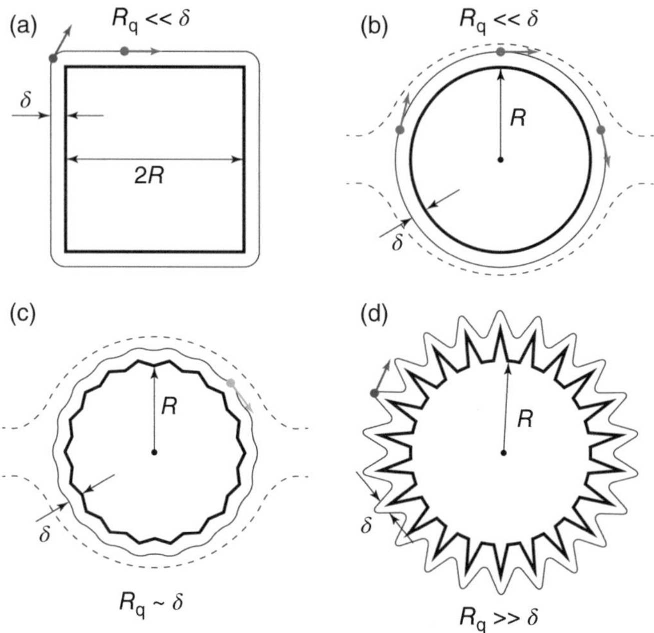

Figure 7.1 Schematic illustration of how surface roughness and sharp corners affect the estimates of maximum velocity derivatives in the high-frequency limit, where $\delta \ll R$. In (a) and (d) both velocity derivatives present in the Navier–Stokes equation are estimated using δ, while in case (b) the tangential velocity derivative is estimated using R and the Laplacian using δ. In (c), where $R_q \simeq \delta$, reliable estimates of the tangential derivative prove difficult, but a smooth cross-over between cases (b) and (d) is expected. Reproduced with permission from Schmoranzer *et al.* (2019), used under CC BY 4.0 (https://creativecommons.org/licenses/by/4.0).

In the limit $\delta_n \ll R$, if no turbulence is present in either component of He II, the superflow will be potential and the normal component exhibits laminar viscous flow inside the boundary layer, remaining approximately potential outside. Moreover, if δ_n is smaller than the typical radius of curvature of the oscillator surface, the surface may be described as consisting of many planar elements, and the velocity profile within the boundary layer is given by the solution to Stokes' second problem (an oscillating plane). In laminar flow around such a body, the average energy dissipation per unit time is given by (Landau and Lifshitz, 1987)

$$\langle \dot{E} \rangle = \frac{1}{2}\frac{\eta}{\delta_n} \oint |\Delta u_{L0,t}|^2 dS = \frac{1}{2}\frac{\eta}{\delta_n} \oint \alpha_L^2 u_{L0}^2 dS, \tag{7.3}$$

where $\Delta u_{L0,t}$ is the difference between two local velocity amplitudes projected tangentially to the surface – that of the potential flow just outside the boundary layer and that of the surface element of the body. Then α_L is the local flow enhancement factor relating this velocity difference to the (local) velocity amplitude u_{L0} of the surface element in question; that is, $|\Delta u_{L0,t}| = \alpha_L u_{L0}$. Integrating over the entire surface of an oscillator, we get

$$\langle \dot{E} \rangle = \frac{\alpha \xi U_p^2 S_r}{2} \frac{\eta}{\delta_n}, \tag{7.4}$$

where U_p is the maximum velocity amplitude along the surface of the resonator (peak velocity). The dimensionless quantity of order unity, $\xi = \oint u_L^2 dS / \left(S_r U_p^2\right)$, describes the velocity profile along the resonator, and an effective surface area $S_r \geq S$ may be used to account approximately for surface roughness. The integrated flow enhancement factor α is defined from $\alpha\xi = \oint \alpha_L^2 u_{L0}^2 dS / \left(S_r U_p^2\right)$. We note that for a smooth rigid oscillator this becomes $\alpha = \oint \alpha_L^2 dS/S$. For example, for a sphere, $\alpha_L = 3/2 \sin(\theta)$, with the angle θ measured from the direction of the flow, and $\alpha = 3/2$. Similarly, for a cylinder oriented normally to flow, we have $\alpha_L = 2\sin(\theta)$ and $\alpha = 2$. We emphasize that the above derivation is valid for all cases described in Fig. 7.1, as the length scale relevant to viscous drag is always δ_n.

Using the peak velocity U_p, it is possible to model a given mode of the resonator as a 1D linear harmonic oscillator, as done for a tuning fork (Blaauwgeers *et al.*, 2007). This leads to the definition of a (net) dissipative force amplitude:

$$F = \frac{2\langle \dot{E} \rangle}{U_p} = \alpha \xi S_r U_p \sqrt{\frac{\eta \rho_n \omega}{2}}. \tag{7.5}$$

We note that this definition is valid only in the 1D model and does not offer a direct measure of the drag forces per unit area discussed above. In analogy with steady flow, we define the dimensionless drag coefficient related to the normal component as

$$C_D^n = \frac{2F}{A \rho_n U_p^2} = \frac{2\alpha \xi S_r}{\rho_n A U_p} \sqrt{\frac{\eta \rho_n \omega}{2}}, \tag{7.6}$$

where A is the cross-sectional area in the direction perpendicular to the flow. In accordance with the principle of dynamical similarity, the drag coefficient can be expressed in terms of the Donnelly number, and in the case of laminar flow, we get a universal scaling law,

$$C_D^n = \Phi/\mathrm{Dn}, \tag{7.7}$$

where $\Phi = 2\alpha\xi S_r/A$ is determined purely by the geometry of the oscillator. In the laminar case, the drag due to the normal component is thus fully described by laws of classical fluid dynamics. For turbulent flows, where only the normal component contributes to the nonlinear drag, a unique function $C_D^n(\mathrm{Dn})$ is expected if Dn is the only governing parameter in the dynamical equation (Fig. 7.1(a,d)). Any departure from this function must then signify an instability occurring in the superfluid component. If the superfluid component becomes turbulent at some critical velocity U_C, we expect a marked increase in the drag coefficient C_D^n above the dependence measured in a classical fluid.

We have already described in Chapter 1 the Donnelly–Glaberson instability leading to the production of quantized vorticity in the superfluid, related to self-reconnections of seeding vortex loops. The related critical velocity is expected to scale as $U_C \propto \sqrt{\kappa\omega}$ (Hänninen and Schoepe, 2010). Hence, it is convenient to define a reduced dimensionless velocity $\hat{U} = U/\sqrt{\kappa\omega}$. To facilitate a hydrodynamic description of the drag forces originating in the superfluid component, we again follow Schmoranzer *et al.* (2019) and define the superfluid drag coefficient as

$$C_D^s = \frac{2F}{A\rho_s U^2} = \frac{2F}{A\rho_s \kappa\omega \hat{U}^2}. \tag{7.8}$$

The situation is very different at velocities sufficiently above the critical values in the developed turbulent drag regime. In the latter case the normal and superfluid components are expected to be coupled due to the mutual friction force and, together, contribute to the pressure drag. In this situation, the classical definition of the drag coefficient is applicable, and we have $C_D = 2F/\left(A\rho U^2\right)$, where the total density $\rho = \rho_n + \rho_s$ is used. It is expected that in coupled turbulent flows, C_D tends towards a temperature-independent constant value of order unity (Blažková *et al.*, 2009; Skrbek and Vinen, 2009).

The total energy contained in the resonator is given as $E = m_{\rm eff} U_p^2/2$, defining the effective mass of the resonator, $m_{\rm eff}$. For a quasi one- or two-dimensional resonator oscillating perpendicular to its large dimension(s) – such as a thin cantilever, beam, or membrane – it follows that $m_{\rm eff} = \xi m$, where m is the actual mass of the resonator. In such a case, it is convenient to define a fluidic quality factor Q_f as (Schmoranzer *et al.*, 2019)

$$\frac{1}{Q_f} \equiv \frac{\langle \dot{E} \rangle}{\omega E} = \frac{\alpha S_r}{m}\sqrt{\frac{\eta\rho_n}{2\omega}} = \frac{\alpha\rho_n S_r \delta_n}{2m}, \tag{7.9}$$

this being independent of the velocity of the oscillator in the laminar regime. This expression differs from the one given by Ekinci *et al.* (2010) by the explicit inclusion of the flow enhancement factor α. We note that this factor is related to the potential flow outside the boundary layer and is necessary for correctly recovering the analytical solutions obtained for the drag force acting on an oscillating sphere or cylinder. Some oscillators, such as the vibrating wire and the torsionally oscillating disc permit analytical solution of the Navier–Stokes equations to obtain the required drag coefficient. Approximating the vibrating wire as an infinite smooth cylinder, one obtains $C_D^n = 4\pi/\mathrm{Dn}$ (see, e.g., Landau and Lifshitz, 1987). For a smooth disc $C_D^n = 2/\mathrm{Dn}$ and for a quartz tuning fork $C_D^n \simeq 4.67/\mathrm{Dn}$ (Schmoranzer *et al.*, 2019).

This theoretical description of high-Stokes-number oscillatory flow of He II in the temperature range, where He II displays the two-fluid behavior,[2] has been observed and verified experimentally by Schmoranzer *et al.* (2019) with several different oscillators: vibrating wire resonators, tuning fork, double-paddle, and torsionally oscillating disc. Figure 7.2 illustrates the behavior for two of these devices: (i) the quartz tuning fork with prong length $L_{\mathrm{f}} = 3.50\,\mathrm{mm}$, tine thickness (parallel to the direction of motion) $T_{\mathrm{f}} = 90\,\mu\mathrm{m}$, and width $W_{\mathrm{f}} = 75\,\mu\mathrm{m}$, its two prongs separated by $D_{\mathrm{f}} = 90\,\mu\mathrm{m}$, driven at two different flexural resonant modes – the fundamental resonance at 6.5 kHz and the first flexural overtone at 40.0 kHz; and (ii) the vibrating wire resonator consisting of a semi-circular 40 μm loop of superconducting NbTi wire with a leg spacing of 2 mm. For low velocities *a universal viscous drag scaling* was observed for the drag coefficient $C_{\mathrm{D}}^{\mathrm{n}}$ in terms of the Donnelly number Dn. Moreover, the same prefactor for the laminar scaling is displayed for the fundamental mode and overtone of the tuning fork, and agrees well with the theoretically derived value of 4.67 for this particular fork. In the case of the vibrating wire resonator, the prefactor for the laminar scaling is somewhat larger than one would expect for a smooth cylinder – most likely due to the surface roughness and excrescences left on the wire during the ablation of its insulation, as well as variations in the cross section of the wire caused by the extrusion process, effectively increasing the surface area. The universal viscous drag scaling $C_{\mathrm{D}}^{\mathrm{n}} \propto \mathrm{Dn}$ breaks down upon reaching the critical velocity for the generation of quantized vorticity (Donnelly–Glaberson instability) in the superfluid component. By increasing the velocity further, the normal and superfluid velocity fields become more and more tightly coupled by the action of the mutual friction force and, at very high velocities, He II effectively behaves as a single component fluid. As seen in the left panels of Fig. 7.2, classical universal scaling $C_{\mathrm{D}} = 2F/\left(A\rho U^2\right)$, where C_{D} is of the order unity, is indeed observed, as in classical single-component fluids.

Following Schmoranzer *et al.* (2019), we examine the transition to nonlinear drag regime in more detail. We are interested in determining which type of instability occurs first upon increasing oscillation amplitude: a classical instability of the normal component or the multiplication of remanent quantized vortices in the superfluid component. We remove from the measured drag force the part that is linear with velocity, keeping only the nonlinear contribution. Such a quantity needs to be normalized and plotted against parameters relevant to either component in

[2] We note that the two-fluid model description of independent and coupled oscillatory flows of the normal and superfluid components was first considered by Donnelly and Penrose (1956) in an attempt to explain the experimentally observed crossover between two regimes of U-tube oscillations. Although the notion of quantized vortices and their role for the mutual friction force were not yet widely appreciated, their approach was capable of formally explaining the existence of the two observed decay regimes, assuming that at low velocity the two fluids move independently and their motion becomes gradually coupled upon reaching some critical velocity, eventually moving together as a single fluid, as in coflow.

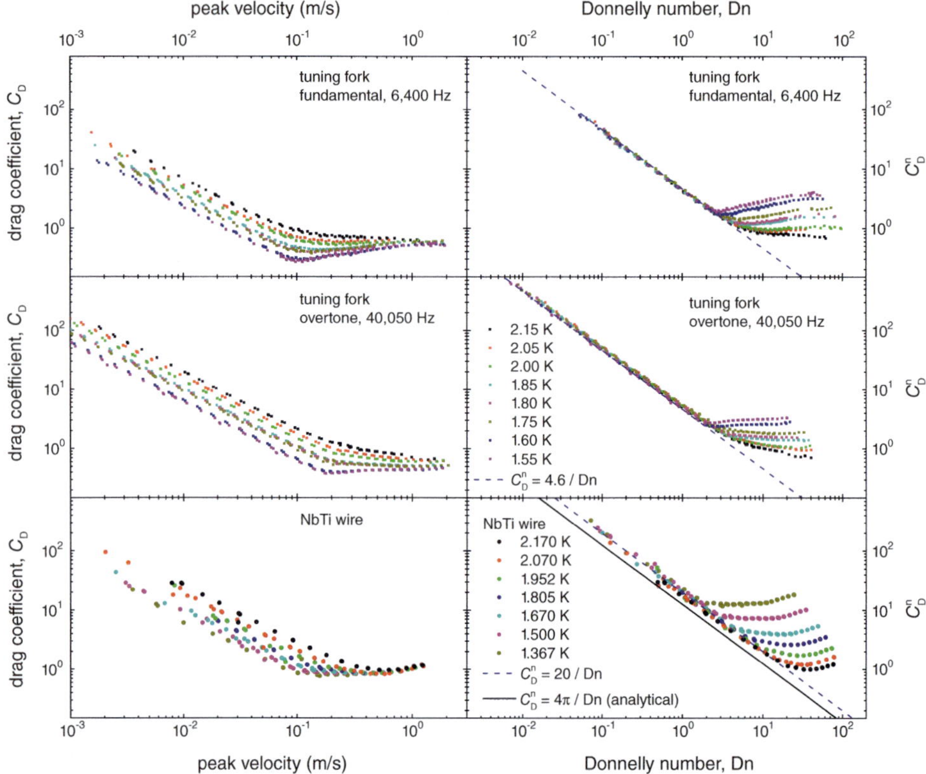

Figure 7.2 (Left) Drag coefficient as function of velocity for the quartz tuning fork (its fundamental and overtone modes) and the vibrating wire resonator. (Right) The corresponding normal fluid drag coefficients as a function of the Donnelly number. Note that (i) despite the difference in the velocity profile and the viscous penetration depth $\propto \omega$, the same prefactor for the laminar scaling is displayed for the fundamental mode and overtone of the tuning fork; and (ii) the prefactor for the laminar scaling is somewhat larger than one would expect for a smooth cylinder in the case of the wire. Reproduced with permission from Schmoranzer *et al.* (2019), used under CC BY 4.0 (https://creativecommons.org/licenses/by/4.0).

order to deduce the nature of the instability first detected. It seems particularly advantageous to use the quantity $1 - \Phi/\left(\mathrm{C}_\mathrm{D}^\mathrm{n}\mathrm{Dn}\right)$ in a plot against Dn to describe the action of the normal component and, analogously, $1 - \phi/\left(\mathrm{C}_\mathrm{D}^\mathrm{s}\hat{\mathrm{U}}\right)$ against $\hat{U}$ for the superfluid component; see Eq. (7.8). These definitions guarantee that the result is always close to zero in laminar flow and approaches unity as the nonlinear drag begins to dominate.

The resulting plots are shown in Fig. 7.3 for two tuning forks and the vibrating wire resonator, with each oscillator displaying a different behavior. We consider the instability occurring at a given departure from the linear drag, which must be above

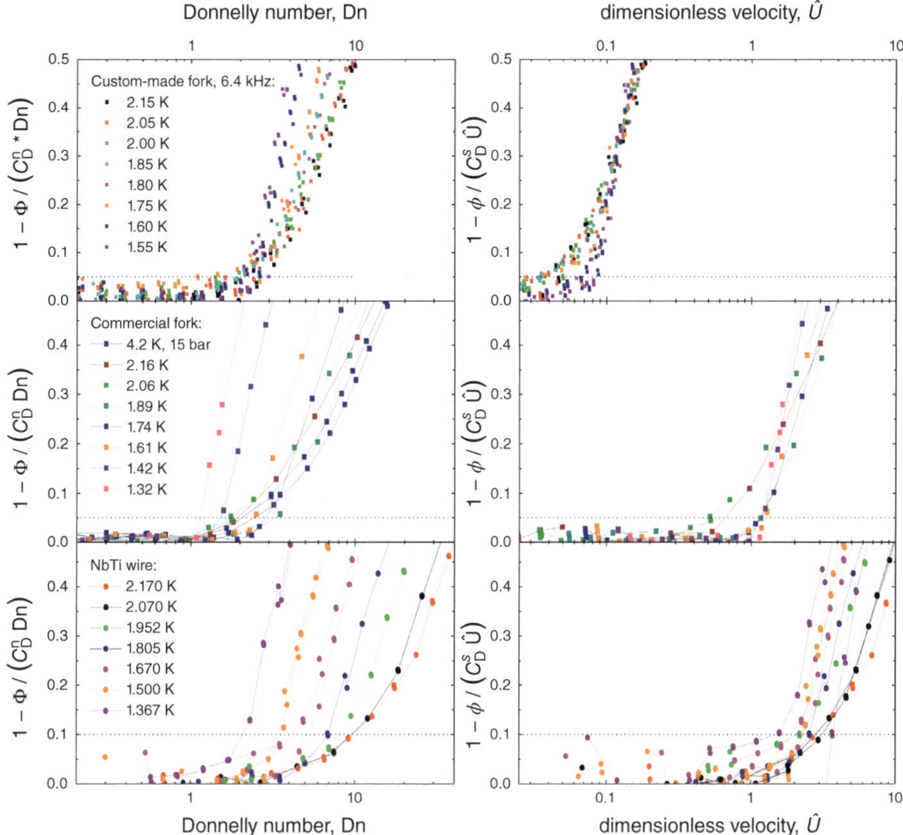

Figure 7.3 Turbulent instability analysis for tuning forks and the vibrating wire resonator. (Left) Nonlinear drag normalized using the normal component properties versus the Donnelly number. (Right) Nonlinear drag normalized by superfluid component properties versus nondimensional velocity $\hat{U}$. Reproduced with permission from Schmoranzer *et al.* (2019), used under CC BY 4.0 (https://creativecommons.org/licenses/by/4.0).

the experimental noise level in the data acquired in laminar flow. To understand the results, it is useful to consider two aspects: (i) the magnitude and relative spread of critical values of either Dn or $\hat{U}$ when crossing the given threshold; (ii) the rate at which the nonlinear drag sets in. In the top two panels of Fig. 7.3, the custom-made fork shows a notably lower spread in Dn than in $\hat{U}$, signifying that Dn is likely to be the correct parameter governing the (classical) instability in a larger part of the temperature range investigated. On the other hand, the vibrating wire resonator (bottom two panels) displays a rather well defined critical value of $\hat{U}$, while showing significant spread in Dn (except for the two highest temperatures, for which the critical values of Dn coincide), giving evidence of a Donnelly–Glaberson type of instability in the superfluid component. The commercial tuning fork (middle panels) shows a clear crossover: At temperatures below 2.0 K the instability is governed by $\hat{U}$, while at higher temperatures it is determined by Dn. It is interesting to note that whenever the instability is determined by $\hat{U}$, the onset of nonlinear drag is sharper. A crossover between a classical and quantum instability might be present in the other two oscillators as well, but is not as pronounced as for the commercial tuning fork.

In summary, which instability (i.e., classical hydrodynamic instability of laminar flow of the normal component or Donnelly-Glaberson instability in the superfluid component) occurs first depends both on the geometry of the oscillator and temperature. A crossover between these instabilities could occur thanks to the steep temperature dependence of the kinematic viscosity of the normal fluid. Upon increasing oscillation amplitude, either instability can live on its own until eventually it serves as a trigger for the other, mediated by the mutual friction force or by pressure forces. At high velocities, both fluids are tightly coupled in the vicinity of the oscillator and He II behaves as a single-component quasi-classical fluid.

7.3 Oscillating Spheres, Wires, and Grids in He II

There is a long history of experiments on spheres oscillating both torsionally (e.g., Benson and Hollis-Hallet, 1956) and transversally (e.g., Luzuriaga, 1997) in superfluid ^{4}He, but we shall focus here on the systematic and pioneering work of Jager *et al.* (1995) and Niemetz *et al.* (2002, 2004), as described in Chapter 4. Besides the recent results described in Section 7.2, a rich variety of experiments performed over several decades of investigations have revealed interesting results that are only partly understood. Here we describe some of them, obtained in He II.

Typical results showing the dependence of oscillation amplitude and drag coefficient on the driving force are shown in Fig. 7.4. There is a region at low drives where the response is proportional to the drive, indicating a laminar flow, but at higher drives this linear response is replaced by the proportionality to the square of the velocity, indicating a turbulent flow regime. At temperatures lower than about 1 K, while performing the amplitude sweeps of the drive the transition from laminar to turbulent response is accompanied by significant hysteresis. At the lowest temperature, this region of hysteresis is replaced by one in which neither the laminar state nor the turbulent state appears to be stable, so that the response switches intermittently between the two states.

We note that such an intermittent switching between laminar and turbulent states has been observed with various mechanical oscillators, e.g., with tuning forks (Bradley *et al.*, 2014).

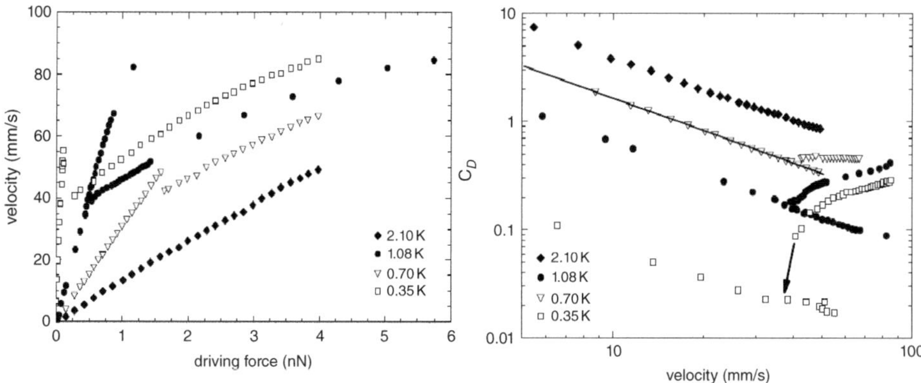

Figure 7.4 (Left) Typical results showing the dependence of velocity amplitude on driving force for the oscillating sphere in the experiment of Schoepe's group (Jager *et al.*, 1995) at four temperatures. Note that the laminar drag coefficient at 0.7 K is higher than that at 1.08 K. This reflects the fact that the laminar drag force displays a maximum in the temperature range below 1 K associated with a crossover from the hydrodynamic to the ballistic drag regime. (Right) The same results plotted as the drag coefficient against the velocity amplitude. The solid line is a linear fit in the laminar regime at $T = 0.7$ K.

A variety of interesting experiments with vibrating wire resonators in ^{4}He, relevant to the transition to both classical and quantum turbulence, has been reported over many decades of research. In particular, the vibrating wire resonators (Morishita *et al.*, 1989), as well as spheres (Jager *et al.*, 1995) and grids (Charalambous *et al.*, 2006), clearly show the transition from hydrodynamic behavior at temperatures above about 1 K to ballistic behavior at millikelvin temperatures. Vibrating wire resonators in ^{4}He have been thoroughly studied, especially by the Lancaster Group (reporting the history-dependent behavior [Bradley *et al.*, 2011b] as well as two critical velocities [Bradley *et al.*, 2009a]), and by Yano *et al.* (2005, 2006, 2007) and Hashimoto *et al.* (2007).

We now describe some results obtained at Osaka with a rather smooth, thin wire of diameter 2.5 μm at a frequency of 610 Hz. Note that the flow of He II due to such a vibrating wire cannot at any temperature be classified as a high Stokes number flow, and so the analysis discussed above is not applicable. If the wire was surrounded initially by helium in the normal phase, and if the helium is then pumped through the λ-transition and cooled rapidly to 1.4 K, the initial response of the wire indicated very strong damping, even at low drives, with very strong hysteretic effects. Repeated scanning of the drive, however, eventually leads to the steady and reproducible response with no hysteresis, exhibiting a transition from a laminar regime to a turbulent drag regime upon reaching a wire velocity of about

30 mm/s. Alternatively, the wire can be directly immersed, slowly and carefully, into the superfluid phase. In this case, the critical velocity was significantly higher. However, such a high critical velocity is not stable, and repeated amplitude sweeps of the drive lead to its reduction to about 50 mm/s, again with no significant hysteresis.

The behavior of a wire at much lower temperatures has also been studied by both the Lancaster and Osaka groups. Again, there is a transition that corresponds to a laminar regime changing to a turbulent drag regime, but now with considerable hysteresis, as just discussed for a sphere. We note that, in the flow region identified as laminar, the response is not proportional to the drive. This is due to the fact that a substantial part of the damping in this temperature region occurs by a nonlinear internal friction in the wire.

The Osaka group has repeated the measurements at 35 mK, taking greater care in filling the region round the wire with helium (Hashimoto *et al.*, 2007). The authors placed the wire inside a small cell, which was then filled very carefully and slowly through a small (0.1 mm) pinhole at a very low temperature. In this case, the response remains laminar up to the largest velocity that could be imposed, greater than 1 m/s.

These observations strongly suggest that transition to turbulence in bulk ^{4}He depends on the extrinsic vortex nucleation. Indeed, direct and convincing experimental evidence for the role of remanent vortices has been provided by Goto *et al.* (2008), based on experiments with two vibrating wire resonators. Both were placed in a fluid at a very low temperature and filled very slowly through a small orifice. Sometimes it was found that for one of the vibrating wire resonators (but not for the other) the critical velocity was high enough (greater than 1 m/s) that it could not be observed. Confirmation that this high critical velocity was associated with the absence of any attached remanent vortex was provided by exciting the second wire above its (rather small) critical velocity, which led rapidly to a large reduction in the critical velocity for the first wire. The reason is that the excitation of a second wire, to which a few remanent vortices are attached, leads to the emission of a beam of vortex rings (Nago *et al.*, 2011), which leads to the generation of turbulence around the first wire.

7.4 Multiple Critical Velocities

We stress that the hydrodynamical description that we have presented is applicable only for oscillatory high-Stokes-number two-fluid flows of He II, i.e., above about 1 K. By lowering the temperature, the hydrodynamic regime ceases to hold due to a steep increase of the mean free path of quasiparticles (mostly phonons) composing the normal fluid. Detailed description of ballistic propagation and scattering of quasiparticles off the moving bodies, i.e., ballistic drag regime, is beyond the scope of this book. We comment only briefly on some peculiarities in the transition to turbulent drag regime observed in the superfluid He II at very low temperatures. Here, a number of experimental studies using vibrating wires (Bradley *et al.*, 2005a), grids (Nichol *et al.*, 2004), and tuning forks (Bradley *et al.*, 2009b; Garg *et al.*, 2012) have reported the existence of more than one critical velocity of hydrodynamic origin. Recently the Prague and Lancaster groups presented convincing evidence for three distinct hydrodynamic critical velocities and proposed an explanation linking all the observations of oscillatory flows in the limit of zero temperature to a single framework (Schmoranzer *et al.*, 2016).

The first critical velocity, connected mostly to frequency shifts rather than changes in the drag force, is associated with the formation of a number of quantized vortex loops near the surface of the oscillator (Nichol *et al.*, 2004), possibly forming a thin layer, which affects the coupling to the fluid via the hydrodynamic added mass. This first critical velocity is hardly observable in the two-fluid regime above 1 K.

The second critical velocity is related to the quantized vorticity propagating into the bulk of the superfluid, either in the form of emitted vortex loops or, eventually, as a turbulent tangle. It is always accompanied by a marked increase in the drag force and usually hysteresis (detectable with amplitude sweeps). We stress that it is this critical velocity which we discussed above in relation to the experiments performed in the hydrodynamic regime above 1 K.

For completeness, we mention the third and higher critical velocity of hydrodynamic origin, above which the drag coefficient starts to grow towards unity even at mK temperatures (the drag coefficient is of order 10^{-1} or 10^{-2} at velocities exceeding the second, but not the third, critical velocity). The third critical velocity was observed to be high, always above 1 m/s (Schmoranzer *et al.*, 2016).

7.5 Second Sound as Oscillating Turbulent Counterflow

We have already discussed in Chapter 4 that quantum turbulence in He II can be generated by high-amplitude second sound. This is a special form of quantum turbulence, which we shall refer to as *ac counterflow turbulence*, as second sound represents just counterflow oscillations of the normal and superfluid components. Kotsubo and Swift (1989, 1990) were the first to generate this form of quantum turbulence using the cylindrical second sound resonator with the ends of somewhat larger diameter than the center. Longitudinal second sound in the resonator was generated externally, via a superleak, by a bellows driven by a stationary superconducting magnet, which exerts pressure on a superconducting plate attached to the moving end of the bellows. The authors measured the ac critical velocity and found it to be the same as in dc counterflow measurements; the critical velocity

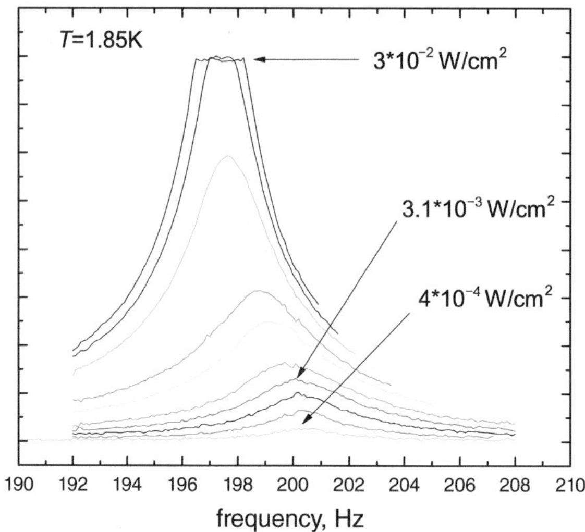

Figure 7.5 Amplitude versus frequency dependence of the first harmonic of second sound at temperature 1.85 K. Upon reaching the critical counterflow velocity in the second sound drive, the amplitude grows no further, owing to generation of quantum turbulence in the vicinity of second sound antinode. Reprinted from Chagovets (2016) with permission from Elsevier.

was independent of the resonator diameters and displayed a square-root-frequency dependence.

The signature of quantum turbulence appearing in the resonator is the flat top frequency dependence of the second sound amplitude. This feature was later observed also by Chagovets (2016), see Fig. 7.5, and Midlik *et al.* (2021) when generating quantum turbulence by second sound thermally, applying an ac power to a heater at one side of the resonator, and detecting the second sound amplitude by a sensitive thermometer at the opposite side.

Experiments of Kotsubo and Swift (1989, 1990) and Chagovets (2016) did not directly investigate the steady-state vortex line density of second sound generated counterflow turbulence or its temporal decay. The Prague group (Midlik *et al.*, 2021) therefore added to a resonator (3.2 cm long square tube of 1 cm side), a pair of second sound sensors perpendicular to its longer dimension, in the middle of its length. This modification opened the possibility for localizing, by using the fundamental as well as higher longitudinal second sound resonances, the appearance of quantum turbulence first in the vicinity of antinodes of the second sound standing modes. Moreover, employing low-amplitude second sound harmonics in the perpendicular direction allowed a study the dynamics of generation and decay of vortex line density upon switching on and off the strong longitudinal thermal drive, delivered by a function generator to an ac resistive heater. The voltage signal on the heater has the form of a harmonic wave function $V_0 \cos(\omega t)$, V_0 being the signal amplitude. The actual time-dependent heat power supplied to the heater is $P \propto 1 + \cos(2\omega t)$, resulting in both ac and dc components of the supplied power. The ac component is used for the generation of a second sound standing wave, i.e., the oscillatory counterflow with frequency doubling that supplied by the generator.[3]

Midlik *et al.* (2021) determined, using two different approaches, the ac critical velocity that leads to the generation of quantum turbulence in the antinodes of the longitudinal second sound. These data, together with those on critical velocities from all oscillatory counterflow experiments just described, allowed the analysis, within the framework of the two-fluid model of He II, of the underlying physics of transition to turbulence in oscillatory flows of He II for high Stokes numbers, along the lines discussed earlier in this chapter.

For low velocities in either coflow or counterflow, only viscous drag is offered by the normal fluid, obeying a universal scaling law in terms of the Donnelly number (boundary layer–based Reynolds number), defined above as $\mathrm{Dn} \equiv (\delta_n \rho_n U)/\eta$. Upon exceeding a certain critical value of the Donnelly number, Dn_{cr}, the normal component undergoes a transition akin to the classical case, subsequently triggering the generation of quantized vortices in the superfluid component. The corresponding critical velocity of the normal component is denoted as $u_{n,cr}$.

On the other hand, quantized vorticity in the superfluid component may be triggered via the Donnelly–Glaberson instability. Indeed, Hänninen and Schoepe (2008, 2010) have argued that the onset of quantum turbulence in oscillatory flows of superfluid helium is universal and can be derived from a general argument based on the "superfluid Reynolds number." The critical velocity scales as $u_{s,cr} \propto \sqrt{\kappa\omega}$, with only the numerical prefactor depending weakly on the geometry of the oscillating object because the flow velocity near the surface of the object may differ from the velocity amplitude elsewhere. A more detailed analysis (Hänninen and Schoepe, 2008) derived from the dynamics of the turbulent state gives the criterion

$$u_{s,cr} \approx 8\kappa\omega/\beta, \tag{7.10}$$

where the numerical factor β is about unity and depends on the mutual friction parameters: For example, $\beta = 1$ below 1 K, $\beta = 0.95$ (at 1.3 K), 0.89 (at 1.6 K), and 0.79 (at 1.9 K). This trend implies a slow *increase* of $u_{s,cr}$ by about 10% and is in fair agreement with experimental results obtained over a wide temperature range, from below 0.4 K up to 1.9 K, for a 100 μm diameter sphere oscillating at 236 Hz. It must be noted that virtually the same approach was used to analyze the turbulent

[3] The dc component of the heat power establishes a steady thermal counterflow, but the exact geometry of such a counterflow is unknown because the channel is closed, but not leak-sealed. The authors therefore treated the vortex line density in the center of the resonator, generated at frequencies of longitudinal second sound resonances, simply as a heat flux–dependent background.

instability in oscillatory counterflow without any consideration of an instability in the normal component. This resulted in the observation of a strong and systematic temperature dependence of superfluid critical velocities that could not be explained by this scaling theory (Kotsubo and Swift, 1990, 1989).

Which of the two instabilities – classical hydrodynamic instability of laminar flow of the normal component or the Donnelly–Glaberson instability in the superfluid component – occurs first depends on both the geometry of the oscillator and the temperature, which determines the dynamic viscosity of He II and the densities of the two components. In experiments on flow due to mechanical resonators, the comparison of the two criteria for the transition is straightforward: In a coflow, the velocities of the normal and superfluid components are practically identical. Even in thermal counterflow, noting that $u_{n,cr} = \rho_s/\rho_n u_{s,cr}$, we can find a common dimensionless parameter. For this purpose, the superfluid critical velocity $u_{s,cr}$ may be converted to an *effective critical Donnelly number*

$$\mathrm{Dn}_{\mathrm{cr,eff}} = \frac{\rho_s}{\rho_n}\frac{\delta_n \mathrm{u}_{s,cr}}{\nu_n(\mathrm{T})}, \qquad (7.11)$$

where $\nu_n(T)$ is the kinematic viscosity of the normal fluid. Unlike the true critical Donnelly number describing the classical instability $\mathrm{Dn}_{\mathrm{cr}}$, the critical value of $\mathrm{Dn}_{\mathrm{cr,eff}}$ is no longer expected to be constant. In fact, requiring a constant value of the correct critical parameter, $u_{s,cr}$, also requires $\mathrm{D}_{\mathrm{cr,eff}}$ to be a function of temperature. However, $\mathrm{Dn}_{\mathrm{cr,eff}}$ will be independent of the frequency of oscillations, as both $u_{n,cr}$ and $u_{s,cr}$ have the same frequency dependence of either critical velocity as $\propto f^{1/2}$.

The critical Donnelly numbers from all available experiments are shown in Fig. 7.6. It is remarkable that, within the experimental accuracy, in the temperature range from 1.2 K to 1.7 K, different experiments such as the mechanically driven second sound (Kotsubo and Swift, 1989, 1990) and thermally driven counterflow (Chagovets, 2016; Midlik *et al.*, 2021) display the onset of the transition to quantum turbulence at the same critical Donnelly number $\mathrm{Dn}_{\mathrm{cr}} = 16 \pm 3$, suggesting that the transition is triggered by the instability in oscillatory laminar flow of the viscous normal component. Above ≈ 1.8 K, however, $\mathrm{Dn}_{\mathrm{cr}}$ decreases as the temperature is increased, in accordance with the effective Donnelly number $\mathrm{Dn}_{\mathrm{cr,eff}}(\mathrm{T})$ given by Eq. (7.11). This behavior is fully explained by the instability in the superfluid component, namely the production of quantized vorticity by means of the Donnelly–Glaberson mechanism. Thus a crossover of two different mechanisms of turbulence generation in oscillatory counterflow is observed: one related to a classical instability of the normal fluid dominating at lower temperatures in the two-fluid regime, and the other purely a consequence of quantized vortex dynamics in the superfluid component dominating at higher temperatures. This observation strongly suggests, perhaps surprisingly, that transition to turbulence in oscillatory

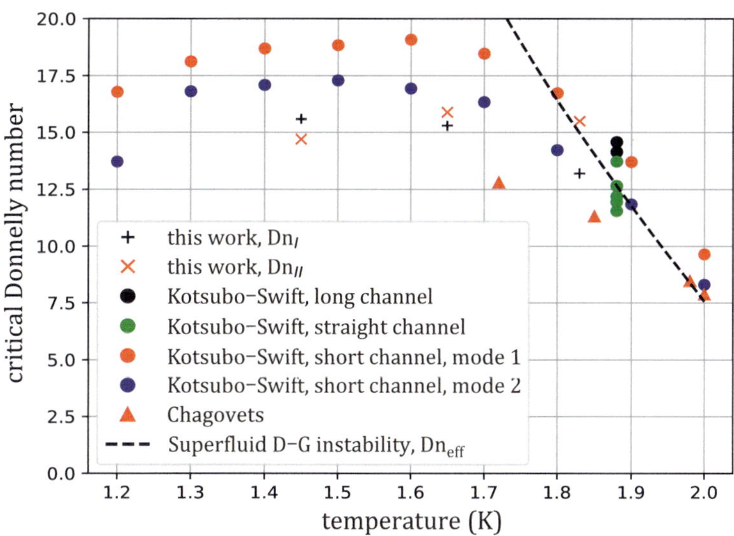

Figure 7.6 Temperature dependence of the critical Donnelly number for oscillatory thermal counterflow as determined from the experiments of Kotsubo and Swift (1989, 1990), Chagovets (2016), and Midlik *et al.* (2021). The dashed line represents an effective Donnelly number calculated for the instability based on the Hänninen–Schoepe criterion, Eqs. (7.10) and (7.11). Reprinted figure with permission from Midlik *et al.* (2021). Copyright 2021 by the American Physical Society.

coflow and counterflow is governed by the same underlying physics, although the crossover occurs in opposite directions for counterflow and coflow, as described in Section 7.2.

We emphasize that the hydrodynamic approach is applicable only in the temperature range where He II displays the two-fluid behavior.

7.6 Summary

In Chapter 6 we saw that in He II, at temperatures above about 1 K, the two-fluid properties make the problem of transition to quantum turbulence in coflows, counterflows, or pure superflows quite complex. Thus, the transition physics is only partly understood. One reason is that, experimentally, it is very difficult to create steady laminar flow of He II, and nearly impossible to generate potential flow of He II in the zero-temperature limit. It is much easier to create oscillatory potential flows using mechanical oscillators of various shapes. In the zero-temperature limit, i.e., in pure quantum flows generated by these oscillating objects, several critical velocities of hydrodynamic origin and intermittent switching among them have been observed.

In the case of oscillatory flows driven at high Stokes number, a fair degree of understanding has recently been achieved because instabilities associated to the normal and superfluid components can be disentangled. In particular, the crossover between instabilities based on the classical Donnelly number and of the (quantum) Donnelly–Glaberson type can be identified in special cases.

8
Turbulence in ^{3}He-B

In superfluid ^{3}He-B at relatively high temperatures (i.e., in terms of T/T_c, where the critical temperature T_c is of order 1 mK, though it depends on pressure), the kinematic viscosity of the normal fluid is large enough that the normal fluid stays still in experimental containers of typical size of 1 cm^3. A completely new class of quantum turbulence becomes possible in which only the superfluid component moves with respect to boundaries. At lower temperatures, including the zero-temperature limit where no normal fluid exists, the superfluid turbulence in ^{3}He-B displays extraordinary features, some of them unique because of its fermionic nature. In this chapter, we consider a few of them.

8.1 NMR Experiments

The relevant nuclear magnetic resonance (NMR) experiments on transition to this type of quantum turbulence in ^{3}He-B have been carried out by the Helsinki ROTA group using rotating cryostats; this work has contributed much, over the years, to our knowledge of the rotating superfluid phases of ^{3}He. Though NMR in ^{3}He-B does not directly measure the number of vortices in rotation, it offers valuable related information.

Most of the relevant experiments have been basically organized as follows. The ^{3}He sample is contained inside a quartz tube of 6 mm in diameter with smooth walls, connected, via a small orifice in the bottom, to the heat exchanger volume on the nuclear demagnetization stage of the rotating cryostat. Three independent superconducting coils produce axially oriented magnetic fields, at both ends of the tube (for the NMR measurements), as well as in the middle, this latter being called the barrier field, which serves to stabilize the superfluid ^{3}He-A-phase in this part of the sample. For a wide range of pressures and temperatures, this experimental configuration allows a stable A/B phase boundary in the cell.

It is important to realize that in rotating ^{3}He-B the critical velocity for intrinsic vortex nucleation is rather high (although still few orders of magnitude lower than in He II) and the vortex-free superflow, the so-called Landau state, can be maintained up to high angular velocities. Other metastable states with an intermediate number of rectilinear vortex lines running through the central part of the sample, up to the equilibrium number N_{eq}, are also stable. On the other hand, the critical velocity for vortex nucleation in the A-phase, which floats just above the B-phase, is small and, for our purposes, can be neglected hereafter. Therefore, the rotating A-phase can always be thought of as containing (nearly) the equilibrium number of vortices mimicking its solid body rotation, as explained in Chapter 2 for a rotating bucket of He II.

Now, if the cryostat is slowly accelerated from zero up to some angular rotation velocity Ω, the following configuration is possible. The B-phase (below the stable A/B boundary) is in the vortex-free Landau state, while, above it, the A-phase is in solid body rotation. Note that in both A and B phases the thick viscous normal fluid always rotates together with the container, providing a unique frame of reference.

It was shown by Blaauwgeers *et al.* (2002) and recently revisited by Eltsov *et al.* (2019) that, at some critical angular velocity Ω_{cr}, the A/B boundary undergoes a shear wave Kelvin–Helmholtz instability. This celebrated instability is well known in classical hydrodynamics corresponding to a spontaneous emergence of interfacial surface waves as a function of the flow velocity parallel to the interface. A typical example is generation of waves on water due to a blowing wind. The Kelvin–Helmholtz instability can be carried over to the inviscid two-fluid dynamics of superfluids to describe the stability of the phase boundary separating two bulk phases A and B of superfluid ^{3}He in rotating flow when the A/B boundary is localized with a magnetic field gradient. Upon reaching the critical velocity of the flow of the superfluid component in the A-phase, rotating on top of stationary superfluid B-phase, the A/B interface becomes wavy and an injection of a few vortices from the A-phase to the B-phase occurs. This injection is the experimental mechanism by which ΔN seeding vortex loops are introduced into the B-phase, which would otherwise remain in the vortex-free Landau state.

This unique vortex nucleation mechanism was used by Finne *et al.* (2003) in the experiment schematically shown in Fig. 8.1. The outcome of this injection falls in one of two categories depending on the temperature. At high temperature, the final number of vortices, N_f, observed in the B-phase equals the number of injected vortices, ΔN. This result requires some transition time, typically a few seconds, until the injected seeding vortices expand, with their ends sliding along the container walls, until a rectilinear vortex array is formed around the rotation axis. At low temperature, however, the expansion of the seeding vortex loops is turbulent and $\Delta N \ll N_f \leq N_{eq}$. This process leads to a total removal of the macroscopic

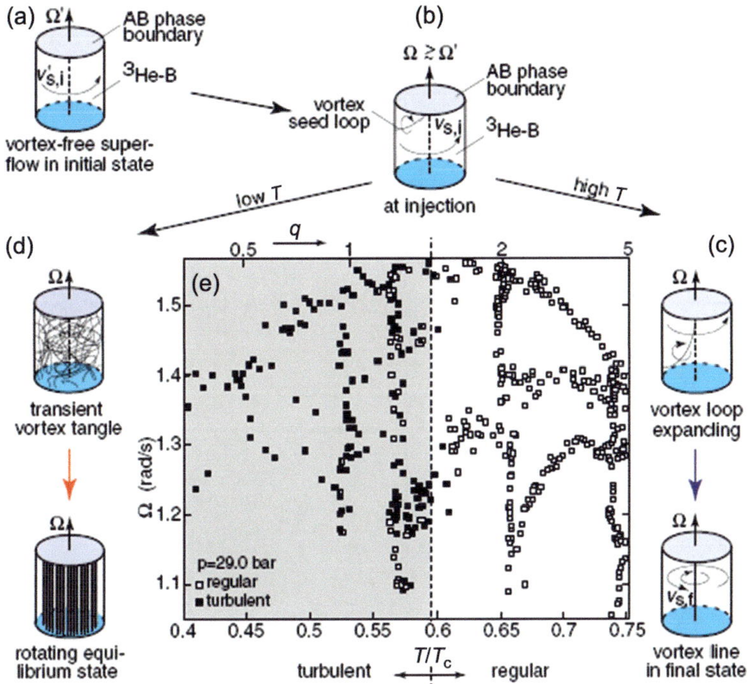

Figure 8.1 Phase diagram of turbulent superflow in ^{3}He-B. The principle of the measurements is as follows. (a) The initial state is vortex-free (Landau state) superflow in the container rotating at angular velocity Ω; when in the rotating frame the normal component is stationary and the superfluid component flows. (b) A few (ΔN) vortex loops are injected and, after a transient period of loop expansion, the number of rectilinear vortex lines N_f in the final steady state is measured. This state is found to fall in one of two categories, (c) or (d). (c) $N_f = \Delta N$, regular mutual-friction-damped loop expansion. (d) $\Delta N \ll N_f \leq N_{eq}$, turbulent loop expansion. This process leads to a total removal of the macroscopic vortex-free superflow (in the rotating frame of reference), as the superfluid component is forced into solid-body-like rotation (on an average) by the formation of a vortex array with the equilibrium number of rectilinear lines, $N_{eq} \approx \pi R^2 2\Omega/\kappa \sim 10^3$. (e) Phase diagram of measured events. From Finne *et al.* (2003).

vortex-free superflow, as the superfluid component is forced into solid-body rotation (on average) by the formation of a vortex array with the equilibrium number of rectilinear lines, $N_{eq} \approx \pi R^2 2\Omega/\kappa \sim 10^3$.

Figure 8.1 shows the experimentally observed phase diagram of turbulent superflow in ^{3}He-B. Each data point monitors the outcome from the injection event in the final steady state as a function of rotation and temperature. Regular loop expansion is marked by open symbols and the turbulent state by filled ones. An abrupt transition at $0.60\,T_c$, independently of Ω, has been found to divide the (X,T) plane

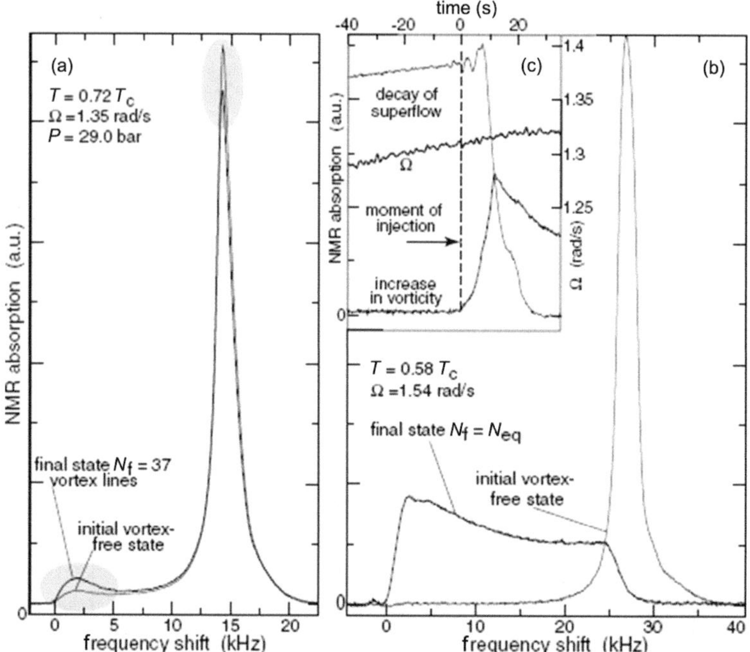

Figure 8.2 NMR absorption spectra before and after vortex loop injection. See text for details. From Finne *et al.* (2003).

vertically into the two types of vortex dynamics. Only within a narrow interval $\Delta T \approx 0.05\,T_c$ around $0.60\,T_c$ can the intermediate behavior be observed.

Figure 8.2 shows how Finne *et al.* (2003) came to the above conclusions from the observed NMR spectra. As already pointed out, NMR in ^{3}He-B does not directly measure the number of vortices in rotation, but the spectral shape reflects the spatial texture of the sample. The texture problem consists of finding the spatial distribution of the rotation axis $\hat{n}$ of the order parameter. The overall NMR absorption signal from the sample can be thought of as consisting of the signal of a number of small volumes (local oscillators) of constant angle β between $\hat{n}$ and the magnetic field (vertical) direction. The normalized frequency shift from the Larmor frequency for each of these oscillators is approximately proportional to $\sin 2\beta$. Hence the overall shape of the observed NMR spectrum depends on the texture. The type of the texture is a result of interplay of orienting interactions among the magnetic field energy, the flow energy, and the surface energy.

It is known that, in a long cylindrical sample of ^{3}He-B at rest, the so-called flare-out texture is established: $\hat{n}$ points up along the rotation axis of the container (i.e., $\beta(0) = 0$), but bends off the axis to reach an angle $\beta(R) = 63°$ at the wall. The influence of rotation on the texture in a cylindrical container is rather complicated, as there are several textural transitions induced by rotation, as described in the original work of Korhonen *et al.* (1990). The observed NMR spectra shown in Fig. 8.2 can be interpreted as follows.

The two spectra on the left, marked (a), represent the vortex-free Landau state, with only minor changes introduced by regular mutual-friction-damped loop expansion above $0.60\,T_c$. The sharp peak at large frequency shift (from the Larmor frequency) is caused by the large-scale superflow present in the rotating frame. When a few rectilinear vortex lines are formed, some absorption from the large peak is shifted into a tiny new peak at small frequency shift. The heights of both peaks change approximately linearly with N_f (in the highlighted regions).

The two NMR spectra on the right, marked (b), show a radical restructuring that occurs below $0.60\,T_c$ when the large-scale superflow is replaced by an equilibrium array of rectilinear vortices.

The inset (c) displays peak heights of the two absorption maxima as a function of time during the transient turbulent period: (i) The sharp peak at large frequency shift decreases, monitoring the decay of the large-scale superflow; (ii) The tiny peak at small frequency shift first increases, reflecting the build up of the high-density vortex tangle. Subsequently it starts to decrease, corresponding to the rarefaction and polarization of the tangle into an equilibrium array.

The Helsinki data clearly shows a surprisingly sharp transition between regular high-temperature processes (marked with open symbols in the phase diagram in Fig. 8.1) and turbulent low-temperature processes (filled symbols) observed at $0.60\,T_c$, with $N_f \sim 1$ above the transition and $N_f \approx 10^3$ below. The main conclusion is that the transition occurs as a function of temperature, with little or no dependence on the maximum superflow velocity $U = \Omega R$, in striking contrast to the situation in classical liquids.

It is interesting to study the transient mechanism of the turbulent low-temperature process in more detail. In He II, due to strong surface pinning the container walls are always rough and covered with a large number of remanent vortices that serve as seeds for rapid transition to turbulence when flow takes place. In ^{3}He-B, on the other hand, the (quartz) container walls are smooth. Further experiments by the Helsinki group (Finne *et al.*, 2006) show that an initially injected ring generates new vortices while interacting with the container wall in a process that is based on computer simulations in Fig. 8.3. Experimentally, this process is observed as slow vortex formation that precedes the more rapid turbulence generation. The precursor process supplies new vortices so that ultimately rapid turbulence will switch on at some location where the loop density has grown sufficiently. This extraordinary mechanism of vortex multiplication is relevant in the case of steady flow of ^{3}He-B with clean smooth surfaces.

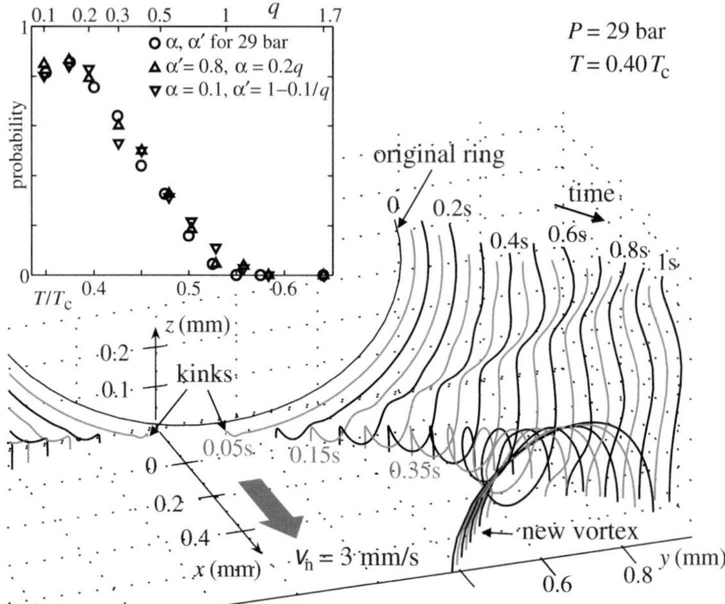

Figure 8.3 As a model of the precursor, the collision of a vortex ring with a plane wall (at $z = 0$) is shown from calculations. A ring of 0.5 mm radius is initially above the wall, with the plane of the ring tilted by $\pi/10$ from the $x = 0$ plane. In the frame of reference of this figure, uniform flow of the normal component at $u_n = 3$ mm/s is applied in the x-direction. Different contours show the ring at intervals of 0.05 s. The sharp kinks at the wall reconnections induce Kelvin waves on the loop. They are able to grow in the applied flow only on one of the two legs formed in the reconnection (here, on the right). The largest wave reconnects with the wall and a new loop is then separated. The conditions for the growth of Kelvin waves are fulfilled in such collisions only in the underdamped temperature regime: The inset on the top shows the probability that a new vortex is created when the original ring is initially placed at $z = 1$ mm with random orientation. This probability depends on the ratio $q = \alpha/(1-\alpha')$, shown on the top axis, rather than on α or α' separately. Reprinted figure with permission from Finne *et al.* (2006). Copyright 2006 by the American Physical Society.

How to theoretically understand the extraordinary transition to turbulent flow occurring in the thick “soup” of the (almost) stationary normal fluid? Let us assume that the quantized vortices in the flow are arranged in such a way that the coarse-grained hydrodynamic equation,

$$\frac{\partial \boldsymbol{u}_s}{\partial t} + \nabla\mu = (1-\alpha')\boldsymbol{u}_s \times \boldsymbol{\omega} + \alpha\hat{\boldsymbol{\omega}} \times (\boldsymbol{\omega} \times \boldsymbol{u}_s), \tag{8.1}$$

obtained by Sonin (1987) from the Euler equation, after averaging over vortex lines and written in the frame of reference of the normal fluid, provides a sufficiently accurate description of the superflow. The symbols α and α' are the mutual friction coefficients of ^{3}He-B experimentally determined by Bevan *et al.* (1997), $\boldsymbol{\omega}$ is the course-grained vorticity, and $\hat{\omega}$ is a unit vector in the direction of ω. In order to apply this equation, we must bear in mind that the coarse-grained equation describes with sufficient accuracy the superfluid velocity field on the scale over which the averaging is done.

Since the normal fluid provides a unique frame of reference, we have to deal only with the superfluid velocity u_s. Rescaling the time variable such that $t \to (1 - \alpha')t$ leads to

$$\frac{\partial u_s}{\partial t} + \nabla \mu = u_s \times \omega + q\hat{\omega} \times (\omega \times u_s). \tag{8.2}$$

A theoretical analysis of the fluid dynamical problem based on this equation was performed by Finne *et al.* (2003), Volovik (2003), and L'vov *et al.* (2004), and independently by Vinen (2005). We shall describe the resulting shape of the energy spectra in Chapter 9; here, we restrict our discussion to different aspects of the problem. As was first emphasized by Finne *et al.* (2003), Eq. (8.2) has a remarkable property that distinguishes it from the classical Navier–Stokes equation. In the Navier–Stokes equation, the relative importance of the inertial and dissipative terms is given by the Reynolds number, which depends on the density and viscosity of the fluid, as well as the speed and the geometry of the flow. Here, the role of the Reynolds number is played by the parameter

$$Re_{\text{eff}} = \frac{1}{q} = \frac{1 - \alpha'}{\alpha}, \tag{8.3}$$

which, surprisingly, depends on the temperature but not on the geometry or flow velocity. We stress that the superfluid Reynolds number (as soon as it is large enough, assuming that Eq. (8.2) is a good approximation of the superflow) is not relevant to a consideration of the problem of a flow obeying Eq. (8.2), whose beauty is that one can derive more general conclusions about turbulent flow generated from suitable initial conditions, depending only on a single temperature-dependent parameter $1/q$, regardless of the actual geometry of the flow.

In ^{3}He-B, a wide range of q values is experimentally achievable since q increases with temperature (Bevan *et al.*, 1997). Similar to the Navier–Stokes equation, Eq. (8.2) possesses both laminar ($q \gg 1$) and fully turbulent ($q \ll 1$) solutions.

8.2 Spin-Up of ^{3}He-B

Spin-up is the transient adjustment occurring in the confined fluid when the rotation rate undergoes an impulsive positive change. A simple example is spin-up of a vertically oriented cylindrical container of radius R and height H, filled with incompressible viscous fluid, accelerated from rest to finite angular velocity Ω (Greenspan, 1968). Boundary layers form on horizontal surfaces following the

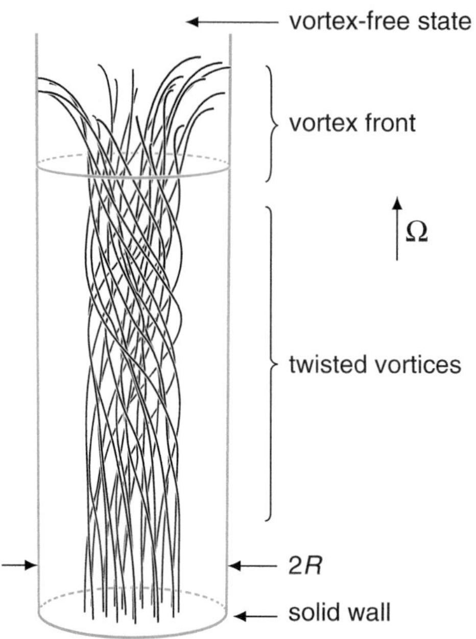

Figure 8.4 The formation of a twisted vortex state by numerical simulation. The vortices have their propagating ends bent to the sidewall of the rotating cylinder. As they expand upwards into the vortex-free state, the ends of the vortex lines rotate around the cylinder axis. The twist is nonuniform because boundary conditions allow it to unwind at the bottom solid wall. Reprinted figure with permission from Eltsov *et al.* (2006). Copyright 2006 by the American Physical Society.

impulsive start. Nonrotating fluid is entrained into these layers, spun-up by viscous action. Eventually, solid body rotation prevails throughout the fluid.

In Chapter 3 we discussed the case of a rotating bucket of He II. We learned that, in the temperature range where He II displays the two-fluid behavior, thanks to the vortex lattice, He II mimics solid body rotation on length scales exceeding the intervortex distance. The outcome is therefore similar to the case of classical viscous fluids. The spin-up process of He II in the zero-temperature limit, however, awaits further studies.

In superfluid ^{3}He-B the spin-up scenario is very different and depends crucially on the temperature. Indeed, by injection of vortex loops into a sample of ^{3}He-B in the Landau state (i.e., the thick normal fluid rotating with the container and the superfluid at rest with no vortices) the Helsinki group of Eltsov and coworkers observed and studied the generation of an interesting twisted vortex state (Eltsov *et al.*, 2006). This state appears spontaneously when vortex lines expand into vortex-free rotating superfluid. An example of turbulent vortex front motion in a rotating cylinder is shown in Fig. 8.4. The vortices form a front, where the ends of the lines bend towards the sidewall. The existence of the front is deduced from both simulation and experiment. The average superfluid velocity in the vortex front is the average of the velocities of the vortex state and the vortex-free state. It follows that the average angular velocity of the front is half that of the container's. The vortex front therefore lags behind the vortex array giving rise to the twisted vorticity. As the temperature drops below $0.45\,T_c$, sustained turbulence appears at the front, profoundly affecting the vortex dynamics (Eltsov *et al.*, 2007). Careful experimental studies supported by numerical simulations showed that a vortex front propagating into a region of vortex-free ^{3}He-B changes its nature from the laminar state to the turbulent state as the temperature decreases, providing direct measurement of the dissipation rate in turbulent vortex dynamics (Eltsov *et al.*, 2007).

Further measurements and numerical simulations of vortex dynamics (Hosio *et al.*, 2011) showed that at low temperatures the density of the propagating vortices falls well below the equilibrium value, i.e., the superfluid rotates at an angular velocity, Ω_s, that is smaller than that of the container, Ω. The superfluid angular velocity Ω_s was measured using NMR by detecting the precession frequency of the vortex bundle behind the front, using the experimental arrangement shown in Fig. 8.5. This study provides the first evidence of the decoupling of the superfluid from the container reference frame in the zero-temperature limit.

A few further details are in order. The experiment starts with the container at rest, with a vortex-free nonrotating superfluid. The container is then set into rotation at an angular velocity Ω, and a turbulent front forms at the rough surfaces of the heat exchanger and starts to move upwards along the cylinder, bringing the superfluid behind it into rotation. The axial front velocity u_f is determined from the flight time between two pickup coils of NMR spectrometers, shown in Fig. 8.5. The energy dissipation rate can be inferred by considering the free-energy difference of the superfluid before and after the front. The rate of the angular momentum exchange with the bounding walls can be determined either directly from the rotation Ω_s of the superfluid behind the front or indirectly from the dependence of u_f on Ω.

To explain the observed behavior, Hosio *et al.* (2013) found that two separate effective friction parameters have to be introduced for the energy and angular momentum transfers. By comparing the measurements with a phenomenological model of turbulent front propagation, the authors found that these parameters differ by two orders of magnitude. This difference suggests a new physical effects in superfluid dynamics in the zero-temperature limit. While quantum turbulence still provides substantial energy dissipation, the angular momentum exchange between the superfluid and its container vanishes gradually with dropping temperature, and the drag almost disappears in the zero-temperature limit.

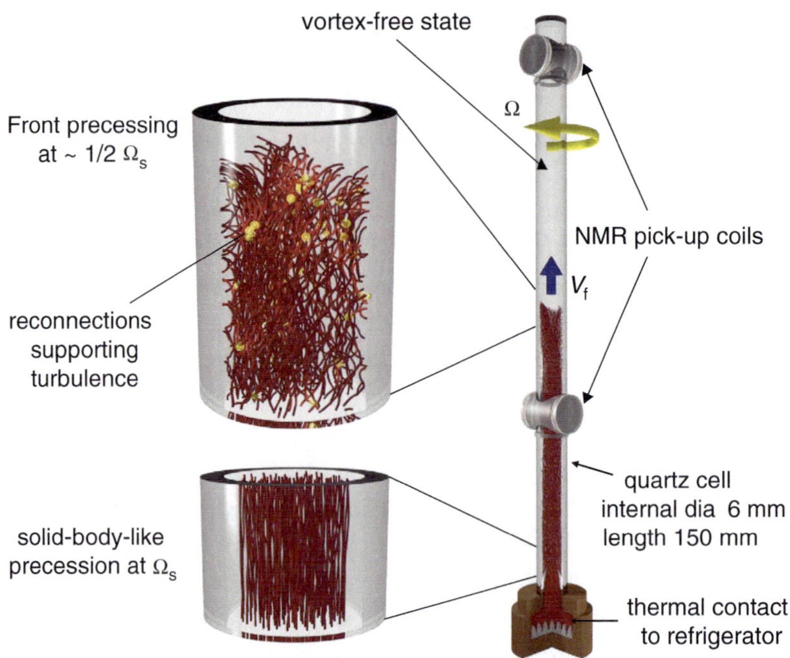

Figure 8.5 Turbulent vortex front in a rotating cylinder. The front moves axially upward and rotates azimuthally with respect to the cylinder. The motion is detected with two NMR pick-up coils, which are 9 cm apart. In the front, where the vortices bend perpendicular to the side wall, the angular velocity of the superfluid (and also the precession frequency of the vortices) changes from zero to $\Omega_s \lesssim \Omega$. The difference in the average precession frequencies between the front and the vortex bundle behind it forces reconnections (yellow dots) and turbulence at the front, as seen in the zoomed view on the top left. The differential precessions also wind the vortex lines behind the front to a twisted configuration. This snapshot of the ensuing vortex configuration comes from numerical simulations at $0.27\,T_c$. Reproduced with permission from Hosio *et al.* (2013).

8.3 Vibrating Wires and Grids in ^{3}He-B

There are several interesting studies of the flow of superfluid ^{3}He-B created by vibrating wires. At relatively high temperatures the wire suffers a large damping due to the large viscosity of the normal fluid. At submillikelvin temperatures, the damping disappears, and studies similar to those previously described for He II become possible. These studies are, however, more difficult to interpret because the critical velocity for generation of turbulence in ^{3}He-B is similar to that for pair-breaking velocity (i.e., destroying superfluidity), and the two effects cannot be easily

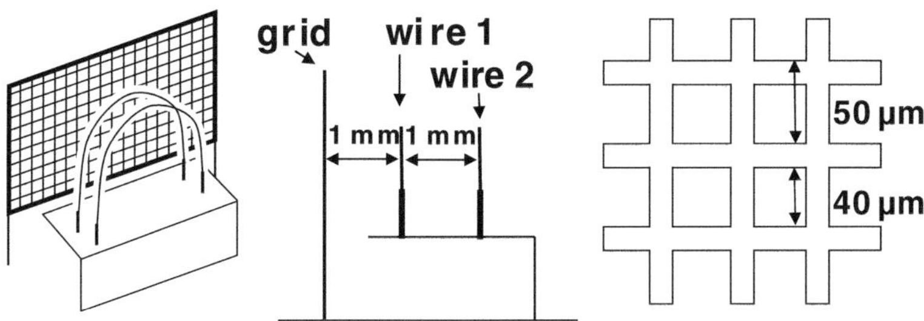

Figure 8.6 Schematic of the vibrating grid and the geometry of the Lancaster experiment (including the two detector wires) used by Bradley and coworkers in superfluid ^{3}He-B at very low temperatures.

disentangled (Bradley *et al.*, 2013). Fortunately, the Lancaster group have perfected the Andreev scattering technique (Fisher *et al.*, 2001) for the detection of quantum turbulence generated either by the same vibrating wire or by a second one. At these very low temperatures ^{3}He-B is almost a pure superfluid, and turbulence consists of a tangle of interacting vortex lines moving in a sparse background of ballistic thermal excitations. The presence of vortex lines is determined by the decrease in the thermal damping of a vibrating wire resonator, thanks to Andreev reflection of thermal quasiparticle excitations, as explained in Chapter 4.

Even though all experiments at submillikelvin temperatures are technically difficult, a relatively simple and efficient way of generating quantum turbulence in ^{3}He-B at microkelvin temperatures is to make a small grid to oscillate back and forth. The experimental arrangement of the first vibrating grid experiment, utilizing the Lancaster style nuclear cooling stage (Bradley *et al.*, 2004), is shown in Fig. 8.6. The grid consists of a mesh of fine copper wires spaced 50 µm apart, leaving 40 µm square holes. Facing the grid are two vibrating wire resonators, made from 2.5 mm diameter loops of 4.5 µm NbTi wire and positioned 1 mm and 2 mm from the grid, which act as detectors. An additional wire resonator (not shown in Fig. 8.6) is used as a background thermometer. By driving vibrations of the grid above a certain value, it was possible to observe the transition to quantum turbulence (Bradley *et al.*, 2005b). While the detailed processes are no doubt complex, the current view is that the vibrating grid emits vortex rings more frequently with increasing grid velocity. At some critical velocity the density of rings becomes sufficiently high that they can no longer avoid each other, resulting in collissions. Subsequent reconnection brings more complex shapes, and a cascade of further reconnections leads rapidly to quantum turbulence. This scenario is backed by computer simulations of Fujiyama *et al.* (2010), as illustrated in Fig. 8.7.

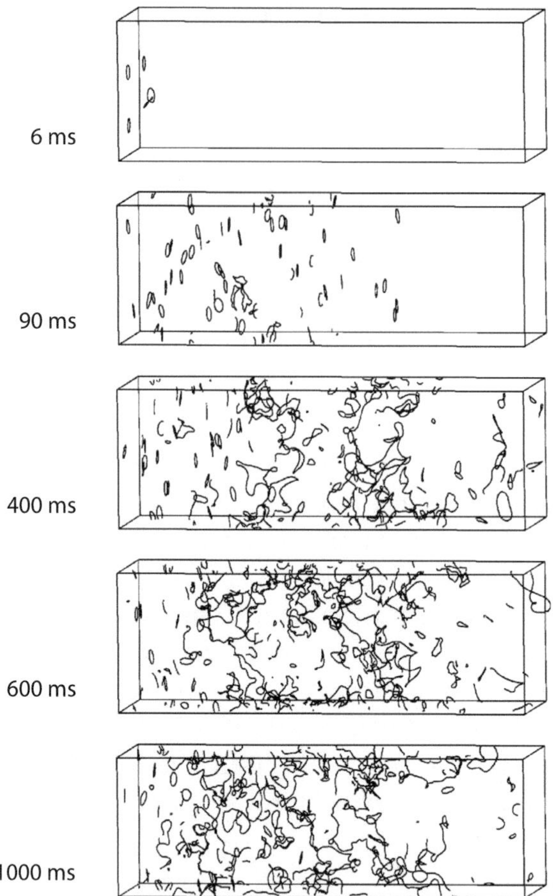

Figure 8.7 Simulation of quantum turbulence formation by an oscillating grid (Fujiyama *et al.*, 2010). Each frame shows the vortex configuration at the labeled times. Rings injected from the left quickly collide and recombine producing a vortex tangle that evolves on longer time scales. From Fujiyama *et al.* (2010), used under CC BY 3.0 (https://creativecommons.org/licenses/by/3.0).

8.4 Summary

While in He II the transition to pure superfluid turbulence is in most cases associated with extrinsic vortex nucleation, i.e., the vortex tangle grows from the seeding remanent vortices that are nearly always present, both intrinsic and extrinsic vortex nucleations are possible in ^{3}He-B.

At nonzero temperatures a transition to a completely new class of quantum turbulence occurs, in which only the superfluid component moves with respect to the boundaries. In this specific case, the role of the Reynolds number is played by a combination of mutual friction parameters, which approximately scales as the inverse temperature; the transition therefore depends on the temperature rather than on the flow speed and the geometrical size, in striking contrast with classical turbulence.

9

Spectra and Structure Functions in Quantum Turbulence

We are still far away from understanding turbulence in its full glory, and the progress that has been made so far has come by combining theoretical inquiries with specific experimental configurations. Homogeneous and isotropic turbulence is an idealized version of turbulence in which the statistical properties are invariant under translations and rotations, and is thus amenable to analytical progress and physical understanding of problems such as dispersion and scale-to-scale energy transfer. An even simpler version of turbulence is obtained if, besides neglecting boundaries, turbulence is in a statistical steady state (or stationary state in time) because of an externally applied forcing that balances energy dissipation. Taylor (1935) suggested that this idealized form of turbulence can be achieved in a wind tunnel, a fruitful idea that the low-temperature physics community has followed up with the construction of several cryogenic wind tunnels, such as the *Toupie* liquid helium wind tunnel developed by Roche *et al.* (2007).

In this chapter we will introduce and discuss important characteristics of homogeneous and isotropic quantum turbulence in the form of spectral distributions of the turbulent kinetic energy in both the superfluid and the normal fluid components over a wide range of scales. Conceptually, the simplest situation occurs in the zero-temperature limit, where the normal fluid is absent. At nonzero temperature, as already stated multiple times, in the framework of the two-fluid model we may expect turbulence to occur in He II either in the superfluid component or the normal component, or both, with the velocity fields coupled by mutual friction. A complex interplay between normal and superfluid energy spectra takes place, depending on the temperature and the specific character of the turbulent flow. In turbulent ^{3}He-B, the thick stationary normal fluid provides a unique laboratory frame of reference (in which the normal fluid is essentially stationary), but the mutual friction, acting on all length scales, shapes the turbulent spectrum in the superfluid component.

Before discussing these aspects in superfluids, we remind the reader of the basic properties of energy spectra in classical turbulent flows.

9.1 Classical Kolmogorov Scaling

In classical turbulent flows, the state of turbulence is usually quantified by a dimensionless parameter, the Reynolds number

$$Re = \frac{UD}{\nu}. \tag{9.1}$$

The Reynolds number is a measure of the ratio of the orders of magnitude of inertial to viscous terms in the Navier–Stokes equation at a typical large length scale D – for example, the size of the flow channel or the size of the body around which the fluid flows. The quantity U is the characteristic velocity of the flow; in some cases U is interpreted as the root mean square (rms) velocity of turbulent fluctuations (any mean flow is subtracted away using the so-called Reynolds decomposition applied to the flow under consideration). From the Navier–Stokes equations, Eq. (5.1), we have

$$\frac{|(\boldsymbol{u} \cdot \nabla)\boldsymbol{u}|}{|\nu \nabla^2 \boldsymbol{u}|} \sim \frac{U^2/D}{\nu U/D^2} = \frac{UD}{\nu}. \tag{9.2}$$

The Reynolds number can also be interpreted as a dimensionless velocity with length and time measured in units of D and D^2/ν, respectively. The larger the inertial term (compared to the viscous term), the larger the dimensionless velocity and the more intense the turbulence.

Turbulent flows contain scales of varying lengths (or, colloquially, eddies of many sizes), so it is natural to ask how the properties of turbulence are distributed, in a statistical sense, over these length scales. Turbulent spectra provide one type of answer. To manage the task better, we consider the simplest case of steady, homogeneous, isotropic turbulence; by steady we mean that the macroscopic observables fluctuate around well-defined mean values. To maintain this turbulence in the steady state compensating for viscous dissipation, suitable forcing must be applied continuously at some large length scale. An example is turbulence in a wind tunnel away from boundaries, for which the forcing could arise from the drag on a grid of bars.

The phenomenology of steady, homogeneous, isotropic turbulence is based on Richardson's idea of the *energy cascade*. Let r be the size of an eddy. By nonlinear effects, large eddies ($r \approx D$) continuously break up into smaller eddies, which break up into further smaller eddies, and so on, until, at the dissipation length scale η, viscous forces become as important as inertial forces (so the effective Reynolds number at that scale is unity), and the kinetic energy of the eddies is transformed into heat by viscous action. *Developed turbulence* is characterized by the coexistence of eddies over a wide range of length scales; the subset of that range, defined loosely by $D \gg r \gg \eta$, is the *inertial range*. In the inertial range, details of the injection of energy (taking place at scales $r \approx D$) and viscous dissipation (taking place at scale $r \approx \eta$ called the Kolmogorov or dissipation length) can both be ignored and the statistical properties of the eddies are presumed to be universal: The only quantity of importance is the energy transfer across scales, which is equal to the rate at which energy is being pumped at scale D, and also what eventually manifests as the energy dissipation (see below). This immediately leads to the result that the scale of dissipation is

$$\eta = Re^{-3/4} D. \tag{9.3}$$

The Reynolds number thus also effectively measures the separation of the two important length scales, D and η.

Since the turbulence is assumed to be isotropic, it can be characterized by the one-dimensional energy spectrum $\Phi(k)$; starting from the kinetic energy per unit mass, Fourier transforming the velocity field $\boldsymbol{u}$, and applying Parseval's theorem, we have

$$E = \frac{1}{V} \int_V \frac{1}{2} \boldsymbol{u}^2 dV = \int_0^\infty \Phi(k)\, dk, \tag{9.4}$$

where V is the volume occupied by turbulence, and the wavenumber $k = |\boldsymbol{k}|$ (the magnitude of the three-dimensional wavevector $\boldsymbol{k}$) can be interpreted as the inverse eddy size ($k = 2\pi/r$) at the scale r. The quantity

$$\epsilon = -\frac{dE}{dt} \tag{9.5}$$

(of dimensions m^2/s^3) is the rate of kinetic energy dissipation (per unit mass). Assuming that in the inertial range ϵ is independent of k and that $\Phi(k)$ is independent of viscosity, a simple dimensional argument leads to the celebrated Kolmogorov $k^{-5/3}$ law (also known popularly as K41),

$$\Phi(k) = C\epsilon^{2/3} k^{-5/3}, \qquad \text{in} \quad k_D \ll k \ll k_\eta. \tag{9.6}$$

Here, $k_D = 2\pi/D$, $k_\eta = 2\pi/\eta$ are the wavenumbers corresponding to the large length scale D and the dissipation length scale η, and C is a dimensionless constant of the order unity. The scaling $\Phi(k) \sim k^{-5/3}$ of Eq. (9.6) is one of the most important results of turbulence theory and describes the energy distribution over the length scales in the inertial range: Large eddies (large r, small k) contain more energy than small eddies (small r, large k). The essence of K41 is that, as long as the inertial range satisfying the inequality of Eq. (9.6) exists, the $-5/3$ scaling holds even if the large scales are not isotropic and homogeneous. The requirement is essentially that the scales within the inertial range be statistically isotropic and homogeneous. This character of K41, called local homogeneity and isotropy, is what makes the Kolmogorov scaling very useful and applicable in practice.

The Kolmogorov spectral form has been verified to various levels of satisfaction in both experiments and numerical simulations, but it is worth pointing out an important difference between experiments and numerics. In simulations, three-dimensional (3D) snapshots of the velocity field at a fixed time are Fourier analyzed and ensemble averaged. In experiments, it is not always practical to determine the velocity field on a 3D mesh; instead, a probe (such as hot-wire anemometer) measures the velocity at a single fixed position for a sufficiently long time. To determine the wavenumber spectrum in an experiment, it is therefore necessary to make the further assumption that, as the mean flow advects the eddies past the velocity probe, the eddies remain unchanged; this is the so-called *Taylor's frozen field hypothesis*. More precisely, let $f = f(x, y, z, t)$ be a property of the flow $\boldsymbol{u} = (u_x, u_y, u_z)$. If f does not change as it moves with the flow, then

$$\frac{df}{dt} = \frac{\partial f}{\partial t} + u_x \frac{\partial f}{\partial x} + u_y \frac{\partial f}{\partial y} + u_z \frac{\partial f}{\partial z} = 0. \tag{9.7}$$

Therefore, if (for simplicity) the flow is entirely in the x-direction, $df/dt = -u\partial f/\partial x$, and fluctuations in t can be related to fluctuations in x; in other words, the (frequency) spectrum measured by the probe can be related to the (wavenumber) spectrum of the theory.

9.2 Kolmogorov Spectrum in He II

9.2.1 Experimental Results

Quantitative agreement with the classical Kolmogorov energy spectrum was demonstrated experimentally by Maurer and Tabeling (1998), who generated turbulence in liquid helium by using two counterrotating propellers, as shown in Fig. 4.2. They measured local pressure fluctuations using a small pressure tube (Pitot tube), which was described in Chapter 4.

The striking results shown in Fig. 9.1 are that (a) there are no detectable changes in the form of the spectrum over the whole observed temperature range; (b) over the range of relevant wavenumbers detected by the total-head tube (limited mainly by its size due to organ pipe resonance) the spectrum is of the same Kolmogorov form within measurement accuracy. Indeed, Fig. 9.1 shows the spectrum at $T = 2.2\,\mathrm{K}$ in classical viscous He I, where the Kolmogorov scaling is expected. The middle and bottom curves exhibit the same classical scaling in He II at $T = 2.03\,\mathrm{K}$ and $T = 1.4\,\mathrm{K}$. At the latter temperature, the normal fluid fraction is only $\rho_n/\rho \approx 7\%$. With so little normal fluid left, the Kolmogorov spectrum must arise also from the superfluid, as stated by Nore *et al.* (1997).

Moreover, the deduced numerical value of the Kolmogorov constant was $C \cong 1.5$, well within the range of accepted classical values (Sreenivasan, 1995) $C = 1.62 \pm 0.17$. Even the intermittency corrections (more details on intermittency will be described later in this chapter) deduced by further analysis remained independent

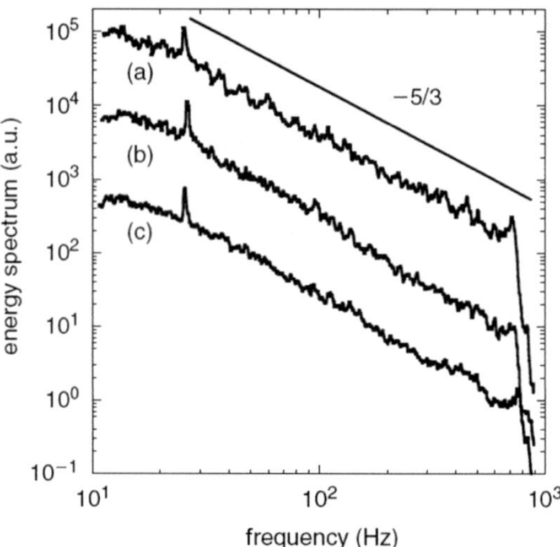

Figure 9.1 Turbulent flow of liquid helium confined between counterrotating discs at temperatures: (a) 2.3 K; (b) 2.08 K; (c) 1.4 K. Measured energy spectrum Φ vs. frequency f. The solid line is a guide to the eye to demonstrate the clear $\Phi \sim f^{-5/3}$ dependence. For further details, see text. Reproduced with permission from Maurer and Tabeling (1998).

of the fraction of the normal to superfluid components. A simple explanation of this result is that *on scales to which this experiment had access* (see later in this chapter), the two fluids move with the velocity field that is identical to that expected in classical turbulence.

9.2.2 Numerical Results

When discussing numerical results on the energy spectra in the inertial range of scales of a turbulent superfluid, it is important to keep three important wavenumbers in mind: k_D, which was already defined, as well as $k_\ell = 2\pi/\ell$ and $k_\zeta = 2\pi/\Delta\zeta$. They correspond, respectively, to the box size D, the average intervortex spacing ℓ (which we refer to as the *quantum length scale*, arising from the discrete, quantized nature of vorticity as it manifests itself at scales $r \lesssim \ell$), and the numerical resolution $\Delta\zeta$ along vortex lines (for the vortex filament model); for the GPE the corresponding high frequency cutoff is based on the size of the vortex core a_0.

The first (qualitative) evidence for the classical Kolmogorov energy spectrum in a quantum fluid was obtained numerically by Nore *et al.* (1997) who solved the Gross–Pitaevskii equation (GPE) in a periodic domain for a decaying Taylor–Green flow. Since the GPE describes a compressible fluid, care must be taken in identifying the energy spectrum that should be compared to the spectrum predicted by the incompressible Navier–Stokes equation. Nore *et al.* (1997) remarked that the

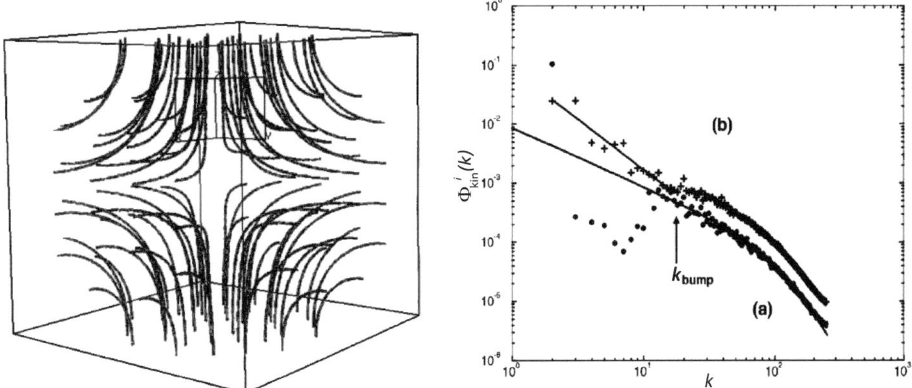

Figure 9.2 Numerical simulation of Nore *et al.* (1997) using the GPE model. (Left) Vortex lines in the initial Taylor–Green configuration at $t = 0$. (Right) Spectrum of the incompressible kinetic energy $\Phi^{i}_{\text{kin}}(k)$ vs. wavenumber k at (a) $t = 0$ and (b) $t = 5.5$. The least-square fit of (b) in the range $2 < k < 16$ gives $\Phi^{i}_{\text{kin}}(k) \sim k^{-1.70}$, claimed to be in qualitative agreement with the $k^{-5/3}$ Kolmogorov scaling. Here, $k_{\text{bump}} = k_{\ell}$ is the wavenumber corresponding to the average distance between vortices. Reprinted figure with permission from Nore *et al.* (1997). Copyright 1997 by the American Physical Society.

GPE conserves the total energy, which can be decomposed into kinetic, internal, and quantum energy contributions, as described in Section 5.1.4. The kinetic energy can be further decomposed into compressible and incompressible parts, E^{c}_{kin} and E^{i}_{kin}, which are separately Fourier transformed as in Eq. (9.4) to identify the compressible and incompressible kinetic energy spectra $\Phi^{c}_{\text{kin}}(k)$ and $\Phi^{i}_{\text{kin}}(k)$:

$$E^{c}_{\text{kin}} = \int_{0}^{\infty} \Phi^{c}_{\text{kin}}(k)\, dk, \qquad E^{i}_{\text{kin}} = \int_{0}^{\infty} \Phi^{i}_{\text{kin}}(k)\, dk. \tag{9.8}$$

The quantity that should be compared to the energy spectrum of K41 theory, Eq. (9.6), is thus $\Phi^{i}_{\text{kin}}(k)$; see Fig. 9.2.

Nore *et al.* (1997) found that, initially, E^{i}_{kin} is large (corresponding to the Taylor–Green vortex configuration) and $E^{c}_{\text{kin}} \approx 0$ (no sound waves). During the decay, E^{c}_{kin} increased, while E^{i}_{kin} decreased. Nore *et al.* noticed that when the energy dissipation was the largest, the spectrum of the incompressible kinetic energy, $\Phi^{i}_{\text{kin}}(k)$, acquired a distribution that is roughly consistent with the $\Phi(k) \sim k^{-5/3}$ Kolmogorov scaling for length scales larger than the typical intervortex distance. It must be stressed, however, that since the GPE describes a condensate, Nore's results must refer to the zero-temperature limit when applied to superfluid helium. More direct evidence for Kolmogorov scaling will be presented in the coming sections.

9.2.3 Further Numerical Results on the Energy Spectrum in Quantum Turbulence

The finding that the GPE model (Nore *et al.*, 1997) yields the inertial spectrum with the $k^{-5/3}$ Kolmogorov scaling was confirmed by complementary numerical work using the vortex filament model at $T = 0$ performed by Tsubota and collaborators (Araki *et al.*, 2002b,a). The results are shown in Fig. 9.3. Note that the spectrum becomes shallower for $k > k_\ell$, in agreement with the k^{-1} scaling of isolated straight vortex lines. Additionally, starting from a different initial condition consisting of staggered vortex lines, Baggaley *et al.* (2012b) also obtained the $-5/3$ law.

Just as for computing the energy spectra in turbulent Bose–Einstein condensates, Kobayashi and Tsubota (2005) used the GPE in a periodic domain (see Fig. 9.4) for computing a harmonically trapped condensate (Kobayashi and Tsubota, 2007). In the former, turbulence developed from a random-phase, constant-density initial condition; in the latter, turbulence was created by simultaneous rotation of the condensate about two perpendicular axes. In both cases, artificial dissipation was applied at scales of the order of the healing length ξ and shorter, resulting in a more convincing Kolmogorov scaling than in Nore *et al.* (1997).

All numerical results presented above refer to the $T = 0$ limit, but the Kolmogorov spectrum, as we have seen in Fig. 7.2, is also observed in the two-fluid regime. This regime has been studied using the vortex filament method (VFM) by Baggaley *et al.* (2012d) and Sherwin-Robson *et al.* (2015), who explored various strategies to represent the turbulent normal velocity field for the frozen Navier–Stokes turbulence, an ABC flow (representing localized regions of strong vorticity), and the synthetic model of time-dependent turbulence (Osborne *et al.*, 2006) given by

$$u_{\rm n}(\boldsymbol{r}, t) = \sum_{m=1}^{M} \left(\boldsymbol{A}_m \times \boldsymbol{k}_m \cos(\phi_m) + \boldsymbol{B}_m \times \boldsymbol{k}_m \sin(\phi_m)\right), \tag{9.9}$$

where $\phi_m = \boldsymbol{k}_m \cdot \boldsymbol{r} + f_m t$, $\boldsymbol{k}_m$ are wavevectors and $f_m = \sqrt{k_m^3 \Phi_{\rm n}(k_m)}$ are angular frequencies. The random vectors $\boldsymbol{A}_m$, $\boldsymbol{B}_m$, and $\boldsymbol{k}_m$, which take into account the periodicity of the computational domain (Wilkin *et al.*, 2007), are chosen so that the normal fluid's energy spectrum $\Phi_{\rm n}(k)$ obeys the Kolmogorov spectrum in the inertial range $k_{\rm D} \le k \le k_{\rm M}$ (where $k_{\rm D} = 2\pi/D$ and D is the size of the computational domain) and k_M determines the Reynolds number via $Re = (k_M/k_{\rm D})^{4/3}$. This synthetic turbulent flow is solenoidal, time-dependent, and compares well with Lagrangian statistics observed in experiments and in numerical simulations of Navier–Stokes turbulence. In all cases, it was found that the superfluid energy spectrum displayed the classical Kolmogorov scaling in the expected hydrodynamic range $k_{\rm D} < k < k_\ell$. An example is shown in Fig. 9.5.

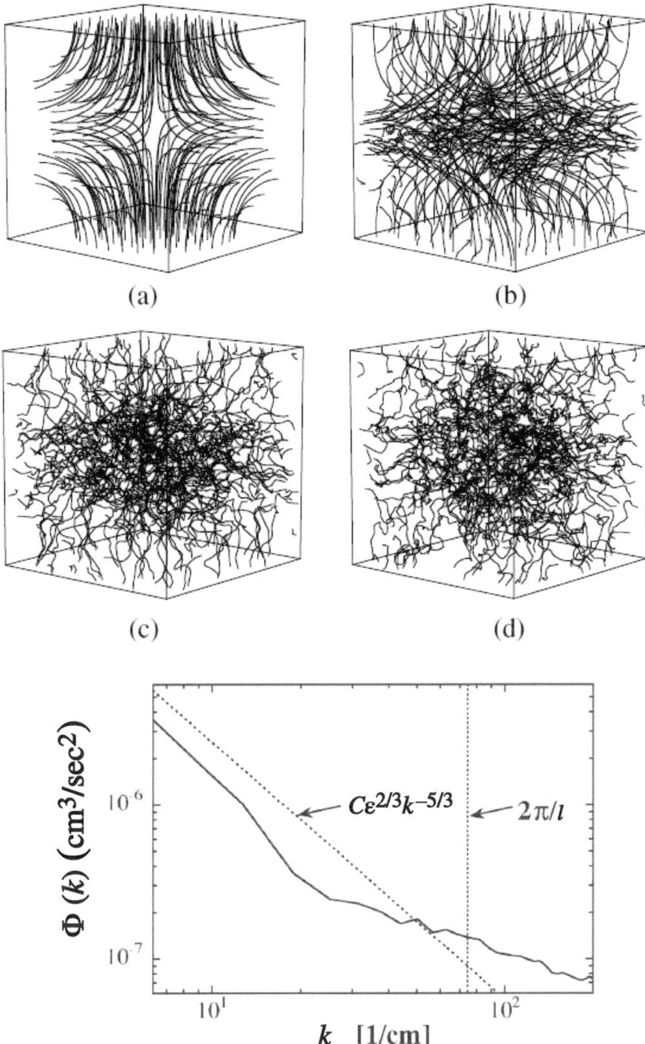

Figure 9.3 Numerical VFM simulation illustrating the temporal decay of the initial Taylor–Green vortex configuration at (a) $t = 0$, (b) $t = 30$ s, (c) $t = 50$ s, and (d) $t = 70$ s. Numerical simulation of energy spectrum at $t \approx 70$ s, as in panel (d). The bottom panel shows the energy spectrum $\Phi(k)$ vs. the wavenumber k computed at $t = 70$ s. The dashed line denotes the Kolmogorov $k^{-5/3}$ scaling. The vertical line marks the wavenumber $k_\ell = 2\pi/\ell$ corresponding to the intervortex distance. Parameters: $D = 1$ cm, $L = 151.9\,\text{cm}^{-2}$, $\Delta\zeta = 0.0183$ cm. Reproduced with permission from Araki *et al.* (2002a).

9.3 Origin of Kolmogorov Scaling in Quantum Turbulence

We have shown that there is plausible evidence for a Kolmogorov spectrum in superfluids. But what is its physical origin? In ordinary viscous fluids, it is generally

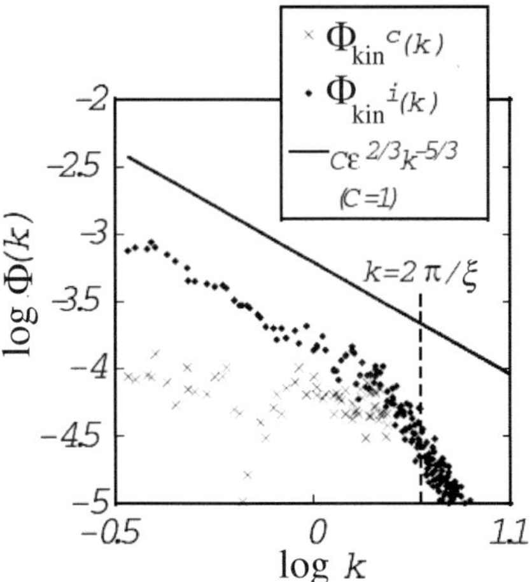

Figure 9.4 Numerical simulation by Kobayashi and Tsubota (2005) of the GPE model in a periodic domain with small-scale dissipation. Spectra of the incompressible and compressible kinetic energy, $\Phi^{i}_{\text{kin}}(k)$ and $\Phi^{c}_{\text{kin}}(k)$, respectively, defined in Eq. (9.8), plotted vs. wavenumber k (sample averaged over many initial conditions). Notice that $\Phi^{i}_{\text{kin}}(k) \sim k^{-5/3}$ for sufficiently small k. The vertical dashed line indicates the wavenumber that corresponds to the size of the vortex core. Reprinted figure with permission from Kobayashi and Tsubota (2005). Copyright 2005 by the American Physical Society.

thought that the smaller scales are produced by vortex stretching. The mechanism is the following. Consider a thin vortex tube of cross section S. Let z be the coordinate along the tube and $\boldsymbol{u} = (u_x, u_y, u_z)$ the velocity. Taking the curl of the incompressible Navier–Stokes Eq. (5.1) and neglecting viscosity, we have

$$\frac{D\boldsymbol{\omega}}{Dt} = (\boldsymbol{\omega} \cdot \nabla)\boldsymbol{u}, \tag{9.10}$$

where $D/Dt = \partial/\partial t + \boldsymbol{\omega} \cdot \nabla$ is the convective derivative. Since the vorticity is mainly in the z direction, $\boldsymbol{\omega} \approx (0, 0, \omega)$, the z-component of Eq. (9.10) is

$$\frac{D\omega}{Dt} = \omega \frac{du_z}{dz}. \tag{9.11}$$

Therefore, if the instantaneous z-component of the velocity increases along the tube ($du/dz > 0$), then the vorticity increases with time. According to the Kelvin circulation theorem, the circulation

$$\Gamma = \int_S \boldsymbol{\omega} \cdot \hat{\boldsymbol{n}} dS \tag{9.12}$$

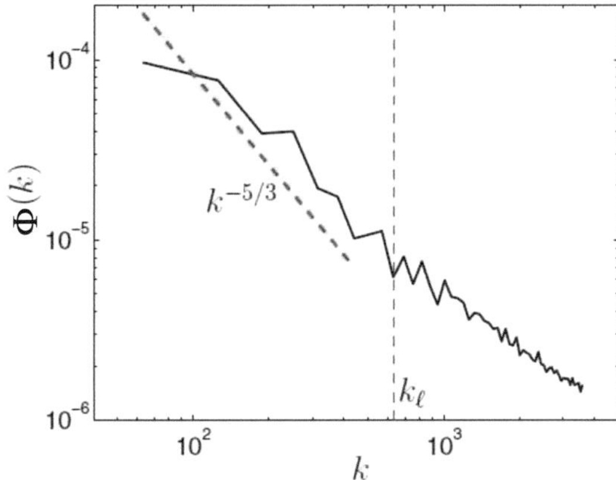

Figure 9.5 Numerical simulation of Baggaley *et al.* (2012d) using the VFM at $T = 1.9\,\mathrm{K}$: Superfluid energy spectrum $\Phi(k)$ vs. wavenumber k. Turbulence in the superfluid component is driven by synthetic turbulence of the normal fluid; see Eq. 9.9. The vertical dashed line marks k_ℓ, the wavenumber corresponding to the average intervortex distance. The red dashed line represents the $k^{-5/3}$ slope. Reprinted figure with permission from Baggaley *et al.* (2012d). Copyright 2012 by the American Physical Society.

is constant along the tube (where $\hat{\boldsymbol{n}}$ is the outward unit vector); for a thin tube, $\Gamma \approx \omega S$ is constant, and an increase of ω requires that the tube becomes thinner in order to conserve the angular momentum.

In a quantum fluid, however, both circulation and the core size of an isolated vortex line are held fixed by quantum mechanics, and classical vortex stretching seems impossible. The answer to the puzzle is that quantized vortex lines may form bundles, and the individual vortices in the same bundle can move closer to each other, stretching the bundle and increasing the coarse-grained vorticity. The idea is schematically illustrated in Fig. 9.6. It must be stressed that this figure is only an idealized cartoon: The bundles may be only a statistical alignment effect, in the sense that, in practice, a partial polarization of the vortex lines appears to be sufficient for displaying the Kolmogorov type of quantum turbulence.

Figure 9.7 shows a vortex tangle generated by numerical simulation (Baggaley *et al.*, 2012c). By computing the coarse-grained vorticity, it is possible to decompose the total vortex configuration (left) into two parts: a coherent part (top right) that consists of bundles of vortex lines oriented in the same direction, and an incoherent part (bottom right) in which the vortex lines are randomly oriented. Figure 9.8 shows that the Kolmogorov spectrum arises from the vortex lines that are polarized, not the random vortex lines, confirming the intuition of Fig. 9.6.

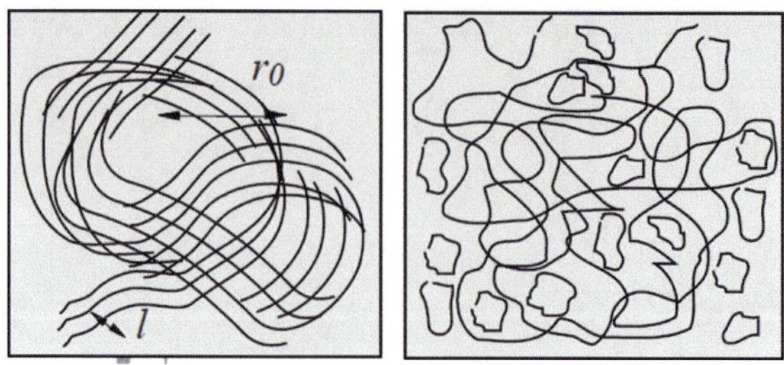

Figure 9.6 A schematic drawing of vortex line geometry leading to two types of quantum turbulence: (Left) Kolmogorov-like turbulence in which vortex lines are arranged in bundles and (Right) Vinen-like turbulence in which vortex lines are arranged randomly. Reproduced with permission from Volovik (2004).

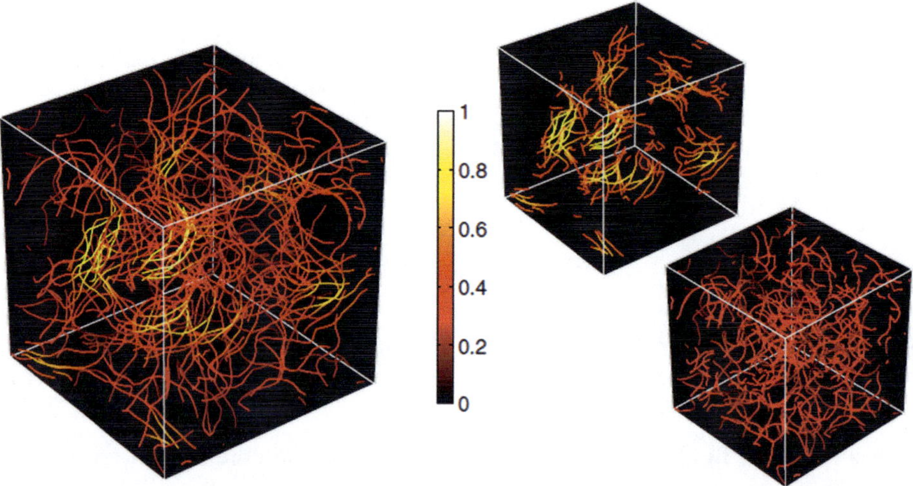

Figure 9.7 (Left) A snapshot of quantum turbulence. The vortex lines are colored according to the local magnitude of the coarse-grained vorticity field $\boldsymbol{\omega}_{\mathrm{s}}$; red lines correspond to small coarse-grained vorticity, yellow lines to large coarse-grained vorticity. (Right) The same snapshot but split into locally polarized vortex lines (top, $\omega(\boldsymbol{s}) > 1.4\,\omega_{\mathrm{rms}}$), which form bundles, and unpolarized vortex lines (bottom, $\omega(\boldsymbol{s}) < 1.4\,\omega_{\mathrm{rms}}$), which are in random directions to each others. Here, ω_{rms} is the rms value of the coarse-grained superfluid vorticity. Reprinted figure with permission from Baggaley *et al.* (2012c). Copyright 2012 by the American Physical Society.

9.4 The 4/5-law in Quantum Turbulence

For statistical investigation of classical turbulence, one usually uses structure functions, which are defined as

$$S_n(r) = \langle [(\boldsymbol{u}(\boldsymbol{x}+\boldsymbol{r}) - \boldsymbol{u}(\boldsymbol{x})) \cdot \hat{\mathbf{e}}]^n \rangle, \tag{9.13}$$

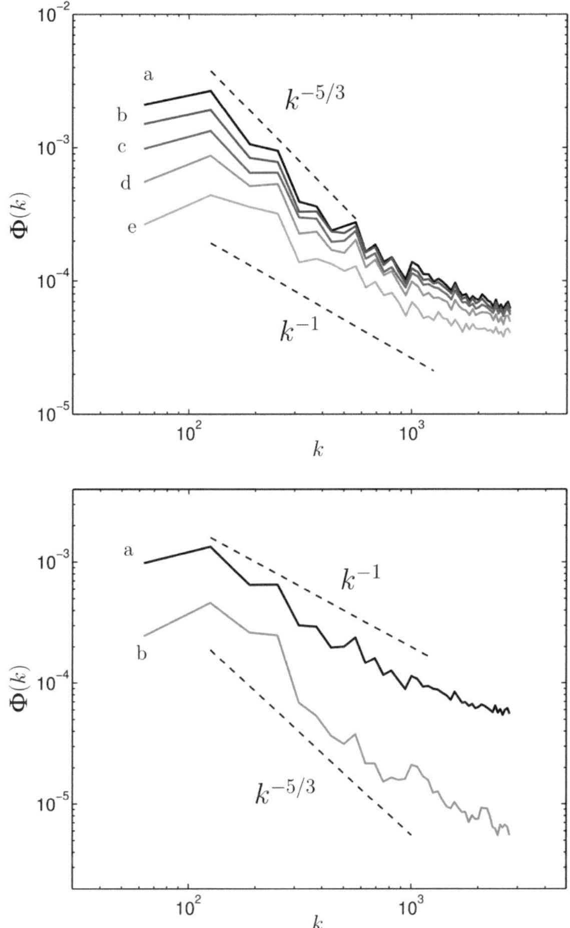

Figure 9.8 Energy spectra $\Phi(k)$ vs. wavenumber k corresponding to Fig. 9.7. (Top) The upper solid line (a) is the energy spectrum of the flow induced by all the vortex lines; lower curves correspond to the flow induced by vortex lines with coarse-grained vorticity below the following thresholds: (b) $\omega_s < 1.7\,\omega_{\rm rms}$, (c) $\omega_s < 1.4\,\omega_{\rm rms}$, (d) $\omega_s < 1.2\,\omega_{\rm rms}$, and (e) $\omega_s < \omega_{\rm rms}$. (Bottom) Energy spectra corresponding to vortex lines with coarse-grained vorticity below (a) and above (b) corresponding to the threshold value $1.4\,\omega_{\rm rms}$. The dashed lines display the k^{-1} and $k^{-5/3}$ scalings. Reprinted figure with permission from Baggaley *et al.* (2012c). Copyright 2012 by the American Physical Society.

where $r = |\boldsymbol{r}|$ is the separation distance and $\hat{\mathbf{e}}$ is the unit vector in a prescribed direction. In particular, the longitudinal and transverse second-order structure functions $S_2^{\|}(r)$ and $S_2^{\perp}(r)$ are respectively defined by choosing $\hat{\mathbf{e}}$ in the direction parallel to $\boldsymbol{u}$ or perpendicular to it; both quantities scale as $r^{2/3}$, consistent with the $-5/3$ power law for the energy spectrum.

Often cited as the only exact result of classical fully developed turbulence (i.e., for asymptotically large *Re*), the Kolmogorov 4/5-law (Kolmogorov, 1941a) states that within the inertial range of scales the third-order longitudinal velocity structure function is given by

$$S_3^{\parallel}(r) = (-4/5)\epsilon r, \tag{9.14}$$

where we remind the reader that ϵ is the rate of energy dissipation per unit mass. It is therefore interesting to test its validity in mechanically generated quantum turbulence, as it was done in Grenoble by Salort *et al.* (2012a). Experimentally, they utilized a He II wind tunnel, operated down to 1.56 K. The authors used the experimental fact (Salort *et al.*, 2010) that, keeping the same mean-flow velocity above and below the transition, ϵ does not change when the superfluid transition is crossed. In order to estimate ϵ, they used He I velocity recordings, since He I is a classical fluid, where the Kolmogorov 4/5-law is known to be valid and used that value to compensate the third-order velocity structure function obtained in He II. In this way, they observed a plateau for nearly half a decade of scales, corresponding to the resolved inertial range of the turbulent cascade. This may be viewed as the first experimental evidence that the 4/5-law is valid in He II, which was further backed up by high-resolution simulations of the Landau–Tisza two-fluid model down to 1.15 K.

9.5 General Shape of 3D Energy Spectra

The shape of the 3D energy spectrum for turbulence in classical viscous fluids is well known for the locally homogeneous and isotropic case. We already know that, neglecting intermittency corrections (see Section 9.9), for high enough Reynolds numbers, the spectral energy density in the inertial range of scales follows the Kolmogorov scaling with a roll-off exponent of $-5/3$; see Eq. (9.6). On the low k end, the spectrum is naturally truncated by the size of the container, D ($k_D = 2\pi/D$), reflecting the simple fact that eddies larger than the system cannot exist. The low k end takes the form Ak^m; assuming the validity of the Birkhoff–Saffman invariant, we have $m = 2$. On increasing k, there is a broad maximum in $\Phi(k)$ that corresponds to energy containing eddies. Most of the turbulent energy resides at scales around ℓ. In steady-state 3D turbulence, the energy is injected at the outer scale $\approx \ell$ and, by the Richardson cascade, transmitted without loss in k-space to the dissipation wavenumber, and becomes dissipated by the viscosity around $k_\eta = \left(\epsilon/\nu^3\right)^{1/4}$, where ν is the kinematic viscosity. This result is determined by dimensional reasoning.

Let us consider the simplest case of quantum turbulence, i.e., the pure superfluid turbulence consisting solely of a tangle of quantized vortex line in the zero-temperature limit where there is no normal fluid. The superflow exists down to smallest length scale, which is the size of the vortex core. There is no viscosity, so the notion of a viscous dissipation length scale does not apply. We have already introduced another physical length scale – the quantum length scale – defined by the mean intervortex distance that could be estimated as $\ell = 1/\sqrt{L}$, where L stands for the vortex line density. The corresponding quantum wavenumber is $k_\ell = 2\pi/\ell$.

As mentioned in Chapter 1, vorticity of the superfluid component is quantized in terms of the circulation quantum $\kappa = 2\pi\hbar/M$ where M is the mass of a superfluid particle (^{4}He atom in He II, two ^{3}He atoms in superfluid ^{3}He-B phase). In analogy with classical viscous flow, the flow of the superfluid component can be characterized by the superfluid Reynolds number

$$Re_s = \frac{UD}{\kappa}. \tag{9.15}$$

The Feynman criterion $Re_s \sim 1$ gives the velocity at which it becomes energetically favorable to form a quantized vortex although it does not mean, due to nucleation barrier, that vortices are always created upon exceeding this criterion. The limit $Re_s \gg 1$ is equivalent to a vanishing Planck constant, $\kappa \propto \hbar \to 0$, so that the vorticity becomes a continuous variable, as in the classical case. The very existence of the quantum length scale is thus a purely quantum mechanical effect.

The quantum length scale has a clear physical meaning. It qualitatively divides the scales in quantum turbulence into large scales, where quantum restrictions in the form of quantization of circulation do not play any role (or, in other words, the granularity of quantum turbulence does not matter), and small scales, where quantum restrictions are essential. For $k > k_\ell$, the spectral energy density $\Phi(k)$ ought to depend, in addition to $\epsilon = -dE/dt$ and k, also on κ. A dimensional argument similar to that of Kolmogorov requires the spectral density in the wavenumber space k to take the general form

$$\Phi(k) = C\epsilon^{2/3}k^{-5/3}f\left(\frac{\epsilon}{k^4\kappa^3}\right) = C\epsilon^{2/3}k^{-5/3}f\left(\frac{k}{k_\ell}\right). \tag{9.16}$$

9.6 Vorticity Spectrum

In classical turbulence, it is known that the vorticity spectrum that corresponds to the Kolmogorov scaling of the energy spectrum has a $k^{1/3}$ dependence that peaks near the dissipation length η. In other words, most of the vorticity is contained in the smallest eddies. The natural question is whether there is an analogous effect in quantum turbulence.

Baggaley *et al.* (2012b) showed that this is indeed the case. Although the superfluid velocity is irrotational outside vortex lines, the presence of a Kolmogorov spectrum means that there are hydrodynamical scales $k_D \ll k \ll k_\ell$ in which the

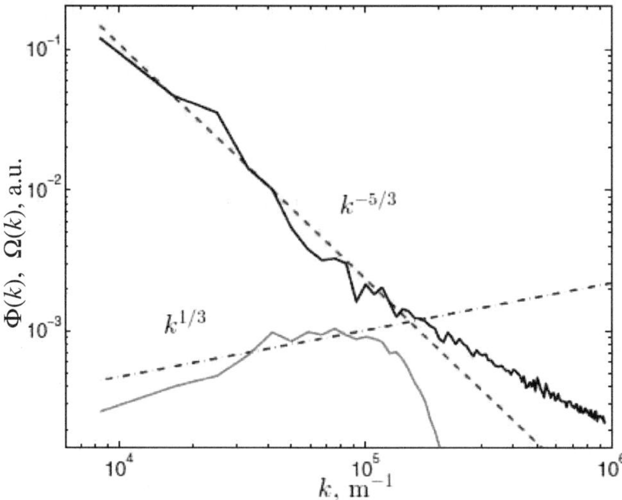

Figure 9.9 Spectra of the superfluid energy spectrum, $\Phi(k)$, and of the coarse-grained vorticity, $\Omega(k)$, plotted vs. wavenumber k, obtained by Baggaley *et al.* (2012b) using the vortex filament method at $T = 0$. Note that both energy and vorticity spectrum are consistent with the classical $k^{-5/3}$ and $k^{1/3}$ scalings, respectively, in the hydrodynamic range of length scales $k_D < k < k_\ell$ indicated by the dashed lines. Reproduced with permission from Baggaley *et al.* (2012b).

"add up" vortex lines make the velocity field, creating flow structures of sizes larger than ℓ. To highlight the rotational motion created by these structures, Baggaley *et al.* (2012b) also coarse-grained the vortex lines using Gaussian kernels, thus constructing a "superfluid vorticity" field (they computed the spectrum of such vorticity field and found that it increases with k in the region $k_D \ll k \ll k_\ell$), peaking at k_ℓ, with a dependence that is consistent with the classical $k^{1/3}$ scaling of the vorticity corresponding to the $k^{-5/3}$ scaling of the energy spectrum. See Fig. 9.9.

9.7 Beyond the Kolmogorov Phenomenology

The Kolmogorov (1941a) theory of homogeneous and isotropic turbulence assumes that classical turbulence in viscous fluids is self-similar within the inertial range and can be described by universal statistics that depend on only one relevant parameter, the energy flux across scales, which is equal to the dissipation rate $\epsilon = -dE/dt$. Then the velocity at scale r may be estimated as $(\epsilon r)^{1/3}$ and velocity structure functions of order n, defined in Eq. (9.13), follow the simple scaling (Frisch, 1995)

$$S_n(r) \sim (\epsilon r)^{\zeta(n)} = (\epsilon r)^{(n/3)}, \tag{9.17}$$

i.e., the scaling exponents are given by $\zeta(n) = n/3$. Real turbulent flows do not fulfill this expectation, and a phenomenon termed turbulent intermittency (Sreenivasan and Antonia, 1997) intervenes. Physical manifestation of intermittency is typically observed as sudden high-intensity bursts of velocity increments that occur more commonly than one would expect for self-similar turbulence. The lack of uniformity of all transport processes in space and time, in real space as well as wavenumber space, and an increasing propensity for this behavior with increasing control parameter in the problem – the Reynolds number in this instance – is a property shared by all strongly nonlinear systems.

In an attempt to account for this behavior, Kolmogorov in his 1962 paper, known as K62 (Frisch, 1995), used the analogy of random breaking of stones with that of eddies, and assumed Gaussian statistics of $\ln(x)$. Other models have also been constructed (see Meneveau and Sreenivasan, 1987, She and Leveque, 1994, and Sreenivasan and Antonia [1977]). Unfortunately, neither K62 nor other phenomenological models of multi-scaling (intermittency) can so far be derived directly from the Navier–Stokes equation (but see a recent attempt by Sreenivasan and Yakhot [2021]).

A natural question therefore arises whether this universality can be extended to quasi-classical flows of He II – for a review, see L'vov and Pomyalov (2017). We shall classify in more detail various regimes of quantum flows in Chapter 12, but provide here a simple definition of a quasi-classical turbulent flow of He II as the flow that is forced at a large scale by some mechanical means, such as by counterrotating propellers or by towing a grid of bars through a stationary sample of He II. The quasi-classical behavior (above 1 K) is typically thought to arise from the coupling of the normal and superfluid velocity fields through the action of the mutual friction force (Vinen, 2000; Vinen and Niemela, 2002).[1]

From the experimental point of view, we already discussed the striking results obtained by Maurer and Tabeling (1998), shown in Fig. 9.1, that there are no detectable changes in the form of the spectrum over the whole range of observed temperature and that the intermittency "corrections" to K41 deduced by subsequent analysis (Salort *et al.*, 2012a) remain classical and independent of the fraction of the normal to superfluid components.

The Kolmogorov scaling and classical intermittent behavior have been further confirmed in a variety of quasi-classical He II flows generated and detected in more recent experiments. Roche and collaborators have built on the basis of Tabeling's original von Kármán design a configuration of two counterrotating discs, a cryogenic wind-tunnel, and a pressurized circulator cooled through a heat exchanger, as shown to the left, middle, and right panels of Fig. 4.2, respectively. All three flows

[1] As we shall see later in this chapter, the form of the turbulent spectra is very different in flows other than the quasi-classical type of quantum turbulence.

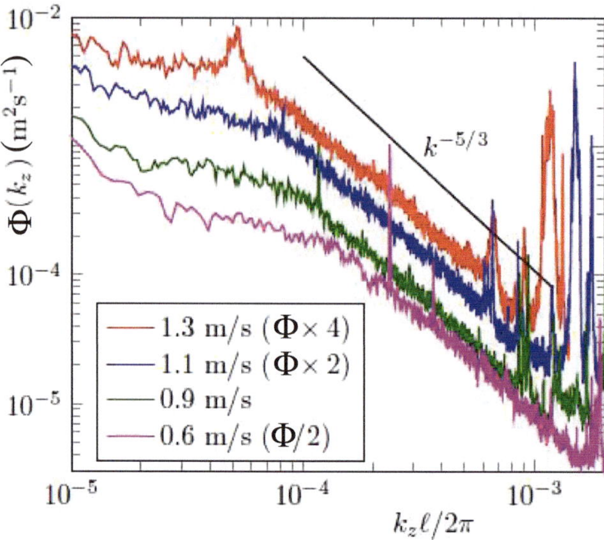

Figure 9.10 Experimental energy spectra (Roche *et al.*, 2007; Barenghi *et al.*, 2014a) for various mean flow velocities at T = 1.55 K. Different scaling factors have been introduced for clarity (see legend). From Barenghi *et al.* (2014a).

are driven by centrifugal forces provided by propellers; the last two are pressurized to avoid cavitation. The resulting Kolmogorov spectra are shown in Fig. 9.10. In addition, the most recent Grenoble measurements of Rusaouen *et al.* (2017a) in the wake of a disk in the two-fluid region of superfluid ^{4}He found no appreciable temperature dependence in intermittency corrections. This means that, within experimental accuracy, *in the scale range to which all these experiments had access*, the two fluids indeed move with the velocity field that is identical to that expected in the classical turbulent case.

On the other hand, very recent local velocity measurements have been performed in the SHREK corotating von Kármán cell shown in Fig. 4.3, using a hot-wire, a miniature Pitot-like sensor (see Fig. 4.13), and a cantilever anemometer, shown in Fig. 4.14. Salort *et al.* (2021) chose flow velocities large enough for developed turbulence to settle, but small enough for the dissipative scales to be approached. Differences between He I and He II velocity spectra are reported at the lowest accessible velocities (0.31 and 0.50 rad/s), but not at the higher one (1.57 rad/s); this result is not yet understood. When comparing the measured spectral differences the authors found more energy in He II than in He I in the dissipative scales that were resolved. This experimental observation is compatible with the prediction of Salort *et al.* (2011a) that superfluid kinetic energy accumulates over a range of mesoscales at the bottom of the inertial cascade in turbulent He II. According to this interpretation, the observed increase of the spectral density signifies the absence of an efficient dissipative mechanism in the superfluid component compared to the normal fluid component of He II.

Generally, it is clear that in quasi-classical He II flows the coupling of the normal and superfluid velocity fields cannot be complete. It must necessarily break down near the quantum length scale of the flow due to the quantisation of circulation in the superfluid component (Khomenko *et al.*, 2015; Babuin *et al.*, 2016). Indeed, it has been observed that even quasi-classical flows of He II display nonclassical statistics when probed by visualization at sufficiently small length scales (La Mantia *et al.*, 2016). Decoupling at smaller scales would necessarily lead to additional dissipation by the mutual friction, which would break the scale-invariance of the inertial range and thus lead to intermittency. As the strength of mutual friction is temperature dependent, it is reasonable to expect temperature dependence also in the turbulent statistics in those scales – in particular, temperature-dependent intermittency corrections.

L'vov's group in Rehovot considered the problem of intermittency in quasi-classical flows of He II theoretically. They adopted the coarse-grained Hall–Vinen–Bekarevich–Khalatnikov (HVBK) equations and performed comprehensive numerical simulations via their Sabra-shell model. Boue *et al.* (2013, 2015) made a detailed comparison of classical and superfluid turbulent statistics and found that the scaling exponents $\zeta(n)$ in quantum He II turbulence are the same as in classical turbulence for temperatures close to the superfluid transition temperature T_λ and for $T \ll T_\lambda$. This is not surprising because of the dominance of the normal component for T close to T_λ. It is also not surprising from previous discussions that the inertial-range intermittency very close to zero temperature, where the superfluid component is large, are similar. At intermediate temperatures, where the densities of the superfluid and normal fluid components are comparable, Biferale *et al.* (2018) identified a flip-flop scenario – a random two-way energy transfer between the two fluid components mediated by mutual friction – and claimed that this mechanism is responsible for the temperature dependence and enhancement of intermittency in quantum turbulence.

A reliable experimental determination of intermittency in He II and its possible temperature dependence requires not only the generation of fully developed turbulence but also flow measurement tools with a spatial resolution comparable to the quantum length scale, ℓ. Varga *et al.* (2018) utilized the Tallahassee He_2^* tracer-line visualization setup (Gao *et al.*, 2015) shown schematically in Fig. 9.11 (a). A mesh grid of 7×7 woven wires was towed by a linear motor to move through a He II sample in a channel $9.5 \times 9.5\,\text{mm}^2$ at a controlled speed up to about 65 cm/s. High-intensity femtosecond laser pulses were sent through the channel and a thin

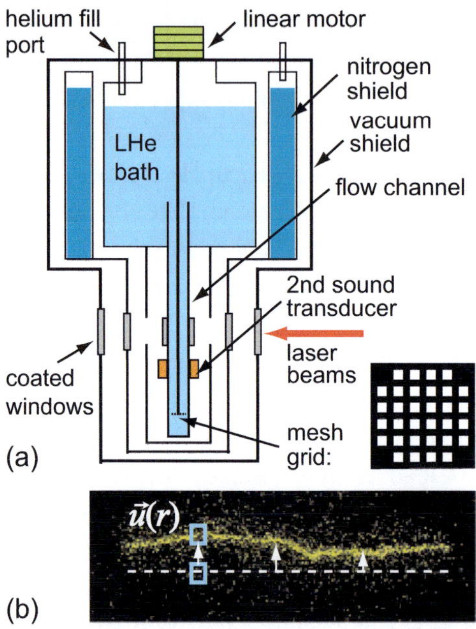

Figure 9.11 (a) Schematic diagram of the experimental setup. (b) A sample image of the He_2^* molecular tracer line. The white dashed line serves to demonstrate the initial location of the trace line for velocity calculations. Reprinted figure with permission from Varga *et al.* (2018). Copyright 2018 by the American Physical Society.

line of He_2^* molecular tracers was created and left to evolve for a drift time t_d of about 10–30 ms before the line was visualized by laser-induced fluorescence. The streamwise velocity $v_y(x)$ was determined by dividing the displacement of a line segment at x by t_d (see Fig. 9.11 (b)). The transverse velocity increments $\delta v_y(r) = v_y(x) - v_y(x+r)$ were evaluated for structure function calculations. Their scaling exponents were obtained by using the so-called extended self-similarity hypothesis (Benzi *et al.*, 1991). This hypothesis states that the scaling of a structure function $S_n(r)$ in the inertial scale range should be equivalent to the scaling of $S_n(r) \propto (S_3(r))^{\zeta_n}$ and allows for significant improvement in behavior, so that the scaling exponents ζ_n can be determined with less ambiguity (Dubrulle, 1994). Figure 9.12 shows transverse velocity structure functions $S_n^{\perp}$ for $n = 1$ to 7, plotted versus $S_3^{\perp}$, displaying distinct power-law scaling in the inertial range of scales. The deduced scaling exponents $\zeta_n^{\perp}$ are shown in Fig. 9.13 as functions of the order n for all temperatures investigated. They closely follow the theoretical prediction of Biferale *et al.* (2018), i.e., temperature-dependent intermittency corrections of the structure function scaling exponents with a maximum deviation from the K41 scaling at about 1.85 K. On the other hand, recent Grenoble measurements of Rusaouen

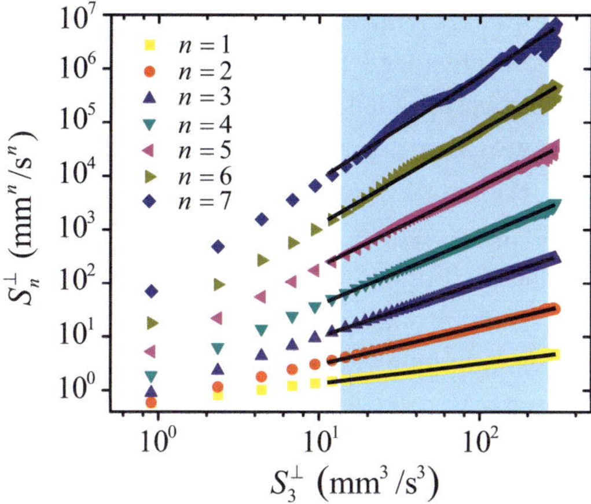

Figure 9.12 Extended self-similarity hypothesis. Transverse velocity structure functions $S_n^\perp$ for $n = 1$ to 7 are plotted vs. $S_3^\perp$. The black lines are linear fits of $\log S_n^\perp$ versus $\log S_3^\perp$ to the data that fall within the inertial range of scales. The particular case shown is for 1.85 K, 300 mm/s grid velocity, and 4 s decay time. Reprinted figure with permission from Varga *et al.* (2018). Copyright 2018 by the American Physical Society.

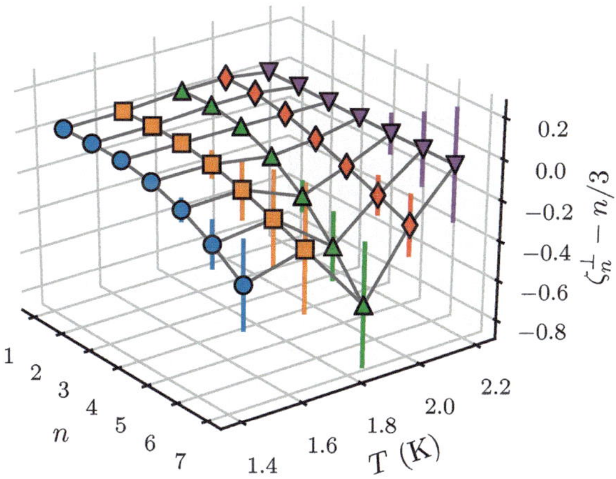

Figure 9.13 Intermittency corrections to the scaling exponents of the transverse structure functions deduced through extended self-similarity for data obtained at 4 s decay time and with grid velocity $v_g = 300$ mm/s. The 3D plot shows the temperature-dependent deviation of scaling exponents from K41 scaling. Reprinted figure with permission from Varga *et al.* (2018). Copyright 2018 by the American Physical Society.

et al. (2017a) in the wake of a disk in the two-fluid region of superfluid ^{4}He found no appreciable temperature dependence in intermittency corrections, so this issue does not appear to be settled at present. In principle, there could be several legitimate reasons why the two experiments show different results. First, the prediction of temperature-dependent enhanced intermittency is explained by Biferale *et al.* (2018) via a flip-flop scenario – a random energy transfer between the normal and superfluid components due to mutual friction. While $^4He_2^*$ molecules solely probe the normal fluid, the cantilever anemometer and pressure probes used in the Grenoble experiment of Rusaouen *et al.* (2017a) may not be able to sense such a flip-flop exchange of energy, if one exists, as they probe both fluids simultaneously. Furthermore, the sizes of the probes used in the Grenoble experiment are typically much larger than the quantum length scale ℓ, perhaps leading naturally to the same intermittency corrections as in classical turbulence. These problems do not seem to be as critical in the Tallahassee scheme of measurement.

9.8 Scaling Laws of Circulation

Another important property of a turbulent flow at different length scales is the circulation of the velocity field $\boldsymbol{u}$, defined in Eq. (3.4) as

$$\Gamma_A = \oint_C \boldsymbol{u} \cdot d\boldsymbol{r}, \tag{9.18}$$

where $A = r^2$ is the area enclosed by the closed loop C of size r. In a classical inviscid fluid, the circulation is a Lagrangian invariant; when viscosity is introduced, however, viscous dissipation smooths out vorticity and allows reconnections, so circulation is no longer conserved. In a quantum fluid, reconnections do occur, as we have seen in Section 3.7, but the circulation assumes only discrete values that depend on the number and orientation of the quantum vortices within C. The circulation is thus an ideal measure to probe similarities and differences between quantum turbulence and classical turbulence at varying length scales.

In classical turbulence, using the same dimensional argument as that leads to the K41 scaling, see Eq. (9.6), one finds that the variance of the circulation behaves closely as $\left\langle |\Gamma_A|^2 \right\rangle \sim A^{4/3}$ in the inertial range, a result observed in both experiments (Sreenivasan *et al.*, 1995) and numerical simulations (Cao *et al.*, 1996; Iyer *et al.*, 2019).

Müller *et al.* (2021) performed numerical simulations of the incompressible Navier–Stokes equation and of the Gross–Pitaevskii equation to compare the scale dependence of the circulation statistics. Firstly, they examined the distribution of the values of the circulation as a function of the loop area A in quantum turbulence. For small A corresponding to loop sizes smaller than the intervortex distance ($r \ll \ell$), the probability that a loop encloses more than one vortex is very small and peaks sharply at $\Gamma_A = 0$. As A is increased, the distribution widens, and develops exponential-like tails for $r \gg \ell$ similar to what is seen in classical turbulence.

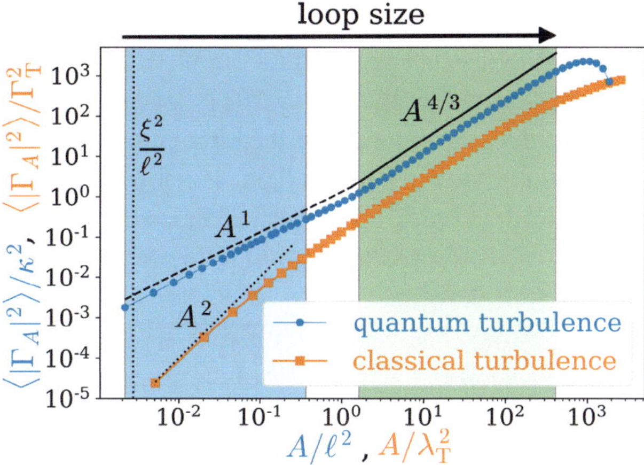

Figure 9.14 Scaling of the circulation in classical and quantum turbulence. The quantity $\left\langle|\Gamma_A|^2\right\rangle$ (in units of the square of the quantum of circulation κ) is plotted vs. the area A enclosed by the loop C (in units of ℓ^2 for quantum turbulence and in units of the square of the Taylor microscale λ_T for classical turbulence). Top (blue) curve: Numerical simulation of the Gross–Pitaevskii equation; the initial condition is an ABC flow. Bottom (orange) curve: Numerical simulation of the Navier–Stokes equation. The left blue region and the right green region represent the quantum length scales and the classical length scales, respectively. Reproduced from Müller *et al.* (2021), used under CC BY 4.0 (https://creativecommons.org/licenses/by/4.0).

Secondly, these authors examined how the variance of Γ_A scales with A; their results are shown in Fig. 9.14. In the limit $r \gg \ell$ (large A), they found that quantum turbulence displays the same $\left\langle|\Gamma_A|^2\right\rangle \sim A^{4/3}$ scaling of the Navier–Stokes equation, in agreement with the K41 scaling of the energy spectrum at such length scales (Cao *et al.*, 1996; Iyer *et al.*, 2019). Indeed, this $\left\langle|\Gamma_A|^2\right\rangle \sim A^{4/3}$ scaling is intermediate between a state of randomly oriented vortices (which would display $\langle|\Gamma_A|^2\rangle \sim A^2$) and a fully polarized tangle (which would correspond to $\langle|\Gamma_A|^2\rangle \sim A$), consistent with the conclusion reached in Section 9.3: The classical K41 scaling in a quantum fluid emerges from the partial polarization of vortex lines that form effective bundles of corotating vortices. In the opposite limit $r \ll \ell$ (small A), the Gross–Pitaevskii equation yields $\left\langle|\Gamma_A|^2\right\rangle \sim A$. This is unlike classical turbulence where the vorticity may be considered constant for a sufficiently small area A, hence $\left\langle|\Gamma_A|^2\right\rangle \sim A^2$ for small enough A.

Iyer *et al.* (2019) found that, from the point of view of scaling, low-order exponents of circulation moments followed K41 but started to show departures for high orders. These departures were weaker than for structure function exponents, and also simpler in the sense that they too varied linearly with the moment order. Iyer *et al.* (2019) argued that circulation behaves like a bifractal, with low-order moments residing on a space-filling set while high-order moments settled on an intermittent fractal set with dimension less than 3. If these results are true, a much simpler intermittency structure is suggested for circulation than for structure functions. Following these results, Müller *et al.* (2021) studied the moments $\langle |\Gamma_A|^p \rangle$ for $p > 2$ to determine the intermittency of the circulation in quantum turbulence and verified that essentially these results hold but also found a new result. For large values of A (in the region of classical length scales) the low-order moments (notably $p = 2$) obey the classical K41 predictions, while the higher order moments that deviate from K41 are well described by the same bifractal model as in classical turbulence (Iyer *et al.*, 2019). The new result is that, for small values of A (in the region of quantum length scales), the moments of the circulation are independent of the order p, which implies a very large intermittency at these scales, unlike classical turbulent flows.

9.9 Energy Spectrum in Superfluid ^{3}He-B

Prior to discussing the peculiarities of the energy spectrum of turbulent superfluid phase ^{3}He-B, it is useful to clarify some practical limits in generating such flows. The most serious practical limitation is that the typical experiment at submillikelvin temperature, using about 1 cm^3 of liquid ^{3}He-B of density about 100 kg/m^3, only allows up to about 1 nW of dissipative power. This is indeed a serious restriction in comparison with laboratory experiments with He II, which at temperatures above 1K tolerate typically up to about 1 W of dissipative power applied to 1 litre of He II of density 145 kg/m^3. As turbulent eddies larger than the box size cannot exist, an important length scale is the size of the turbulent box, D, typically 10 cm in He II and 1 cm in ^{3}He-B experiments; this sets the maximum outer scale of turbulence under study. Another important length scale in classical turbulence, namely the Kolmogorov, or dissipative, length scale η, can be obtained in viscous fluids following the earlier reasoning, as $\eta = 2\pi/k_\eta$, where $k_\eta \approx \left(\epsilon/\nu^3\right)^{1/4}$. Here, ν denotes the kinematic viscosity of the fluid. The dissipative length scale can be understood as the scale below which there is no turbulent motion.[2]

The superflow, however, exists down to smallest length scale, which is the size of the vortex core ξ. There is no classical kinematic viscosity, so the notion of dissipation length scale does not apply. We already introduced another physical length scale defined by the mean intervortex distance that could be estimated based on vortex line density L as $1/\sqrt{L}$. From the point of view of fluid dynamics, it seems sensible to introduce a characteristic quantum wavenumber, $k_Q \approx \left(\epsilon/\kappa^3\right)^{1/4}$ and the corresponding quantum length scale $\ell = 2\pi/k_Q$ (Skrbek *et al.*, 2001). For a

[2] In practice, sub-Kolmogorov scales do exist in classical turbulence (Yakhot and Sreenivasan, 2004), which may be very important for considerations as to the theoretical resolution of simulations and the possibility of singularities, so this statement must be viewed as representing conventional wisdom, which proves adequate for the present level of description.

typical laboratory He II experiment we have $\eta \approx \ell \approx 1/\sqrt{L} \approx 10\,\mu\text{m}$, with D some six orders of magnitude larger. This separation of scales in principle allows for very high Reynolds number experiments in laboratory-sized cryostats (see Eq. (9.3)), up to about 10^8. This is to be compared with turbulent ^{3}He-B, where we arrive at $\eta \approx 10\,\text{cm}$ and $\ell \approx 0.1\,\text{mm}$, with $D \approx 1\,\text{cm}$. It means that the relatively thick normal fluid of ^{3}He-B effectively does not move and, as we have stressed in several previous contexts, provides a unique frame of reference fixed by either stationary or moving turbulent box.

Quantum turbulence of ^{3}He-B (at finite temperatures above $\approx 0.2T_c$) thus takes the form of superfluid eddies, composed of the tangle of quantized vortices by their partial polarization, with mutual friction affecting their dynamics at all relevant scales. Following Vinen (2005), we consider a small volume of superfluid component δV, but much larger than ℓ^3, which moves through the thick stationary normal fluid with the velocity u. There is mutual friction force per unit volume $f = \alpha \rho_s \kappa L u$ acting on it antiparallel to u, where α is the dissipative mutual friction parameter of ^{3}He-B measured by Bevan *et al.* (1997). In order to see how the mutual friction affects the turbulent motion at different length scales, we examine how the Fourier component u_k relates to the motion on the length scale $1/k$. The effect of mutual friction on this Fourier component is described by the linear equation $\dot{u}_k = \alpha\kappa L u_k$. It follows that the mutual friction causes u_k to decay with a time constant given by $\tau_d = 1/(\alpha\kappa L)$, independent of k. In other words, mutual friction gives rise to the decay of an eddy of size D at a rate that is independent of D.

We then argue that the 3D turbulent cascade may exist in the superfluid component of ^{3}He-B only if the turnover time τ of a superfluid eddy of the size D exceeds the decay time of such an eddy due to mutual friction, τ_d. To estimate the turnover time (Vinen, 2005), we use the standard relation used in classical homogeneous and isotropic turbulence $\epsilon = \epsilon_0 u_D^3/D$, where ϵ_0 is of order unity (Sreenivasan, 1984, 1998). It follows that superfluid eddies larger than $D_{\max} \approx \epsilon^{1/2}/(\alpha\kappa L)^{3/2}$ cannot exist, as they would be effectively destroyed by mutual friction within one turnover time τ_D. For typical values for ^{3}He-B experiments assuming $\alpha \approx 1$ (i.e., at about $T/T_c = 0.6$) we arrive at $D_{\max} \approx 2\,\text{cm}$.

Possible 3D superfluid energy spectrum in ^{3}He-B therefore starts from eddies of size not substantially exceeding $D_{\max}$. Its form can be deduced from theory, based on the coarse-grained hydrodynamic equation (Finne *et al.*, 2003; Volovik, 2003) obtained by averaging the Euler equation over vortex lines written in the frame of reference of the normal fluid:

$$\frac{\partial \boldsymbol{u}}{\partial t} + \nabla \mu = \boldsymbol{u} \times \boldsymbol{\omega} + q\hat{\boldsymbol{\omega}} \times (\boldsymbol{\omega} \times \boldsymbol{u}). \tag{9.19}$$

Here, $\boldsymbol{u}$ is the superfluid velocity, $\boldsymbol{\omega}$ denotes the course-grained vorticity, and $\hat{\boldsymbol{\omega}}$ is a unit vector in the direction of $\boldsymbol{\omega}$. As first emphasized by Finne *et al.* (2003),

Eq. (9.19) has the remarkable property that makes it distinct from the ordinary Navier–Stokes equation where the relative importance of the inertial and dissipative terms is given by the Reynolds number, which in turn depends on the geometry of the particular flow under study. Here, the role of the effective Reynolds number is played by the parameter $Re_{\text{eff}} = q^{-1} = (1 + \alpha')/\alpha$, which, via the mutual friction parameters α and α', depends on temperature and not on geometry. A wide range of q values is thus achievable experimentally, with q increasing with temperature in ^{3}He-B (Bevan *et al.*, 1997). For $q \gg 1$, similar to the low-Reynolds-number classical fluid dynamics, the solutions are stable. As q approaches unity, solutions become unstable and for $q \ll 1$ Eq. (9.19) describes fully developed turbulence. For $T \to 0$, Eq. (9.19) becomes the Euler equation with no dissipation term, as mutual friction cannot operate in the absence of the normal fluid.

L'vov *et al.* (2004) and Vinen (2005) have independently shown that, assuming quiescent normal fluid, owing to the action of mutual friction there is strong damping of large eddies, with the result that at low wavenumbers (starting possibly from largest eddies as explained above) the energy spectrum rolls off more rapidly, approaching k^{-3}. Beyond a certain critical value k_{cr}, however, the superfluid energy spectrum gradually recovers its usual Kolmogorov form. Here, the normal and superfluid velocity fields are again nearly independent, as the action of the mutual friction force at these small length scales is negligible. One can also understand this fact on the basis of characteristic times τ_d and τ discussed above. Here, $\tau_d \gg \tau$, so the mutual friction cannot appreciably affect the motion of such small eddies. We again stress that only superfluid eddies exist at all length scales, the normal fluid being at rest due to finite (and large) kinematic viscosity. At small scales, the superfluid Kolmogorov-like cascade exists on its own. These considerations (in the frame of continuum approximation) are justified at length scales larger than ℓ.

At a sufficiently low temperature, similar to He II, a transition to the Kelvin wave cascade takes place at length scales of order ℓ, a phenomenon we shall discuss next.

9.10 Kelvin Wave Cascade

In 2000, an influential oscillating grid experiment (see Fig. 4.9), performed by McClintock and collaborators (Davis *et al.*, 2000), revealed that quantum turbulence generated by an oscillating grid in He II at mK temperatures decays rather quickly. This rapid decay was puzzling. At higher temperatures, the normal fluid provides a route to dissipating the kinetic energy of turbulent vortex lines into heat: Energy can be transferred from the superfluid to the normal fluid by the mutual friction, and then dissipated into phonons and rotons by viscous forces. But this particular experiment was performed at temperatures so small that the normal fluid component was negligible. A second possible route to dissipating the kinetic energy of vortex lines is acoustic: Numerical simulations of the Gross–Pitaevskii equation show that vortex lines radiate sound waves when they accelerate (Leadbeater *et al.*, 2002) and sound pulses when they collide and reconnect (Leadbeater *et al.*, 2001; Zuccher *et al.*, 2012); indeed, the effect that vortices generate sound is well known in classical physics (Howe, 1998).

The explanation of McClintock's experiment in terms of sound emission did not stand up to first quantitative scrutiny. The problem was the following. The sound power that is radiated by vortices depends strongly on the vortex separation ℓ (if we consider the sound radiated by two corotating vortex lines) or the Kelvin wavelength (if we consider sound radiated by a Kelvin wave, which can be reasonably assumed to have wavelength of the order of ℓ); the closer the two vortex lines (or the smaller the Kelvin wavelength), the more rapidly the vortex lines rotate, and more power is radiated. Unfortunately, the vortex line density in McClintock's experiment was not large enough to correspond to the short length scale that would have explained the observed rapid decay of the vortex tangle. Further, stretching of individual vortex lines is not possible; the Kolmogorov cascade therefore cannot shift energy to wavenumbers larger than k_ℓ.

A solution of this puzzle was proposed based on the earlier work of Svistunov (1995), who had argued that the nonlinear interaction of Kelvin waves creates shorter and shorter waves. The idea was that a *Kelvin waves cascade* on a vortex line (analogous to the Kolmogorov cascade of eddies) would transfer energy to length scales shorter than ℓ, short enough that sound can be efficiently radiated away. In this scenario, quantum turbulence at very low temperatures contains two energy cascades: a classical Kolmogorov cascade of eddies (each eddy consisting of many vortex lines) that shifts energy from the injection length scale D (the scale of the grid that generates the turbulence, for example) to the intervortex scale ℓ, and a Kelvin cascade of waves on individual vortices that shifts energy to scales of the order of 10 nm at which Kelvin waves are damped. In He II the damping mechanism is phonon emission (Vinen, 2001). Ultimately, these phonons should thermalize, raising the temperature of the liquid helium (Samuels and Barenghi, 1998).

As in classical turbulence, the final destiny of the flow of energy is to be converted into heat. It is important to stress that the Kelvin wave regime acquires significant importance only at low temperatures (typically below 1 K in ^{4}He), particularly relevant to decay experiments. At higher temperatures, mutual friction damps Kelvin waves, smoothing vortex lines, transferring the energy of the Kelvin waves into the normal fluid: The Kelvin wave cascade is therefore suppressed (Boue *et al.*, 2012).

We now describe the basic physics of the Kelvin wave cascade and the evidence for it. Neglecting the interaction between separate vortex lines for $k\ell \gg 1$, in the absence of normal fluid and friction, superfluid turbulence can be considered as a system of Kelvin waves with different wavevectors interacting with each other on the same vortex. Svistunov predicted that this interaction results in energy transfer toward large k (Svistunov, 1995); this energy transfer was confirmed by numerical simulations in which energy was pumped into Kelvin waves at intervortex scales by vortex reconnections (Kivotides *et al.*, 2001) or simply by shaking the vortex lines (Vinen *et al.*, 2003). The first analytical theory of Kelvin wave turbulence (propagating along a straight vortex line and in the limit of small amplitude compared to wavelength) was proposed by Kozik and Svistunov (2004), who argued that the leading interaction is a six-wave scattering process (three incoming waves and three outgoing waves). They found, under the additional assumption of locality of the interaction, the energy spectrum of Kelvin waves of the form

$$\Phi(k) = C_{\mathrm{KS}}\Lambda\epsilon^{1/5}\kappa^{7/5}\ell^{-8/5}k^{-7/5}, \tag{9.20}$$

where the dimensionless parameter Λ is $\Lambda = \ln(\ell/a_0) \approx 12$ for ^{3}He-B and ≈ 15 for He II, a_0 is the vortex core radius, ϵ is the energy flux in 3D k-space, and C_{KS} is a constant of order unity. Later L'vov and Nazarenko (2010) criticized the assumption of locality and concluded that the spectrum is

$$\Phi(k) = C_{\mathrm{LN}}\Lambda\kappa\epsilon^{1/3}\Psi^{-3/2}k^{-5/3}, \quad \Psi = 4\pi\Phi(k)\Lambda^{-1}\kappa^{-2}, \tag{9.21}$$

where $C_{\mathrm{LN}} = 0.304$ (Boue *et al.*, 2011).

It is important to mention that the 3D energy spectrum $\Phi(k)$ described above can be related to the 1D amplitude spectrum $A(k)$ by $\Phi(k) \sim \hbar\omega(k)n(k)$, where $\omega(k) \sim k^2$ is the angular frequency of a Kelvin wave of wavenumber k, $\hbar\omega(k)$ is the energy of one quantum, and $n(k) \sim A(k)$ is the number of quanta (occupation number); therefore, in terms of the Kelvin waves amplitude spectrum (which is often reported in the literature and can be numerically computed), the predictions of Kozik and Svistunov and of L'vov and Nazarenko are proportional, respectively, to $k^{-17/5} = k^{-3.40}$ and $k^{-11/3} = k^{-3.67}$. The two predicted exponents, -3.40 and -3.67, are close to each other and therefore difficult to numerically distinguish; nevertheless, the VFM simulations by Krstulovic (2012) and by Baggaley and Laurie (2014) observe better agreement with the L'vov–Nazarenko spectrum.

Comparison of the Kolmogorov spectrum and the Kelvin wave spectrum in quantum turbulence at $T = 0$ suggests (L'vov *et al.*, 2007) that the 1D energy transfer mechanism along individual vortex lines is less efficient than the 3D eddy–eddy energy transfer. A pile-up of energy is therefore expected near the transition region $k\ell \approx 1$ between the two cascades (a bottleneck effect). It is not clear how this energy accumulation should manifest in the experiments. Some evidence of flattening out of the 3D energy spectrum in this region is visible in the GPE simulations of Sasa *et al.* (2011) and, more recently, of diLeoni *et al.* (2017); see Fig. 9.15.

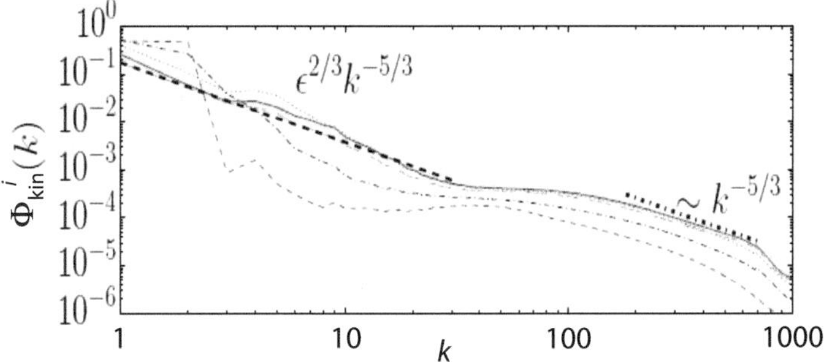

Figure 9.15 Dual cascade: Incompressible energy spectrum $\Phi^i_{\rm kin}(k)$ versus wavenumber k (the horizontal axis ranges from $k = 10^0$ to $k = 10^3$) computed by diLeoni *et al.* (2017) who solved the Gross–Pitaevskii equation on a large 2048^3 grid and succeeded in exciting the Kolmogorov cascade of eddies for $k < k_\ell \approx 79$ and the Kelvin wave cascade for $k > k_\ell$. Notice that the incompressible energy spectrum $\Phi^i_{\rm kin}(k)$ follows the same $k^{-5/3}$ scaling for both cascades, and the plateau from $k \approx 30$ to $k \approx 200$ (hence around k_ℓ) is probably the predicted bottleneck. Reprinted figure with permission from diLeoni *et al.* (2017). Copyright 2017 by the American Physical Society.

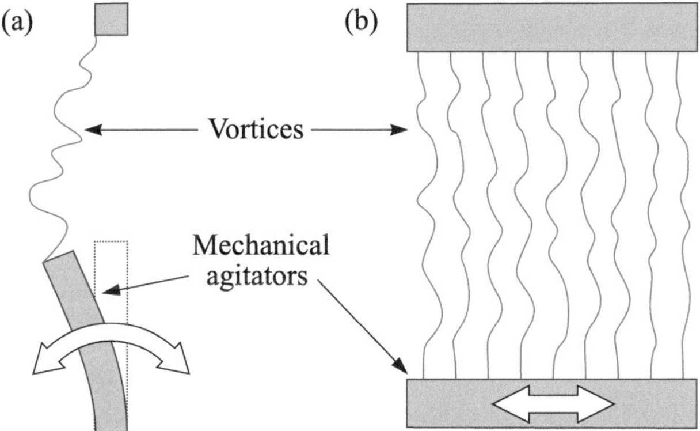

Figure 9.16 Example configurations of vortex lines, agitated to generate Kelvin waves. (a) A single vortex attached to an oscillating device. (b) An array of vortices stretched between parallel plates and agitated by shear or torsional oscillations of the plates. Reproduced with permission from Eltsov and L'vov (2020).

The Kelvin wave spectrum is not yet experimentally confirmed. The most promising way to do it seems to employ recently developed nanomechanical oscillators in a form of a goal post (Kamppinen and Eltsov, 2018; Autti *et al.*, 2020) or plate (Barquist *et al.*, 2020) to drive vortex lines as schematically shown in Fig. 9.16. The drag offered to their motion in superfluid helium in the zero-temperature limit is then related to the dissipation via Kelvin wave cascade. One of the first questions that arises then in the analysis of the experimental results is the relation between the energy flux in the cascade and the amplitude of the Kelvin waves. Eltsov and L'vov (2020) recently provided such a relation based on the L'vov–Nazarenko picture of the cascade discussed above. Remarkably, the amplitude of the waves is predicted to depend on the energy flux very weakly, as the one-tenth power. The authors also show that even a moderate energy flux (corresponding to the amplitude of motion of the agitator of $0.2\,\mu$m) can bring the vortex on the edge of the regime where the turbulence of Kelvin waves may still be considered as weak. Clearly, laboratory confirmation of the Kelvin wave cascade represents a formidable challenge. Indeed, experimental evidence of the cascade has become available only very recently, when Mäkinen *et al.* (2023) excited inertial waves by modulating the rotation of ^{3}He-B and observed a transfer of energy to higher-frequency Kelvin waves.

The Kelvin wave cascade at very low temperatures suffers dissipation at high enough k. In He II it occurs due to phonon irradiation, while in ^{3}He-B via the excitation of the Caroli–Matricon bound states in the vortex cores (Silaev, 2012); note that this dissipation mechanism is not contained in Eq. (9.19) considered above. Very recently, this type of dissipation has been associated with so-called Planckian dissipation, which takes place in quantum systems in equilibrium at temperature T when the relaxation time τ is comparable with $\hbar/T$. Volovik (2022) claimed that the anomalous spectral flow of levels along the chiral brand of the Caroli–de Gennes–Matricon states takes place in the super-Planckian region (for $\tau < \hbar/\Delta E$, laminar flow) and is absent in the sub-Planckian regime (for $\tau > \hbar/\Delta E$, turbulent flow), where ΔE denotes the distance between energy levels in the vortex core.

In short, in ultracold Fermi superfluids, the turbulent energy can be transferred to the ensemble of localized quasiparticles in vortex cores, resulting in a local temperature increase inside them and the subsequent dissipation due to the escape of hot quasiparticles into delocalized states, effectively increasing the amount of the normal fluid, i.e., the temperature. The final destiny of the turbulent energy in both He II and ^{3}He-B is thus the same as in classical turbulence – heating the fluid.

9.11 Summary

Similar to classical turbulence in viscous fluids, energy spectra also provide an important characteristic of turbulent flows of quantum fluids. Due to the two-fluid behavior, there are two energy spectra, one for the normal fluid and one for the superfluid, which, in general, are coupled by the action of the mutual friction force. As in classical turbulent flows, in many mechanically generated quantum flows (with energy input at some large energy containing scale), Richardson cascade operates, giving rise to the existence of an inertial range of scales characterized by the Kolmogorov scaling: At large scales the normal and superfluid velocity fields are coupled by the mutual friction. Thus, the key principle responsible for Kolmogorov-type scaling, namely vortex stretching, operates in those turbulent quantum flows where the tangle of quantized vortices is at least partly polarized (vortex lines are organized in bundles) on length scales exceeding the quantum length scale ℓ, determined by the mean distance between vortex lines. Upon approaching ℓ, the matching of the velocity fields is no longer possible and mutual friction acts, together with viscosity of the normal fluid, as a drain for kinetic energy of the flow. The stretching mechanism cannot operate on an individual quantized vortex. The Richardson cascade therefore cannot operate at scales smaller than, or of the order of, the quantum length scale (mean distance between quantized vortices in the tangle). At low enough temperatures, the turbulent kinetic energy propagates to smaller scales via a Kelvin wave cascade along individual vortex lines and becomes dissipated by phonon emission in He II or by exciting the Caroli–Matricon states in the cores of quantized vortices in ^{3}He-B. Also, as in classical turbulence, intermittency is observed in quasi-classical co-flows of He II. Its enhancement and temperature dependence has been theoretically predicted and experimentally confirmed. In particular, the scaling properties of circulation are remarkably similar towards the low end of the inertial range.

Due to practical reasons of cryogenic restrictions, in experimentally feasible flows of superfluid ^{3}He-B, the thick normal fluid hardly moves and provides a unique frame of reference to study the dynamics of the vortex tangle. This is why superfluid turbulence in a form of a vortex tangle can exist only below a certain critical temperature, where the mutual friction becomes weaker. The inverse temperature therefore serves as an analogue of the Reynolds number. Even below this critical temperature, mutual friction inhibits large superfluid eddies and leads to strong dissipation at large scales, resulting in steeper roll-off exponent for the superfluid spectral energy density, which gradually changes from -3 to the classical $-5/3$ at small scales, where mutual friction ceases to be important.

In this chapter, we discussed only selected quantum flows in order to provide a sample of the rich variety of the normal fluid and superfluid energy spectra and related questions such as the scaling of structure functions and intermittency effects. We shall return to these questions in Chapter 11 when analysing the closely connected temporal decays of various turbulent quantum flows.

10

Quantum Turbulence via Tracer Particles

A great deal of knowledge about classical turbulence depends on our ability to visualize the flow field. It has been known for a long time that seeding a fluid with dye, bubbles, or small particles reveals qualitative details of the flow pattern to the naked eye; an example of historical importance is the 1883 experiment of Reynolds, illustrated in Fig. 1.2. Visualization methods such as particle image velocimetry (PIV), particle tracking velocimetry (PTV) and laser-induced fluorescence based on neutral He_2^* triplet molecules have driven the recent progress in the hydrodynamics of He II to quantitative levels. These visualization methods (mentioned in Chapter 4) were originally developed in the context of classical fluid dynamics, but their implementation at cryogenic temperatures was far from trivial. Nowadays, flow visualization supplements the information about flow patterns provided by computer simulations based on the models introduced in Chapter 5.

Flow visualization typically presents the view of some significant part of the flow at a given time, and, depending on the level of detail one is attempting to visualize, one can obtain temporal evolution as well. Seeing is believing, and nothing builds one's intuition better than flow visualization. For this reason, good statistical measures keep in mind the information from flow visualization. Yet, one must be aware of various limitations in relation to turbulence at high Reynolds numbers. The spatial and temporal resolutions could be constraining (this is also true of single-point measurements); rare events could be missed or confused; what appears to the eye as a dominant feature may not be dynamically the most important part, etc. Thus, care must be taken while the statistical information and flow visualization are interpreted together. In particular, many flow visualization images in He II depend strongly on experimental conditions such as the temperature, the manner in which the He II flow was created, the scale of the experiment, and the scales explored by visualization. It is, in general, hard to make sense of flow visualization images of He II without a good knowledge of the experimental conditions.

The application of classical visualization techniques to turbulent two-fluid hydrodynamics requires particular care. For example, PTV allows the measurement of Lagrangian quantities, e.g., the velocity or the acceleration of small parcels of fluid (material particles). PIV estimates the instantaneous Eulerian streamlines (i.e., the fluid velocity in a section of the flow field) by assuming a single, smoothly varying velocity field. As demonstrated by Duda *et al.* (2014) for the thermal counterflow of He II past a cylinder, this assumption may lead to the detection of spurious vortical structures. Since the tracer particles interact with both superfluid and normal fluid components of He II, and may also become trapped in the quantum vortices, the correct interpretation of PTV or PIV measurements depends on proper understanding of these complicated interactions, which is the subject of this chapter.

10.1 Particle Dynamics in He II

In classical fluid dynamics, given a flow velocity field $\boldsymbol{u}(\boldsymbol{r}, t)$, the trajectory of an infinitesimal parcel of fluid (called a pathline) subject to the initial condition $\boldsymbol{r}_{\rm p}(0)$ at $t = 0$ is the solution $\boldsymbol{r}_{\rm p}(t)$ of the equation

$$\frac{d\boldsymbol{r}_{\rm p}}{dt} = \boldsymbol{u}(\boldsymbol{r}_{\rm p}, t), \tag{10.1}$$

where, as indicated, at every instant t the velocity $\boldsymbol{u}$ is evaluated at the position of the particle, $\boldsymbol{r}_{\rm p}$. But the pathline is only a mathematical idealization. A physical particle, however small, has size and inertia, so its position $\boldsymbol{r}_{\rm p}(t)$ and velocity $\boldsymbol{u}_{\rm p}(t)$ obey *two* equations (unlike the single Eq. (10.1) for the pathline):

$$\frac{d\boldsymbol{r}_{\rm p}}{dt} = \boldsymbol{u}_{\rm p}, \qquad \frac{d\boldsymbol{u}_{\rm p}}{dt} = \frac{\boldsymbol{F}}{m}, \tag{10.2}$$

where m is the mass of the particle. Note that the second equation is just Newton's law for the particle. The identification of the forces $\boldsymbol{F}$ acting on the physical particle is sometimes subtle, and its trajectory is not necessarily a pathline.

The equations governing the motion of a small particle in He II were derived and discussed in detail by Poole *et al.* (2005). They considered a small spherical particle of radius $a_{\rm p}$ and density $\rho_{\rm p}$ moving in the presence of prescribed normal fluid and superfluid velocity fields $\boldsymbol{u}_{\rm n}$ and $\boldsymbol{u}_{\rm s}$ at temperature T. Following the classical *one-way approach* (Maxey and Riley, 1983), Poole *et al.* (2005) assumed that the particle does not significantly affect $\boldsymbol{u}_{\rm n}$, $\boldsymbol{u}_{\rm s}$, or any vortex line present in the flow, and that the particle is not trapped in a vortex line. These assumptions mean that, in the case of turbulent flows, tracer particles do not affect the quantum turbulence they visualize. Poole *et al.* (2005) also assumed that $\boldsymbol{u}_{\rm n}$ and $\boldsymbol{u}_{\rm s}$ change very little over distances of the order of $a_{\rm p}$. This means that, for turbulent superfluid flows, the particle is smaller than the average distance between vortex lines, ℓ, and that, for turbulent normal fluid flows, the particle is smaller than the Kolmogorov dissipation length. In deriving the particle's equation of motion, Poole *et al.* (2005) noticed that in typical experimental conditions the Reynolds number based on the relative velocity between the particle and the normal fluid is quite small,

$$Re_{\mathrm{p}} = 2\rho_{\mathrm{n}} a_{\mathrm{p}} |\boldsymbol{u}_{\mathrm{n}} - \boldsymbol{u}_{\mathrm{p}}|/\eta \ll 1,$$

and hence the viscous drag is linear in this velocity difference $\boldsymbol{u}_{\mathrm{n}} - \boldsymbol{u}_{\mathrm{p}}$ (Stokes' law). They also showed that, in typical experimental conditions, the Basset history force, the Faxen correction, the Magnus lift force, and the shear-induced lift force can be neglected (for details on practical use of PTV in classical fluids, see, e.g., Raffel *et al.*, 2007), and concluded that the particle position $\boldsymbol{r}_{\mathrm{p}}$ and velocity $\boldsymbol{u}_{\mathrm{p}}$ obey the following equations:

$$\frac{d\boldsymbol{r}_{\mathrm{p}}}{dt} = \boldsymbol{u}_{\mathrm{p}}, \tag{10.3}$$

$$\begin{aligned} \rho_{\mathrm{p}}\vartheta\frac{d\boldsymbol{u}_{\mathrm{p}}}{dt} &= 6\pi a_{\mathrm{p}}\eta(\boldsymbol{u}_{\mathrm{n}} - \boldsymbol{u}_{\mathrm{p}}) + \vartheta(\rho_{\mathrm{p}} - \rho)\boldsymbol{g} \\ &+ \rho_{\mathrm{n}}\vartheta\frac{D\boldsymbol{u}_{\mathrm{n}}}{Dt} + C\rho_{\mathrm{n}}\vartheta\left(\frac{D\boldsymbol{u}_{\mathrm{n}}}{Dt} - \frac{d\boldsymbol{u}_{\mathrm{p}}}{dt}\right) \\ &+ \rho_{\mathrm{s}}\vartheta\frac{D\boldsymbol{u}_{\mathrm{s}}}{Dt} + C\rho_{\mathrm{s}}\vartheta\left(\frac{D\boldsymbol{u}_{\mathrm{s}}}{Dt} - \frac{d\boldsymbol{u}_{\mathrm{p}}}{dt}\right), \end{aligned} \tag{10.4}$$

where η is helium viscosity, $\boldsymbol{g}$ is acceleration due to gravity, $\vartheta = 4\pi a_{\mathrm{p}}^3/3$ is the particle volume, and $C = 1/2$ is the added mass coefficient for a sphere; the substantial derivatives of the normal and superfluid velocities are defined, respectively, as

$$\frac{D\boldsymbol{u}_{\mathrm{n}}}{Dt} = \frac{\partial\boldsymbol{u}_{\mathrm{n}}}{\partial t} + (\boldsymbol{u}_{\mathrm{n}}\cdot\nabla)\boldsymbol{u}_{\mathrm{n}}, \qquad \frac{D\boldsymbol{u}_{\mathrm{s}}}{Dt} = \frac{\partial\boldsymbol{u}_{\mathrm{s}}}{\partial t} + (\boldsymbol{u}_{\mathrm{s}}\cdot\nabla)\boldsymbol{u}_{\mathrm{s}}. \tag{10.5}$$

It is instructive to examine some properties of Eqs. (10.3) and (10.4). If the normal fluid and the superfluid are both stationary ($\boldsymbol{u}_{\mathrm{n}} = \boldsymbol{u}_{\mathrm{s}} = 0$) and gravity is present, then, starting from the initial condition $\boldsymbol{u}_{\mathrm{p}} = 0$, after a transient, the particle achieves the terminal velocity

$$u_{\infty} = \frac{2a_{\mathrm{p}}^2 g(\rho_{\mathrm{p}} - \rho)}{9\eta}. \tag{10.6}$$

Therefore, by measuring this sedimentation velocity, one can determine the particle radius (or the average particle radius for a batch of particles).

In some experiments the particles are chosen to have approximately the same density of helium ($\rho_{\mathrm{p}} \approx \rho$) or to be neutrally buoyant, and hence gravity can be neglected and Eq. (10.4) reduces to

$$\frac{d\boldsymbol{u}_{\mathrm{p}}}{dt} = \frac{1}{\tau}(\boldsymbol{u}_{\mathrm{n}} - \boldsymbol{u}_{\mathrm{p}}) + \frac{1}{\rho}\left(\rho_{\mathrm{n}}\frac{D\boldsymbol{u}_{\mathrm{n}}}{Dt} + \rho_{\mathrm{s}}\frac{D\boldsymbol{u}_{\mathrm{s}}}{Dt}\right), \tag{10.7}$$

where the particle relaxation time is

$$\tau = \frac{\rho a_{\mathrm{p}}^2}{3\eta}. \tag{10.8}$$

The tendency of the particles to become trapped in vortex lines is apparent if we consider a buoyant particle initially at rest at distance r_0 to a straight, fixed vortex. At $T = 0$, Eq. (10.7) reduces to

$$\frac{d\boldsymbol{u}_{\mathrm{p}}}{dt} = -\frac{1}{\tau}\boldsymbol{u}_{\mathrm{p}} + \frac{\rho_{\mathrm{s}}}{\rho}(\boldsymbol{u}_{\mathrm{s}} \cdot \nabla)\boldsymbol{u}_{\mathrm{s}}. \tag{10.9}$$

The velocity field of a straight vortex line is $\boldsymbol{u}_{\mathrm{s}} = (0, \kappa/(2\pi r), 0)$ in cylindrical coordinates; thus $(\boldsymbol{u}_{\mathrm{s}} \cdot \nabla)\boldsymbol{u}_{\mathrm{s}}$ has the form of a radial pressure gradient, $\nabla\left(1/r^2\right)$. Using the symbol u_{p} for the radial component of the velocity $\boldsymbol{u}_{\mathrm{p}}$, we find

$$\frac{du_{\mathrm{p}}}{dt} = -\frac{u_{\mathrm{p}}}{\tau} - 2\frac{\beta_0}{\rho r^3}, \tag{10.10}$$

where $\beta_0 = \rho_{\mathrm{s}}\kappa^2/\left(8\pi^2\rho\right)$. Equation (10.10) shows that the particle "falls" radially towards the vortex line, taking approximately the time (Poole *et al.*, 2005) $\tau_{\mathrm{v}} \approx r_0^4/(8\beta_0\tau)$ to travel from the distance $r = r_0$ away from to the vortex to the distance $r = 2a_{\mathrm{p}}$ close to the vortex.

At closer distances, the approximation that the vortex remains fixed will break down due the presence of an image vortex in the particle. The analysis of the particle–vortex interaction at close distance requires a *two-way approach* in which the particle and the vortex affect each other. This can be done by including the particle's dynamics either in the Gross–Pitaevskii equation (GPE) (Winiecki and Adams, 2000; Berloff and Roberts, 2001; Shukla *et al.*, 2018) or in the vortex filament method (VFM) (Kivotides *et al.*, 2008a). Because of the length scales involved, the modified GPE is more realistic for studying the interaction of vortices with small Angstrom-size ion bubbles, while the modified VFM applies better to the micron-size particle tracers that concern us here. The modified VFM shows that, when the particle becomes close, the vortex becomes curved (due to the superfluid displaced by the approaching particle and the interaction with its image inside the particle) and therefore moves until the particle touches the vortex (which thus reconnects with its image). At this point we say that the particle has become "trapped" in the vortex; in reality, the typical particle radius, $a_{\mathrm{p}} \approx 10^{-6}$ m, is about 10,000 times greater than the vortex core radius, $\xi_0 \approx 10^{-10}$ m. The successive evolution of the particle–vortex complex depends on the details of the initial condition (the position and velocity of the particle, the shape of the vortex, the temperature) and may involve the sliding motion of the two end points of the vortex over the particle surface, the generation of Kelvin waves along the vortex line, or the emission of

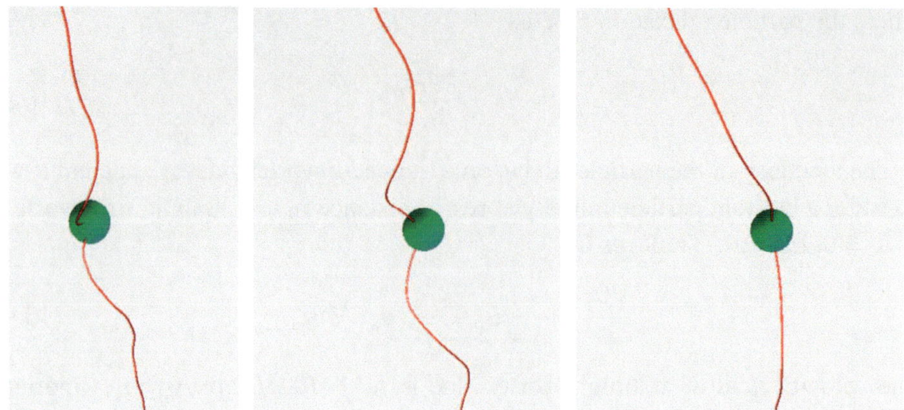

Figure 10.1 VFM calculation of particle–vortex interaction. (Left) The particle, moving from right to left, has just collided with the vortex line, which has reconnected with its image within the particle. (Middle) Kelvin waves on the vortex line are generated and travel away. (Right) Mutual friction and the helium viscosity have damped out the Kelvin waves (the vortex strengthens) and the particle settles down in the trapped configuration. Reprinted figure with permission from Kivotides *et al.* (2008a). Copyright 2008 by the American Physical Society.

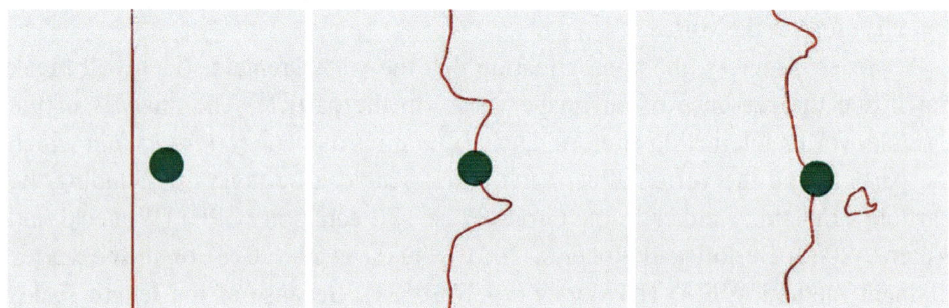

Figure 10.2 VFM calculation of particle–vortex interaction. As in Fig. 10.1, the particle becomes trapped in the vortex, but in this case the relaxation involves the emission of a small vortex ring. Reprinted from Kivotides *et al.* (2008b) with the permission of AIP Publishing.

vortex loops. Typically, the particle will relax in the trapped configuration only if there is sufficient damping, otherwise it may free itself after the collision with the vortex. Examples of computed particle–vortex interactions are shown in Figs. 10.1 and 10.2.

When the particle is trapped on a vortex, the energy of the superfluid is reduced by an amount that is approximately equal to the energy of the displaced superfluid (often called substitution energy), which Poole *et al.* (2005) estimated as

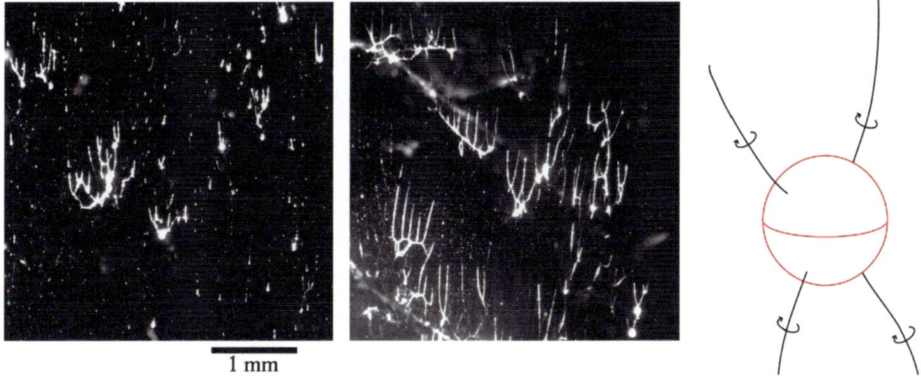

Figure 10.3 Some big particles act as a site for the pinning of many vortex lines, changing their topology (and dynamics) as illustrated in the sketch to the right. From Bewley *et al.* (2006).

$$E_{\mathrm{dis}} \approx \frac{\rho_{\mathrm{s}}\kappa^2 a_{\mathrm{p}}}{4\pi} \ln\left(\frac{a_{\mathrm{p}}}{a_0}\right) \tag{10.11}$$

(the energy per unit length of vortex line times the length of the missing vortex line). For typical tracer particles, $E_{\mathrm{dis}} \ll k_{\mathrm{B}}T$ and thermal fluctuations do not affect the trapping, whereas small ion bubbles trapped in vortex lines are not stable unless the temperature is very low.

When a particle moves within a turbulent vortex tangle, it may be attached to several vortex lines simultaneously, as illustrated in Fig. 10.3. The ratio between the particle radius, a_{p}, and the average distance between vortices, ℓ, is therefore an important parameter. Sergeev *et al.* (2006) developed a phenomenological model of the force on the particle that agrees fairly well with experiments with non-buoyant particles (Zhang and Van Sciver, 2005a) on counterflow turbulence in the presence of gravity.

10.2 Particles in Thermal Counterflow

In Chapter 4 we introduced the PTV visualization method and illustrated that taking movies of small micron-sized particles of frozen hydrogen (as well as deuterium or HD) seeded in the flow led to a Lagrangian picture of the flow. Most of the visualization experiments performed so far can be used for gathering important statistical information of quantum turbulence in thermal counterflow of He II simply because of the ease of its generation. The PTV method also allows us to obtain acceleration statistics of the particles. The velocity and acceleration statistics are both discussed in this chapter.

10.2.1 Particle Velocity Distributions

An example of recorded tracks by Paoletti *et al.* (2008b) is shown in Fig. 4.23, displaying (in the case of moderately small counterflow velocity) that some particles move with the normal fluid while others are presumably trapped into vortex lines or move, on average, in the opposite direction. This is nicely confirmed in Fig. 10.4, which shows local velocity statistics computed from such particle trajectories in vertical counterflow channel with the heat source at its bottom. The probability density function (PDF) of vertical velocities of the particles is therefore bimodal in form, with its peaks associated with the mean normal (up) and superfluid (down) velocity. It is indeed remarkable to watch movies recorded now by several experimental groups that clearly show this extraordinary simultaneous up and down motion of particles in steady-state thermal counterflow.

In practice, the situation described above occurs only for certain subset of the parameter space mainly governed by the temperature and applied heat flux, and is more complex in general. For very low heat flux (depending on the temperature), there are no quantized vortex lines in the flow (except for the remanent ones) and all particles ought to move, under the action of the Stokes drag, with the normal fluid. This is, however, very difficult to observe experimentally, as the effect of the normal fluid drag will be masked by the buoyancy force, caused by the mismatch in densities of seeding particles and He II.

There is therefore only a relatively narrow window in the parameter space within which one can clearly distinguish the particles going up with the normal fluid and those going down with the superfluid. The reason is that in counterflow He II the particles interact with both the normal and superfluid velocity fields simultaneously and may become trapped onto (and de-trapped from) the cores of quantized vortices.

Paoletti *et al.* (2008a) also performed experiments in decaying thermal counterflow. Once the heat is switched off, the vertical velocities collapse to distributions peaked near zero. The PDFs of vertical velocity components derived from all trajectories for times after the heat was turned off are shown in Fig. 10.5. Note the tails of these distributions, which appear as a consequence of rare events, namely trajectories with high, atypical velocities. Paoletti *et al.* (2008a) attributed them to quantized vortex reconnections, described in some detail in Chapter 3.

Their arguments are as follows. Near the reconnection moment, quantized vortices move with velocities much higher than the background mean flow, and based on dimensional arguments the minimum separation distance between reconnecting

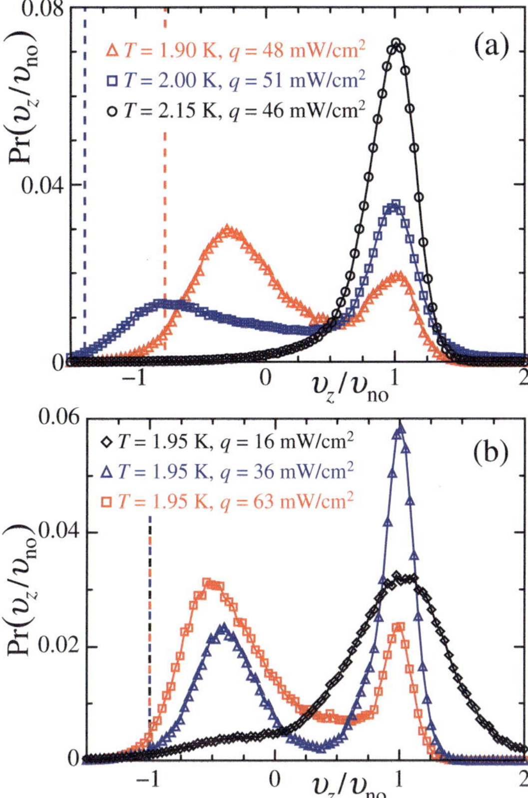

Figure 10.4 An example of velocity statistics in thermal counterflow experiments. All the velocity distributions in the (positive) vertical z-direction of the counterflow are scaled by the observed normal fluid velocity. The predicted (negative) values of the superfluid velocity are shown by vertical dashed lines. (a) Variation of vertical velocity distributions with varying temperature for (approximately) constant heat flux with the experimental parameters given in the legend. (b) Variation of vertical velocity distributions with varying heat flux at constant temperature with the experimental parameters given in the legend. Reproduced with permission from Paoletti *et al.* (2008b).

quantized vortices evolves as $\delta(t) \propto (\kappa|t - t_0|)^{1/2}$, where t_0 is the reconnection moment. Thus, for lengths between the vortex core radius, $\xi_0 \simeq 1$ Å, and the typical intervortex spacing ℓ, they expect the velocities to scale as $u(t) \propto |t - t_0|^{-1/2}$ (see Chapter 3). To model the PDF of the velocity derived from particle trajectories, the authors use the transformation $\mathrm{Pr_v}(u) = \mathrm{Pr_t}[t(u)]|dt/du|$, where $\mathrm{Pr_u}(u)$ is the probability of observing a velocity between u and $u + du$ at any time while $\mathrm{Pr_t}[t(u)]$

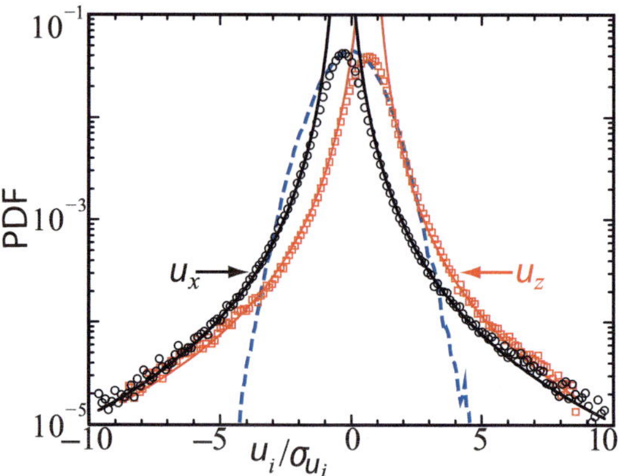

Figure 10.5 Local velocity statistics computed from particle trajectories. Probability distribution function (PDF) of u_x (black circles) and u_z (red squares). The solid lines are fits to $Pr_{(u_i)} = a \mid u_i - \overline{u_i} \mid^3$, where i is either x (black) or z (red) and $\overline{u_i}$ is the mean value of u_i. For comparison, the dashed (blue) line shows the distribution for classical turbulence in water, which has shape close to Gaussian over about five decades in probability. Reprinted figure with permission from Paoletti *et al.* (2008a). Copyright 2008 by the American Physical Society.

is the uniform probability of taking a measurement between t and $t + dt$. This leads to the behavior

$$\mathrm{Pr}_{\mathrm{u}}(u) \propto |dt/du| \propto |u|^{-3}, \tag{10.12}$$

in agreement with Fig. 10.5.

The experimentally observed velocity PDF shown in Fig. 10.5 seemed to contradict other experiments, such as Pitot tube measurements of Maurer and Tabeling (1998), revealing Gaussian distributions of turbulent velocity suggesting a K41 inertial range of scales of the spectral energy density, as observed in classical turbulence in viscous fluids. This puzzle has been addressed both theoretically and experimentally, and the non-Gaussian character of velocity, to which Paoletti *et al.* (2008a) drew attention first, is a genuine characteristic of quantum turbulence, though its nature is better understood now than at that time. On the theoretical/numerical side, Baggaley and Barenghi (2011b) numerically modeled quantum turbulence at $T = 0$ as a tangle of vortex filaments, and concluded that there is no contradiction between the two experiments. The transition from Gaussian to power law velocity PDFs arises from the different length scales that are probed using the two techniques. Figure 10.6 shows the PDFs of the velocity components v_i $(i = x, y, z)$ versus v_i/σ_i calculated by averaging over regions of size $\Delta = 2\ell$ (left) and $\Delta = \ell/6$

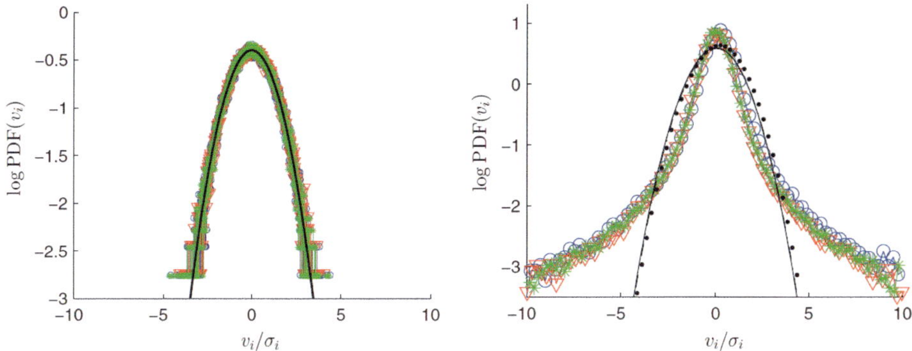

Figure 10.6 PDFs of turbulent velocity components v_i ($i = x, y, z$) obtained by averaging over small regions (left) larger ($\Delta = 2\ell$) and (right) smaller ($\Delta = \ell/6$) than the intervortex spacing. Note the transition from (left) Gaussian to (right) power-law. Green asterisks, blue circles, and red triangles refer, respectively, to x, y, z directions. The solid lines are Gaussian fits with the same mean and standard deviations. Reprinted figure with permission from Baggaley and Barenghi (2011b). Copyright 2011 by the American Physical Society.

(right), ℓ being the average intervortex spacing. The solid lines are the Gaussian fits defined by

$$\mathrm{PDF}(v_i) = \frac{1}{\sqrt{2\pi\sigma_i^2}} e^{-(v_i-\mu_i)^2/(2\sigma_i^2)}, \tag{10.13}$$

where $\mu_i \approx 0$ is the mean and σ_i the standard deviation. Clearly, for $\Delta > \ell$ the PDF is Gaussian, but for $\Delta < \ell$ is has power-law form $\mathrm{PDF}(v_i) \approx v_i^{-b_i}$ where $b_i = 3.2$, 3.2, and 3.1 for $i = x, y, z$, respectively.

On the experimental side, it is important to estimate which range of length scales in quantum flow is probed by the direct visualization of tracer particles. Compared to the vortex core, about $a_0 \simeq 1$ Å in He II (except very close to the λ-point), the size of the particles used for PIV or PTV visualization is typically several orders of magnitude larger, typically of the order of 1 μm, and represents the physically smallest length scale one can access, providing that the images are recorded fast enough so that a particle does not move further than its size between two successive images. The upper length scale limit is determined by the lengths of the particle trajectories that can be experimentally followed, which is typically much less than the outer scale of quantum turbulence; usually this is the size of the container or the device used to generate turbulence, the size of an oscillating object, or the mesh of a grid.

In the experiment, one can usefully change the length scale at which the quantum turbulence is probed by changing the rate at which the consecutive visualization

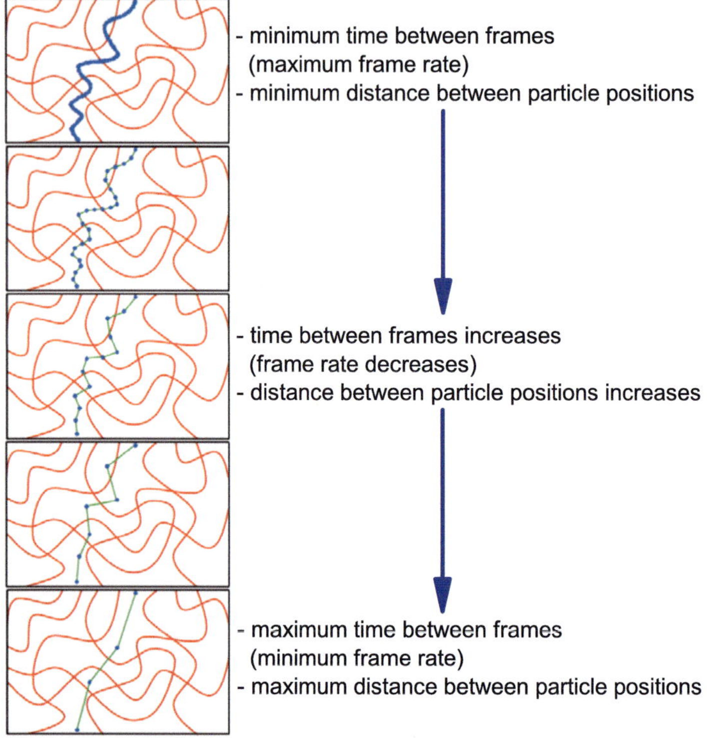

Figure 10.7 A cartoon illustrating the data processing procedure. A (sample) particle trajectory is obtained, at a certain frame rate, by linking (green lines) each relevant particle position (blue •). Each panel (top: maximum frame rate; bottom: minimum frame rate) shows particle positions taken during the same time interval. Each position is recorded in a different image, i.e., each panel can be seen as the sum of several images. Particle positions at a low frame rate can also be obtained, accordingly, by removing particle positions from data sets recorded at a high frame rate. The red lines illustrate the vortex tangle. Note that positions of vortex lines also change between frames; this motion is, however, neglected in the figure for the sake of clarity. As the frame rate decreases, from top to bottom panels, the number of particle positions used to identify the track decreases, while the probed length scale $\ell_{\rm exp}$ increases. Once the particle positions are known, the corresponding velocities or velocity increments can be computed. Reproduced with permission from La Mantia and Skrbek (2014a).

images are taken, or by deliberately selecting only each nth particle position ignoring those in between; see Fig. 10.7 for details. As was shown by La Mantia and Skrbek (2014a) in vertically oriented thermal counterflow, this strategy allowed one to confirm the predictions of the numerical simulations just discussed (Baggaley and Barenghi, 2011b), by using the PTV technique. In thermal counterflow, the quantum length scale can be tuned by the applied heat flux. A suitable quantum length scale

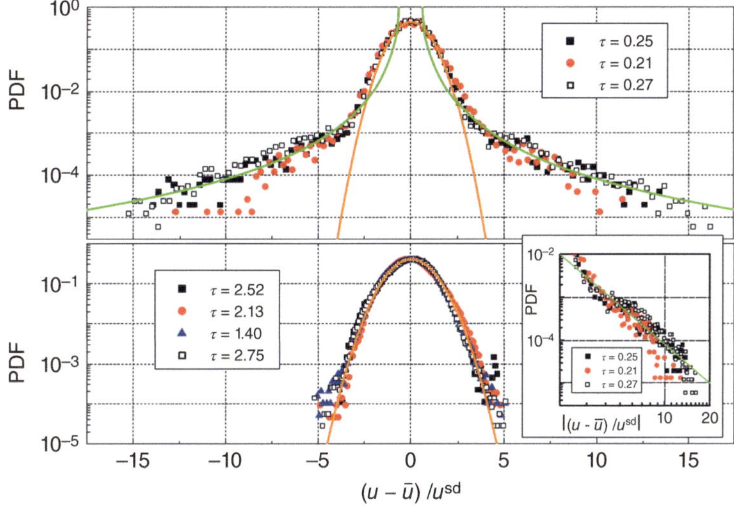

Figure 10.8 Quantum and classical signatures of counterflowing He II PDF of $(u - \overline{u})/u^{\mathrm{sd}}$; $t_1 = 5\,\mathrm{ms}$ (top), $t_1 = 50\,\mathrm{ms}$ (bottom). Tracks with at least five points; the area below the data curves is normalized to 1. ■: images taken at 400 fps, $T = 1.65\,\mathrm{K}$, $q = 487\,\mathrm{W/m^2}$, $\ell = 74\,\mu\mathrm{m}$, $V_{\mathrm{abs}} = 3.72\,\mathrm{mm/s}$; red •: 200 fps, $T = 1.65\,\mathrm{K}$, $q = 490\,\mathrm{W/m^2}$, $\ell = 73\,\mu\mathrm{m}$, $V_{\mathrm{abs}} = 3.13\,\mathrm{mm/s}$; blue ▲: 100 fps, $T = 1.66\,\mathrm{K}$, $q = 492\,\mathrm{W/m^2}$, $\ell = 74\,\mu\mathrm{m}$, $V_{\mathrm{abs}} = 2.09\,\mathrm{mm/s}$; □: 400 fps, $T = 1.77\,\mathrm{K}$, $q = 608\,\mathrm{W/m^2}$, $\ell = 70\,\mu\mathrm{m}$, $V_{\mathrm{abs}} = 3.86\,\mathrm{mm/s}$; orange line: Gaussian fit; green line: power-law fit $0.08|(u - \overline{u})/u^{\mathrm{sd}}|^{-3}$. Inset: Log–log plot of the PDF of $\left|(u - \overline{u})/u^{\mathrm{sd}}\right|$ at $t_1 = 5\,\mathrm{ms}$. Reproduced with permission from La Mantia and Skrbek (2014a).

$\ell = 75\,\mu\mathrm{m}$ was selected, allowing the investigation of Lagrangian dynamics of solid deuterium particles of size $2a_{\mathrm{p}} \simeq \ell/10$ at length scales ℓ_{exp} straddling two orders of magnitude across ℓ.

The PDF of the nondimensional instantaneous velocity $(u - \overline{u})/u^{\mathrm{sd}}$ in the horizontal direction (i.e., the direction perpendicular to the mean counterflow velocity u_{ns}) is plotted in Fig. 10.8, where $\overline{u}$ and u^{sd} are the mean and standard deviation of the velocity u. The length scale ℓ_{exp} that is probed is quantified by introducing the nondimensional time $\tau = t_1/t_2$, where t_1 is the time interval used for the calculation of the velocities along the tracks, and $t_2 = \ell/u_{\mathrm{abs}}$ where u_{abs} denotes the mean particle velocity obtained for the experimental conditions; at the smallest t_1, we have $\ell_{\mathrm{exp}}(\tau = 1) = \ell$. The velocities of particles sufficiently far away from the vortices can explain the observed Gaussian core of the distributions. The use of even smaller particles, which can probe smaller length scales, would lead to narrower cores of the distribution.

Figure 10.9 shows the evolution of the PDF obtained at different τ. As t_1 increases, the PDF changes shape to nearly Gaussian. This evolution can be quantified by

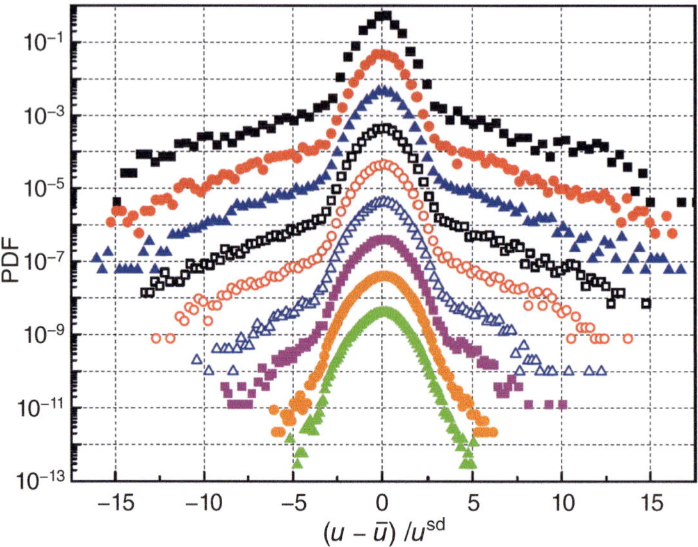

Figure 10.9 Evolution of the PDF of $(u - \overline{u})/u^{\mathrm{sd}}$ with τ, in the direction perpendicular to v_{ns}. Data collected at 400 fps, $T = 1.77$ K, $q = 608$ W/m^2, $\ell = 70\,\mu$m, $u_{\mathrm{abs}} = 3.86$ mm/s. ■: $t_1 = 2.5$ ms, $\tau = 0.14$ (each subsequent data set is shifted down by one decade); red •: $t_1 = 5$ ms, $\tau = 0.27$; blue ▲: $t_1 = 7.5$ ms, $\tau = 0.41$; □: $t_1 = 10$ ms, $\tau = 0.55$; red ○: $t_1 = 12.5$ ms, $\tau = 0.69$; blue △: $t_1 = 20$ ms, $\tau = 1.10$; magenta ■: $t_1 = 25$ ms, $\tau = 1.37$; orange •: $t_1 = 40$ ms, $\tau = 2.20$; green ▲: $t_1 = 50$ ms, $\tau = 2.75$. Reproduced with permission from La Mantia and Skrbek (2014a).

evaluating the flatness of the $(u - \overline{u})/u^{\mathrm{sd}}$ distribution, calculated as the ensemble average of $\left[(u - \overline{u})/u^{\mathrm{sd}}\right]^4$. The flatness of the distribution decreases if plotted vs. τ, i.e., as a function of the time t_1 used to compute the velocities, and reaches the value of 3 – that of a Gaussian distribution – for $\tau \approx 2$, in other words when $\ell_{\mathrm{exp}} \approx 2\ell$. This may be related to the coherent vortex structures of similar size observed in numerical simulations by Baggaley *et al.* (2012b) and experimentally by Rusaouen *et al.* (2017b); it could also depend on the fact that particles are not expected to move between vortices along straight lines.

It needs to be noted that particle dynamics in quantum turbulence is influenced not only by the tangle properties, determined by the applied heat flux (as in thermal counterflow), bath temperature, and boundary proximity of the studied flow region, but also by the particle inertia, as in classical turbulent flows. It significantly affects the way in which the solid particles interact with the vortices (La Mantia and Skrbek, 2014b; Švančara *et al.*, 2018). This will be discussed in connection with particle accelerations in Section 10.2.2.

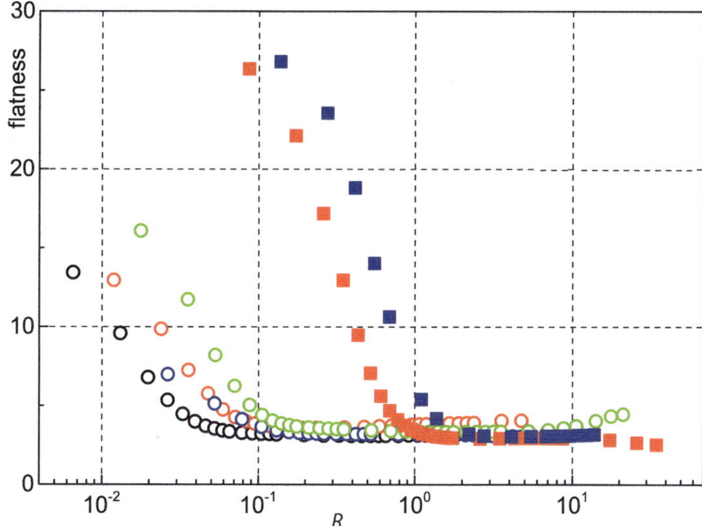

Figure 10.10 Flatness of the horizontal velocity distribution as a function of the scale at which thermal counterflow is probed. Here, R reflects the ratio of the experimentally probed scale, ℓ_{exp}, to ℓ that is determined from the known vortex line density in the bulk. Data series denoted by circles (from left to right: black, blue, red, and green) are counterflow data obtained in the heater proximity using solid deuterium at temperatures 1.24 K, 1.40 K, 1.75 K, and 1.95 K; squares are counterflow data obtained at 1.77 K in the bulk using solid deuterium (blue) and hydrogen (red). Flatness values of the bulk and near the heater approach three of the Gaussian distribution, at different R values. Reprinted figure with permission from Hrubcová *et al.* (2018). Copyright 2018 by the American Physical Society.

Taking the particle inertia into account, Hrubcová *et al.* (2018) calculated the flatness of the horizontal velocity distribution in vertical thermal counterflow, in the bulk and in the proximity of a solid wall. Figure 10.10 plots flatness vs. R – the ratio between the experimentally probed length scale, ℓ_{exp}, and the quantum length scale, the latter determined from the known bulk vortex line density as $\ell = L^{-1/2}$. We see that the flatness values in the bulk and near the heater approach 3, characteristic of the Gaussian distribution, for larger R. Assuming that flatness value approaches 3 for the data probing the flow at quasi-classical scales $\ell_{\text{exp}} \lesssim \ell$, this observation suggests that ℓ is apparently one order of magnitude smaller than that in the bulk for the same temperature and applied heat flux, at least in the velocity range investigated, i.e., of about 2 mm/s – two times larger than for the channel turbulence. In other words, this observation suggests that boundary layers likely exist in quantum flows, at least in thermal counterflow providing convective heat flow, as in the Rayleigh–Bénard convection in classical fluids.

10.2.2 Particle Acceleration Distributions

The research field of Lagrangian acceleration is now well established in classical fluid turbulence. For a review, see Toschi and Bodenschatz (2009); for more recent results from large-scale direct numerical simulations, see Buaria and Sreenivasan (2022). The Prague group utilized the PTV method for investigations of particle accelerations in quantum flows of He II. Using deuterium particles, La Mantia *et al.* (2013) found that in vertical thermal counterflow, at $\ell_{\text{exp}} \approx \ell$, the normalized PDF of the instantaneous acceleration a_z in the vertical direction can be approximated (at least for a range of acceleration values) by the form

$$\text{PDF} = \frac{\exp\left(3s^2/2\right)}{4\sqrt{3}}\left[1 - \text{erf}\left(\frac{\ln\left|a/\sqrt{3}\right| + 2s^2}{\sqrt{2}s}\right)\right], \tag{10.14}$$

where $s = 1$ and $a = (a_z - \overline{a}_z)/a_z^{\text{sd}}$, and a_z and a_z^{sd} are the mean and standard deviation of the dimensional acceleration a_z, respectively. This functional form, associated with a lognormal distribution of the acceleration, has been reported by Mordant *et al.* (2004) in their study on the dynamics of fluid particles in classical turbulence, while the results of Qureshi *et al.* (2007, 2008) refer to inertial particles and are consistent with $s = 0.62$. Note, however, that such a lognormal behavior does not have a clear physical interpretation.

La Mantia and Skrbek (2014b) further showed that experimental results on particle accelerations, similar to those on velocities (La Mantia and Skrbek, 2014a), depend on the length scale ℓ_{exp} at which the stationary thermal counterflow is probed and, consequently, display quantum behavior for $\ell_{\text{exp}} \lesssim \ell$ and classical behavior for $\ell_{\text{exp}} \gtrsim \ell$.

The quantum signatures of velocity and acceleration PDFs of the superfluid particles can be understood simply as follows. The velocity PDF is indeed non-Gaussian; but its tails are $\propto |u|^{-3}$ for a fluid particle even in the absence of vortex reconnections, as in classical systems of vortex points (White *et al.*, 2010), or by noting that close to a quantized vortex, where the superfluid velocity $u_s = \kappa/(2\pi r)$, r is the distance from the vortex core: If the probability $P_u(u)du$ of observing a velocity u is assumed to be proportional to $\int \delta(u - u_s) r dr$, where δ denotes the delta function, it follows that $P_u(u) \propto u^{-3}$ without the need of vortex reconnections. This reasoning can be extended by the assumption that the probability $P_a(a)$ of observing an acceleration a is proportional to $\int \delta(a - a_s) r\, dr$. As the superfluid acceleration $a_s = u_s^2/r$ in the proximity of a quantized vortex, it follows that $P_a(a) \propto a^{-5/3}$ (Rast and Pinton, 2009; Baggaley and Barenghi, 2014). We stress that the influence of the normal fluid velocity field and superfluid velocity field due to other quantized vortices is neglected here, i.e., the necessary condition is $\ell_{\text{exp}} \ll \ell$.

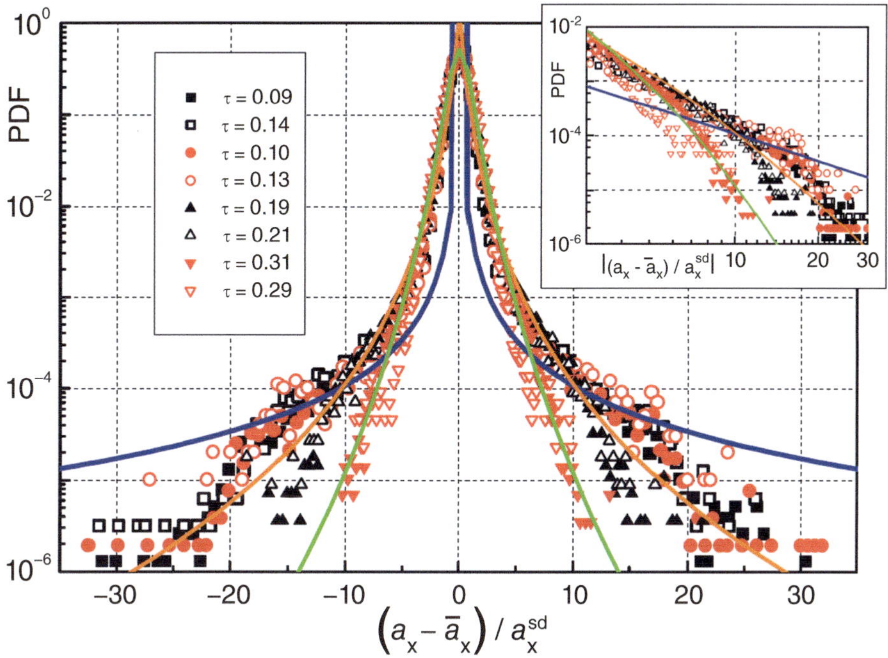

Figure 10.11 PDF of $(a_\mathrm{x} - \overline{a}_\mathrm{x})/a_\mathrm{x}^\mathrm{sd}$. Filled symbols refer to hydrogen particles, while open symbols denote deuterium particles. Black ■: images taken at 400 fps, T = 1.77 K, q = 612 W/m^2, ℓ = 70 μm, u_abs = 2.46 mm/s; black □: 400 fps, T = 1.77 K, q = 608 W/m^2, ℓ = 70 μm, u_abs = 3.91 mm/s; red •: 400 fps, T = 1.66 K, q = 490 W/m^2, ℓ = 75 μm, u_abs = 3.02 mm/s; red ○: 400 fps, T = 1.65 K, q = 487 W/m^2, ℓ = 74 μm, u_abs = 3.76 mm/s; black ▲: 200 fps, T = 1.66 K, q = 489 W/m^2, ℓ = 75 μm, u_abs = 2.90 mm/s; black △: 200 fps, T = 1.65 K, q = 490 W/m^2, ℓ = 73 μm, u_abs = 3.15 mm/s; red ▼: 100 fps, T = 1.66 K, q = 489 W/m^2, ℓ = 75 μm, u_abs = 2.34 mm/s; red ▽: 100 fps, T = 1.66 K, q = 492 W/m^2, ℓ = 74 μm, u_abs = 2.13 mm/s; blue line: power-law fit, $0.005|(a_\mathrm{x} - \overline{a}_\mathrm{x})/a_\mathrm{x}^\mathrm{sd}|^{-5/3}$; orange line: lognormal fit, Eq. (10.14) with $s = 1$ and $a = (a_\mathrm{x} - \overline{a}_\mathrm{x})/a_\mathrm{x}^\mathrm{sd}$; green line: lognormal fit, Eq. (10.14) with $s = 0.62$ and $a = (a_\mathrm{x} - \overline{a}_\mathrm{x})/a_\mathrm{x}^\mathrm{sd}$. Inset: Log–log plot of the PDF of $|(a_\mathrm{x} - \overline{a}_\mathrm{x})/a_\mathrm{x}^\mathrm{sd}|$; symbols as in the main panel. Reprinted figure with permission from La Mantia and Skrbek (2014b). Copyright 2014 by the American Physical Society.

Figure 10.11 shows the nondimensional instantaneous acceleration $(a_\mathrm{x} - \overline{a}_\mathrm{x})/a_\mathrm{x}^\mathrm{sd}$ in the horizontal direction measured in vertical thermal counterflow of He II. At length scales smaller than ℓ, the departure from the lognormal shape is observed at the largest accelerations, when the particles are, on average, closer to the quantized vortices. As shown in Fig. 10.12, this is consistent with the prediction for the tails up to about $20|a_\mathrm{x}^\mathrm{sd}|$, which approximately corresponds to the acceleration of a few micron-sized particles touching a vortex core. Note that the particle size $2a_\mathrm{p}$ has a large influence on the acceleration magnitude as $a \propto (2a_\mathrm{p})^{-3}$.

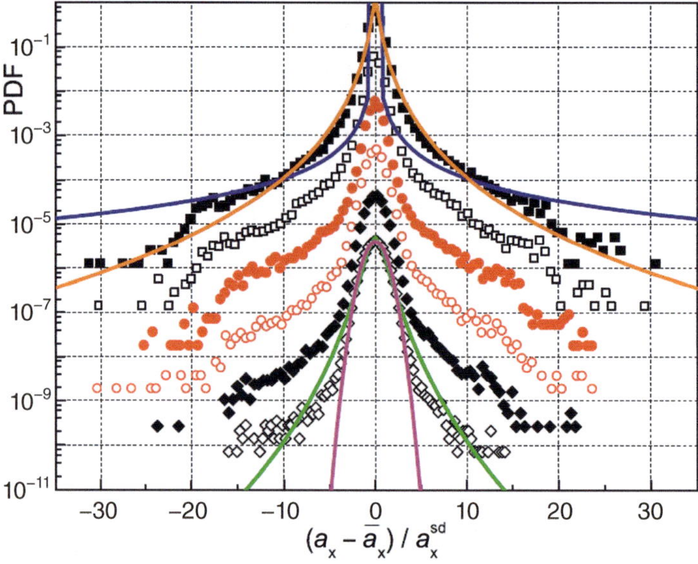

Figure 10.12 Evolution of the PDF of $(a_x - \overline{a}_x)/a_x^{sd}$ with τ. Hydrogen particles, data collected at 400 fps, $T = 1.77$ K, $q = 612$ W/m^2, $\ell = 70$ µm, $V_{abs} = 2.46$ mm/s. Black ■: $t_1 = 2.5$ ms, $\tau = 0.09$ (each subsequent data set shifted down by one decade); black □: $t_1 = 5$ ms, $\tau = 0.18$; red •: $t_1 = 10$ ms, $\tau = 0.35$; red ∘: $t_1 = 25$ ms, $\tau = 0.88$; black ♦: $t_1 = 50$ ms, $\tau = 1.77$; black ◊: $t_1 = 250$ ms, $\tau = 8.83$; blue line: power-law fit, $0.005|(a_x - \overline{a}_x)/a_x^{sd}|^{-5/3}$; orange line: lognormal fit, Eq. (10.14) with $s = 1$ and $a = (a_x - \overline{a}_x)/a_x^{sd}$; green line: lognormal fit, Eq. (10.14) with $s = 0.62$ and $a = (a_x - \overline{a}_x)/a_x^{sd}$ (shifted down by five decades); magenta line: Gaussian fit of the black ◊ data set (shifted down by five decades). Reprinted figure with permission from La Mantia and Skrbek (2014b). Copyright 2014 by the American Physical Society.

10.2.3 Distinctive Features of Particle Motion in Thermal Counterflow

Most PIV visualization studies have been performed so far in thermal counterflow of He II, in channels of constant square cross section. At relatively small heat fluxes, the particles move in both directions from and towards the heater placed at the closed end of the counterflow channel. Specifically, a significant fraction of particles move, on average, towards the heater, in the direction of the superfluid component (Paoletti *et al.*, 2008b; La Mantia, 2016). As the heat flux, $\dot{q}$, increases, the portion of particles moving away from the heater in the normal fluid direction increases, indicating that, at relatively large heat fluxes, the particles tend to roughly follow the normal fluid flow, although the corresponding tracks become less straight than those at smaller $\dot{q}$ values (La Mantia, 2016).

It also turns out that the mean particle velocity in the direction away from the heater is approximately equal to u_n (Paoletti *et al.*, 2008b; Chagovets and Van

Sciver, 2011) or, in the case of larger heat fluxes, to $u_n/2$ (Zhang and Van Sciver, 2005a; Chagovets and Van Sciver, 2011). A similar decrease of the mean particle velocity was obtained in numerical simulations by Kivotides (2008), who attributed it to the stronger interactions between particles and quantized vortices in a more dense tangle.

The particle motion in thermal counterflow visualized by the PTV technique was further investigated by Mastracci and Guo (2018) and Mastracci *et al.* (2019). They found that, at the largest $\dot{q}$ values probed in their experiments in a square counterflow channel with 16 mm sides, the statistical distributions of the particle velocity in the vertical direction were characterized by a single peak centered near $u_n/2$. As the applied heat flux decreased, another peak, centered near u_n, appeared in the vertical velocity distributions and, for even smaller $\dot{q}$ values, the peak at the smaller velocity was centered near u_s, assigned here with a negative sign as the superfluid and normal fluid components move in opposite directions. This behavior indicates that, for small heat fluxes, particles can be trapped onto quantized vortices for relatively long times and can move in the superflow direction with the vortex tangle. This agrees with the numerical findings of Sergeev and Barenghi (2009b). For large heat fluxes, on the other hand, the Stokes drag of the normal component forces most particles to move away from the heater and the particles tended to stay trapped onto vortices for shorter times.

It is interesting to consider the situation of moderate heat fluxes when the vast majority of particles move in the normal fluid direction, away from the heat source, with the corresponding vertical velocity distributions characterized by two peaks centered near $u_n/2$ and u_n. Švančara *et al.* (2021) observed frequent velocity changes along individual particle trajectories and proposed a separation scheme that allows identification of several regimes of motion. The striking observation appears from the time evolution of the vertical position of some particles, as shown in the left panel of Fig. 10.13. Two characteristic slopes can be easily identified, corresponding to the peak velocities u_1 and u_2. If we follow the highlighted trajectory and plot its vertical velocity and acceleration as functions of time, as in the right panel of Fig. 10.13, we observe rapid velocity changes between two roughly constant values that are clearly visible and consistent with the corresponding acceleration changes. This behavior allows one to develop a separation scheme in the velocity–acceleration phase space, shown in Fig. 10.14 for one set of measured data corresponding to the highlighted trajectory in Fig. 10.13. It takes the form of several loops (white points) and areas of higher density of points near the line of zero acceleration, representing the two peak velocities (note that the bivariate PDF is plotted as the color-coded background).

This separation scheme is loosely built on the earlier work of Mastracci and Guo (2018), who, however, separated the motion regimes solely on the basis of the vertical velocity of particles. Švančara *et al.* (2021) divide the two-dimensional

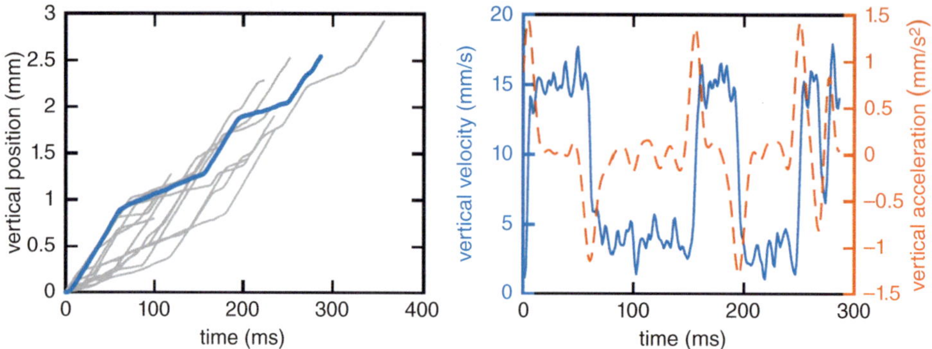

Figure 10.13 (Left) Typical deuterium particle trajectories collected in thermal counterflow at 1.36 K for counterflow velocity 17.3 mm/s. (Right) Vertical velocity (blue solid line) and acceleration (red dashed line) of the particle track that is highlighted in the left panel. From Švančara *et al.* (2021), reproduced with permission.

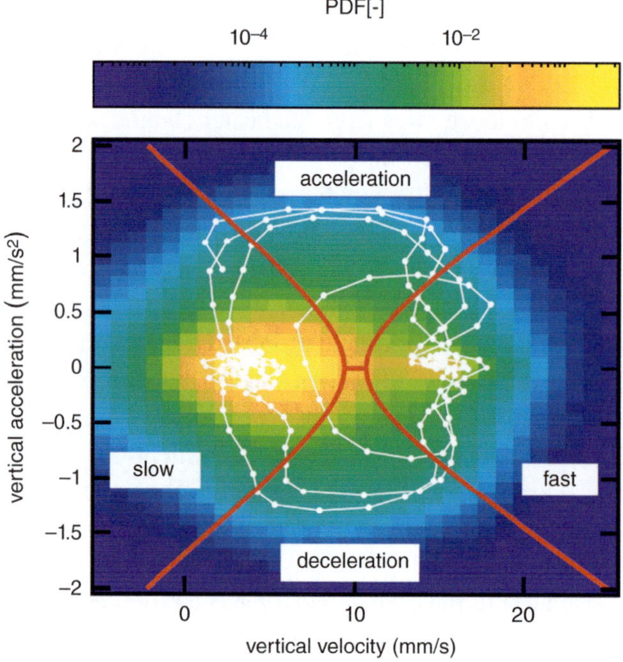

Figure 10.14 An illustration of the separation scheme. (Color-coded map) Bivariate PDF of the measured velocity–acceleration pairs. (White points) Trajectory highlighted in Fig. 10.13. Two hyperbolas and a segment (red lines) divide the phase space into four motion types; see text for details. From Švančara *et al.* (2021), reproduced with permission.

phase space into four subspaces (or motion types), labelled as slow (S), fast (F), acceleration (A), and deceleration (D), and define the respective separating curves as (i) a slow hyperbola, with focus in $[u_1, 0]$ and semiaxes of lengths $2\sigma(u_1)$ and $2\sigma(a_y)$; (ii) a fast hyperbola, with focus in $[u_2, 0]$ and semiaxes of lengths $2\sigma(u_2)$ and $2\sigma(a_y)$; and (iii) a segment between the points $[u_1 + 2\sigma(u_1), 0]$ and $[u_2 - 2\sigma(u_2), 0]$. The values u_i and $\sigma(u_i)$, with $i = 1$ and 2, are obtained from Gaussian fits, and $\sigma(a_y)$ denotes the standard deviation of the vertical acceleration of the particles (the mean value is very close to zero in all cases considered). These hyperbolae are plotted as thick red lines in Fig. 10.14.

The most important outcome, after smoothing out short segments of type A or D above, consisting of only a few points, is the existence of long segments of type F or S. These two regimes can be associated with fast particles moving in the direction of the normal fluid along almost straight tracks, and with slow particles whose erratic upward motion appears to be significantly influenced by quantized vortices. It is important to stress that a single particle can explore both regimes during its motion away from the heat source. The occurrence of very long segments of the same character means that particles can be fast or slow on macroscopic length scales that are appreciably larger than quantum length scale ℓ. Indeed, some fast particles move relatively straight, with velocities close to the normal fluid velocity u_n, while the slow particles seem to be influenced by strong interactions with the vortex tangle (their tracks are considerably more erratic) and their velocity reduced to about $u_n/2$. At the same time, all particles display nonclassical and broad distributions of velocity, indicating the relevance of particle–vortex interactions in both regimes; however, the strengths of particle–vortex interactions in the two cases are appreciably different.

10.3 Lagrangian Pseudovorticity in Quantum Turbulence

Visualization of micron-sized tracer particles is capable of shedding light on an important question of macroscopic vortical structures in turbulent quantum flows. Large-scale vortical structures in mechanically generated flow of superfluid ^{4}He, in the form of vortex rings, have been visualized by Murakami *et al.* (1987) with the help of deuterium particles, and we already discussed the work of Zhang and Van Sciver (2005b) on thermal counterflow past a circular cylinder. Macroscopic vortical structures also appear in classical as well as quantum flows due to various oscillating objects, such as cylinders or tuning forks. Their visualization, using the PTV technique, allows direct comparison of classical and quantum cases.

Such an experimental PTV study using frozen deuterium micron-sized particles was performed by the Prague group. Duda *et al.* (2015) used a rectangular cross section cylinder made of transparent plexiglass, 3 mm high, 10 mm wide, and 30 mm

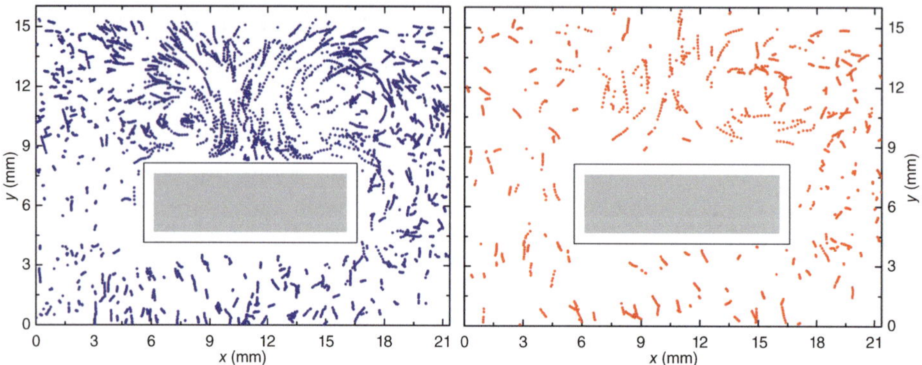

Figure 10.15 Particles trajectories, with at least 5 data points, obtained within the 15° phase interval centered at the lowest position of the cylinder oscillatory cycle; frequency $f = 0.5$ Hz and amplitude $a_{\text{osc}} = 5$ mm. (Left) He II, temperature $T = 1.24$ K. (Right) He I, $T = 2.18$ K. The gray rectangle indicates the cylinder, at the lowest position of the cycle, while the white rectangle denotes the corresponding mask used for data processing. Reprinted figure with permission from Duda *et al.* (2015). Copyright 2015 by the American Physical Society.

long. The laser sheet illuminated the middle part of its length, which was in the horizontal direction. This obstacle was supported at its end; the upper part of the brass support connected to a stainless steel shaft allowed the transmission of the imposed vertical harmonic motion from the crank at the cryostat flange. The cylinder therefore oscillated vertically, perpendicular to its cross-section width, with frequencies between 0.05 Hz and 1.25 Hz and amplitudes of 5 mm and 10 mm, in both He I and He II at temperatures between approximately 1.2 K and 3 K. For various experimental conditions, defined by temperature T, frequency f, and amplitude a_{osc}, images were captured by the camera at 100 fps, and particle positions and trajectories were computed. To take advantage of the periodic character of the imposed motion and to enhance the statistical quality of the data, the calculated particle tracks were phase averaged. More specifically, data obtained within the same phase interval of the periodic motion were merged in such a way that particle trajectories, which depend on time, became functions of the phase interval. Examples of such particle trajectories in He II and He I, shown in Fig. 10.15, clearly demonstrate that large vortical structures exist, but the observed particle trajectories do not allow their direct quantitative investigation.

On the other hand, viscous flows past cylinders, which are relevant to the results reported here, constitute one of the most popular research topics in classical turbulent flows; see, e.g., the review by Williamson (1996). Unsteady flows of classical viscous fluids can often be characterized by the vorticity. This local quantity is calculated from the spatial derivatives of the fluid velocity as

$$\omega_i = \epsilon_{ijk} \frac{\partial u_k}{\partial x_j}, \tag{10.15}$$

where ϵ_{ijk} is the Levi-Civita tensor in three-dimensional space and u_k indicates the kth component of the velocity vector. The problem is that vorticity cannot be computed using the PTV data, such as that shown in Fig. 10.15, due to the fact that the fluid velocity is not known in its entirety at a given time.

In order to quantify the magnitude of the vortices shed behind the cylinder, Duda *et al.* (2015) introduced a new parameter, the Lagrangian pseudovorticity θ, which is related to the flow vorticity and is defined as

$$\theta(\boldsymbol{r}, \varphi) = \left\langle \frac{(\boldsymbol{r}_i - \boldsymbol{r}) \times \boldsymbol{u}_i}{|\boldsymbol{r}_i - \boldsymbol{r}|^2} \right\rangle_{|\boldsymbol{r}_i - \boldsymbol{r}| < R_{\mathrm{M}} \wedge |\varphi - \varphi_i| < \Phi_0}, \tag{10.16}$$

where $\boldsymbol{r}$ denotes the position at which θ is calculated on a chosen mesh that adequately covers the field of view; similarly, φ is the phase of the oscillatory motion at which the computation is performed; $\boldsymbol{r}_i$, $\boldsymbol{u}_i$, and φ_i indicate the position, velocity, and phase of the ith particle, respectively, within the spatial region of radius R_{M}, centered in $\boldsymbol{r}$, and in the phase interval equal to $2\Phi_0$. It must be emphasized that this approach postulates a single velocity field that is due to the visualized particles. Lagrangian pseudovorticity is a meaningful quantity only if it does not depend on the chosen mesh size: In this particular case, the radius R_{M} was set to 5 mm, which corresponds to the approximate diameter of the visualized vortices, and the phase parameter Φ_0 was chosen to be equal to 7.5°, as this allows one to observe adequately the vortical structures generated. See Fig. 10.15. Note that the denominator of Eq. (10.16) is introduced to account for both the number of trajectories included and the linearity of the relevant vector product with respect to the magnitude of $\boldsymbol{r}$. This pseudovorticity consequently has the same dimension as the vorticity, and can be seen as a measure of the vortex strength. It can be shown that, for a large enough number of fluid particles, the parameter θ represents the average flow vorticity ω (Outrata *et al.*, 2021).

Figure 10.16 shows the maps of $\theta(\mathbf{r}, \varphi)$ calculated at the lowest position of the oscillatory motion. The macroscopic vortex pairs shed at the cylinder edges during the cycle are clearly visible. As the imposed oscillation frequency increases, the magnitude of the generated vortices increases as well. Similar results were obtained in normal viscous He I, but the vortical structures observed in He I were somewhat less clearly defined than those in He II.

In order to better quantify the dependence of the magnitude of the shed vortices on experimental conditions, Duda *et al.* (2015) computed for each processed movie the ensemble average of θ^2, i.e., $\langle \theta^2 \rangle$. This parameter retains useful quantitative information for comparing two visualized flow fields with each other (Outrata

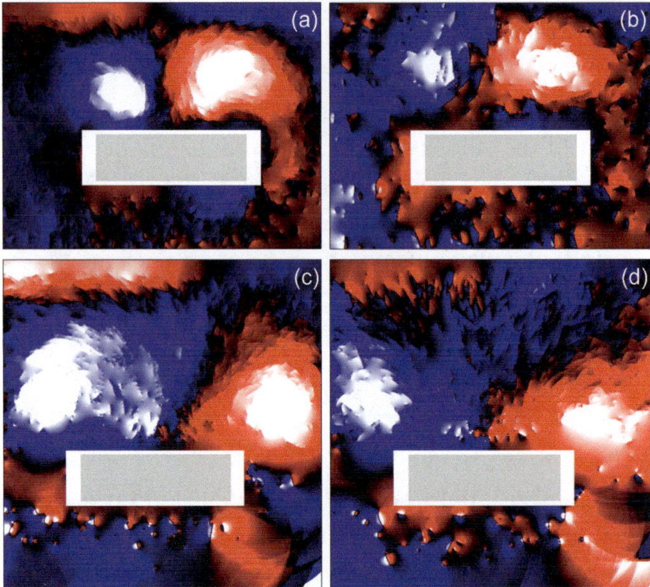

Figure 10.16 Maps of $\theta(\boldsymbol{r}, \varphi)$, calculated at the lowest position of the oscillatory motion; frequency $f = 0.5$ Hz. (a) He II, amplitude $a_{\rm osc} = 5$ mm, temperature $T = 1.24$ K; (b) He I, $a_{\rm osc} = 5$ mm, $T = 2.18$ K; (c) $a_{\rm osc} = 10$ mm, $T = 1.40$ K; (d) He I, $a_{\rm osc} = 10$ mm, $T = 2.18$ K. The gray rectangles indicate the cylinder, while the white rectangles denote the corresponding mask used for data processing. Reprinted figure with permission from Duda *et al.* (2015). Copyright 2015 by the American Physical Society.

et al., 2021) and can be loosely interpreted as a square of the vorticity magnitude relevant to a particular phase of the harmonic motion, coarse-grained over the area of radius $R_{\rm M}$.

The parameter $\langle\theta^2\rangle$ showed no clear temperature dependence in He II but increased with the oscillation frequency; also, as the motion strength increased, the magnitude of the shed vortices increased as well. To quantify this feature, as is customary in fluid dynamics, Duda *et al.* (2015) presented their results on the vortex strength as a function of the suitably defined Reynolds number. In the case of He I, the Reynolds number is defined classically as $Re = 2\pi f a_{\rm osc} D/\nu$, where $2\pi f a_{\rm osc}$ can be understood as the velocity amplitude in the case of harmonic oscillations, D is the cylinder width, and ν denotes the kinematic viscosity of He I tabulated by Donnelly and Barenghi (1998). In the case of He II, they used the definition $Re_\kappa = 12\pi f a_{\rm osc} D/\kappa$, based on effective kinematic viscosity of He II set as $\nu_{\rm eff} \simeq \kappa/6$, where $\kappa/6$ is kinematic viscosity of liquid He just above the transition temperature T_λ (L'vov *et al.*, 2014); see Chapter 11 for a discussion of the relation between $\nu_{\rm eff}$ and κ.

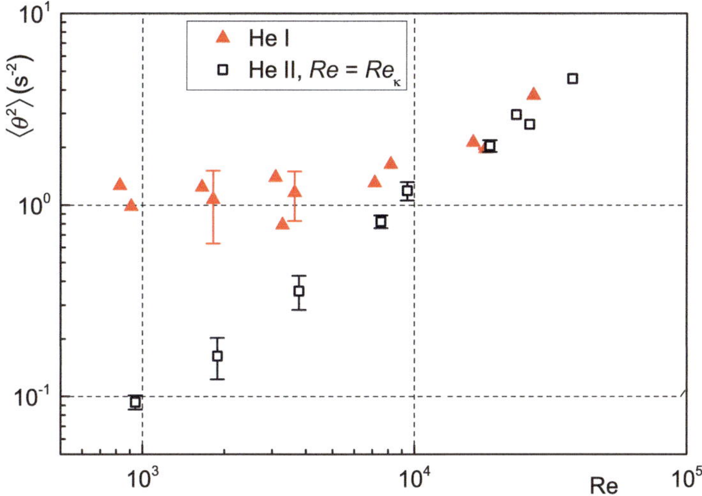

Figure 10.17 The variance $\langle\theta^2\rangle$ as a function of the Reynolds numbers *Re* and Re_κ. (Red triangles) He I. (Open black squares) He II. Each experimental condition is represented by at least 40,000 track points. At each Reynolds number the corresponding average of $\langle\theta^2\rangle$ is plotted; the error bars denote the standard deviation of $\langle\theta^2\rangle$ obtained at that Reynolds number. Reprinted figure with permission from Duda *et al.* (2015). Copyright 2015 by the American Physical Society.

The striking result of this study, shown in Fig. 10.17, is that the parameter $\langle\theta^2\rangle$ behaves differently in He I and He II, as a function of the Reynolds number. This is in apparent contrast with the commonly held notion that mechanically forced flows of He II at large enough length scales are similar to comparable viscous flows. The existence of two distinct branches, for Reynolds numbers smaller than approximately 10^4, therefore calls for plausible physical explanation. The most likely reason is again based on the length scale at which the flow is probed.

Duda *et al.* (2015) argued that the experiment probes length scales larger than the relevant flow scale only for Reynolds numbers larger than about 10^4. In other words, the two branches displayed in Fig. 10.17 might possibly arise from the fact that, at Reynolds numbers smaller than 10^4, the experimentally probed scale is smaller than the dissipative flow scale in He I, and the quantum length scale in He II. While in He I, below the Kolmogorov length scale, the fluid motion is dissipated into heat by the action of the finite viscosity, quantum flows of He II exist all the way down to the Å scale, the size of the cores of quantized vortices. Quantum restrictions on the superfluid motion are the most likely reason why, for Reynolds numbers smaller than 10^4, the parameter $\langle\theta^2\rangle$ behaves differently in He II from that in He I.

The concept of Lagrangian pseudovorticity was recently used by Švančara and La Mantia (2019) and by Outrata *et al.* (2021) to study by PTV visualization the behavior of turbulent vortex rings in He II. The rings of diameter about 5 mm were generated thermally by the application of a short thermal pulse to a heater located at the sealed bottom of a counterflow tube placed in the He II bath. The vortex rings appeared above the nozzle of the tube and moved upwards with velocities of about 10 mm/s. The ring features were quantitatively assessed and, within the temperature range of the experiment, $1.28 \leq T \leq 1.95$ K, their behavior was remarkably similar to that of turbulent vortex rings moving in a viscous fluid. This is in accordance with the notion that coflowing He II behaves classically if probed at classical length scales exceeding ℓ.

10.4 Streaming Flow of Liquid ^{4}He due to a Quartz Tuning Fork

It is well known that high-frequency oscillations in a classical viscous fluids cause steady secondary flows, called streaming flows; that is, the time average of a fluctuating flow often results in a nonzero mean at time scales much longer than the oscillation period; see, e.g., the review by Riley (2001). The specific features of these flows depend on the type of oscillator and the vibration characteristics. The so-called streaming cells are usually characterized by the presence of closed loops in the proximity of the oscillator as a consequence of mass conservation.

The most extensively studied type of classical streaming flow is that generated by a circular cylinder of radius R oscillating transversally in a viscous fluid with frequency f and amplitude $a_{\text{osc}} \ll R$, resulting in a characteristic velocity $U = 2\pi f a_{\text{osc}}$. For the corresponding Reynolds number $Re = UR/\nu \gg 1$, the steady flow pattern consists of four symmetric cells, where the fluid flows towards the cylinder in the direction parallel to the oscillations and away from it in the perpendicular direction. The thickness δ_{dc} of these cells depends on the so-called *streaming Reynolds number*, $Re_{\text{stream}} = Ua_{\text{osc}}/\nu = U^2/(2\pi\nu f)$, and, if $Re_{\text{stream}} > 1$, it is appreciably larger than the Stokes layer thickness δ_{ac} and of the order of $Re_{\text{stream}}^{-1/2}$. Consequently, the fluid outside the inner streaming cells flows is in the opposite direction, i.e., towards the cylinder in the direction perpendicular to the oscillations, and away from it in the parallel direction.

Kim and Troesch (1989) studied, numerically and experimentally, the streaming flow generated by a cylinder of square cross section vibrating in water and found that for $Re_{\text{stream}} > 1$ the outer flow is characterized by jets parallel and perpendicular to the oscillations, while suction jets are directed towards the square corners, resulting in eight cells instead of four in the case of circular cylinder.

An interesting question is if streaming flows exist in complementary quantum flows. Duda *et al.* (2017) answered this question experimentally, by visualizing the flow due to a rapidly oscillating quartz tuning fork, in both normal He I and superfluid He II, by following the flow-induced motions of relatively small particles suspended in the liquid. The particles employed in this study were roughly spherical deuterium particles of radius (5 ± 2) μm. Several sets of movies were recorded, at four different temperatures, in the range 1.22–2.22 K, with a driving force F up to 41 μN, where the simultaneously measured turbulent drag indicated turbulent flow in both He I and He II.

The basic features of the observed streaming flows (see Fig. 10.18), common in He I and in He II, have been characterized by Duda *et al.* (2017) as follows: There are strong outward streaming jets, both above and below each fork prong, perpendicular as well as parallel to the direction of oscillations. Due to mass conservation, the outward jets are compensated by suction jets from the sides, at an angle of about 45°. Additionally, between the prongs, the neighboring suction jets merge, creating a single jet, parallel to the outward streaming jets. If this pattern is compared with the outer streaming flow generated by a square cylinder vibrating in water (Kim and Troesch, 1989), a close resemblance is found if account is taken of the differences in oscillator geometries.

This claim was substantiated by analysing Lagrangian pseudovorticity introduced in Section 10.3. For a fast oscillating fork, however, the amplitude and phase cannot be resolved and therefore the dependence of θ on the phase ϕ can be dropped, i.e., $\theta(\boldsymbol{r}, \phi) \approx \theta(\boldsymbol{r})$. In the limit $R_{\mathrm{M}} \to 0$, for a large enough number of fluid particles, $\theta(\boldsymbol{r})$ would represent the flow vorticity. It is important to choose R_{M} in such a way that the calculated maps of $\theta(\boldsymbol{r})$ are physically sound. The bottom panel of Fig. 10.18, with $R_{\mathrm{M}} = 0.4$ mm (equal to the prong height), gives useful additional information characterizing the streaming flow pattern. Since camera frame rate was about ten times lower than the fork oscillation frequency, the experiment could not measure the oscillatory flow occurring in the Stokes layer, which is at least an order of magnitude smaller than the spatial resolution employed. The streaming flow shown in Fig. 10.18 is very similar to the outer streaming pattern due to a square cylinder vibrating in water.

Perhaps surprisingly at first sight, the outer streaming patterns observed in He II, a quantum liquid, appear to be very similar to those seen in He I, which is a classical fluid. Duda *et al.* (2017) show, however, that the length scales probed in He II are larger than the quantum length scale of the flow, for which the mechanically forced coflowing He II is expected to behave as a single quasi-classical fluid possessing an effective kinematic viscosity. As a consequence of the dynamical locking of the two components of superfluid ^{4}He, as discussed in previous chapters, the large-scale properties of turbulent He II flows cannot be distinguished from those obtained in turbulent flows of viscous fluids.

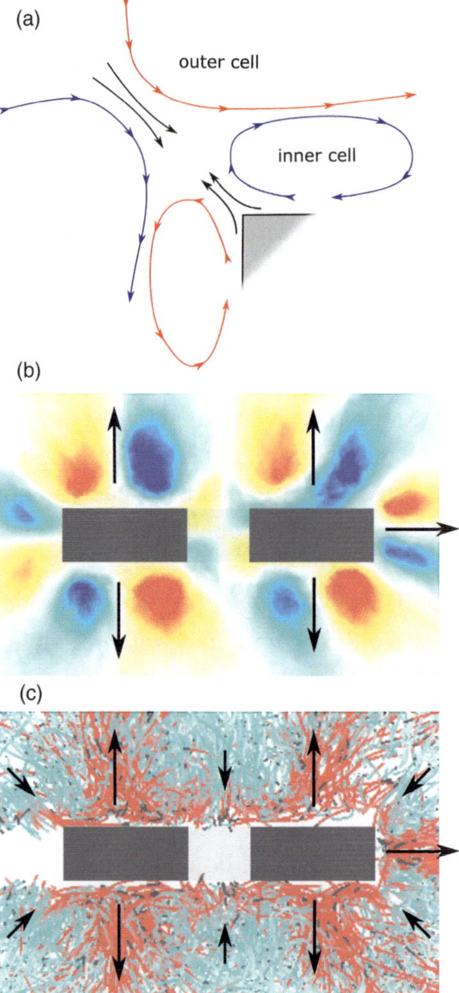

Figure 10.18 (a) Schematic view of the streaming flow features in the proximity of the fork's prong left corner (the fork oscillates in the horizontal direction). (b) Lagrangian pseudovorticity map calculated from all the data obtained in He II, with driving force $F = 41\,\mu\text{N}$. (c) Corresponding particle trajectories, color coded by velocity magnitude (higher velocities in red). The strong outgoing jets are shown with black arrows, as the weaker incoming jets, orientated along the axes of the prong corners. Reprinted figure with permission from Duda *et al.* (2017). Copyright 2017 by the American Physical Society.

10.5 Small-Scale Similarity in Quantum Turbulence

A good starting point for this section is a general statement of La Mantia *et al.* (2016) that "... properties of natural systems often depend on the scale at which they are investigated. Systems that appear similar at a certain scale may look entirely different at another scale, leading to the possibility of discovering hidden connections between apparently dissimilar phenomena." We now discuss how this general principle applies to the study of quantum turbulence.

La Mantia *et al.* (2016) visualized the flow-induced dynamics of relatively small particles in various thermally and mechanically driven flows of He II, which appear distinctly different if experimentally probed at scales ℓ_{exp} that are larger than ℓ. Still, if one probes length scale ℓ_{exp} smaller than ℓ, the tails of the particle velocity distributions (which indicate the occurrence of rare events of large magnitude) are found to be nearly identical. This experimental result supports the long-held expectation that, at small enough length scales, the dynamics of quantized vortices do not depend on the type of imposed large-scale flow. More specifically, three different quantum flows have been probed experimentally at length scales covering two orders of magnitude above their quantum length scale ℓ:

(i) thermal counterflow in the bulk, far enough away from fluid boundaries;
(ii) thermal counterflow in the proximity of a wall, approximately ten times closer to the fluid boundary than in the previous case;
(iii) flow due a mechanically oscillating cylinder.

As we have repeatedly pointed out, the main difference between thermally and mechanically driven flows of He I is that, in thermal counterflow, the normal and superfluid components flow in opposite directions on the average, while, for mechanically generated flows, the two components are locked together at large enough length scales, and move on average in the same direction. However, experimental results show that, at small enough flow scales this difference disappears and that the particle motions, when statistically described, appear to be influenced only by the quantized vortex dynamics, independently of the imposed mean flow. This can be also interpreted as an experimental confirmation of the numerical results reported by Sergeev and Barenghi (2009b), which show that quantized vortices have the tendency to change topology, i.e., to reconnect, when particles get close enough to them. This speaks to certain type of universal behavior inherent in flows with quantized vortices.

This last observation is reinforced by the fact that, at large length scales, the particle velocity distributions are nearly Gaussian (La Mantia and Skrbek, 2014a) as observed in viscous flows, regardless of the imposed flow type. It is also in accord with the behavior of flatness of the nondimensional particle velocity distribution $(u - \overline{u})/u^{\text{sd}}$, where u is the instantaneous dimensional velocity, $\overline{u}$ and u^{sd} denote, respectively, the mean and standard deviation of u, as a function of the ratio ℓ_{exp}/ℓ: For $\ell_{\text{exp}} > \ell$ its value is around 3, that of the Gaussian distribution, while for $\ell_{\text{exp}} < \ell$, the velocity flatness depends on the type of flow. This is linked to the fact that the flatness of a statistical distribution is mainly determined by large-magnitude events that are also rare. For this reason, the relatively short experimental records at small scales available so far can be treated mainly as indicative. Indeed, it is possible that for substantially larger data sets the universal flatness behavior might be observed for $\ell_{\mathrm{exp}} < \ell$ as well when particle motions are influenced solely by the vortex tangle dynamics, characterized by the power-law tails for the distributions of both velocity and acceleration.

In spite of the paucity and special nature of data available so far, the flatness of the velocity distribution at various length scales can be useful for estimating the quantum length scale of the flow ℓ. We reason that it should be of order of the smallest scale at which the distribution flatness becomes approximately equal to 3. Thus, visualization studies are capable of providing information such as the vortex line density $L \approx \ell^{-2}$, usually in support of that same information obtained by other means – such as the second sound attenuation to be discussed in Section 10.6.

As a consistency check, La Mantia and Skrbek (2014a) calculated the flatness values for complementary flows of He I and found that the particle velocity distributions do not depend on the scale at which the turbulent flows are probed but have quasi-Gaussian form at all scales, including those smaller than the dissipative scale. These statistical properties of such scales have not been explored fully in classical viscous flows mainly because (i) such small scales are difficult to probe experimentally while also keeping the Reynolds numbers high; and (ii) the flow behavior at scales smaller than the dissipation length is not expected to be universal. Dedicated investigations addressing this issue in classical turbulence are few; for some recent development, see Khurshid *et al.* (2018) and Buaria and Sreenivasan (2020).

The properties of quantum turbulence just discussed can be linked to small-scale universality observed in classical turbulent flows of viscous fluids – see, e.g., Sreenivasan and Antonia (1997) and Schumacher *et al.* (2014) – as it departs from the pioneering work of Kolmogorov (1941a,b). It is generally accepted that in classical turbulence the departure from a purely Gaussian distribution of the velocity difference is due to intermittency effects originating from rare events of large magnitude. Quantized vortex reconnections may provide such large-magnitude events in quantum turbulence and, consequently, the existence of the observed power law tails in the velocity (and velocity difference) distributions could be similar to intermittency effects in classical viscous flows.

10.6 Flight Crash Events in Quantum Turbulence

Xu *et al.* (2014) have shown that particles probing turbulent flows of viscous fluids tend to gain energy less quickly than they lose, at all probed scales. It is argued that this is a consequence of flight-crash events meaning that particles decelerate on average faster than they accelerate, which provides clear evidence, from a Lagrangian viewpoint, that classical turbulent flows are time irreversible in that the energy put into the system is eventually dissipated by the action of the fluid viscosity (Pumir *et al.*, 2016). We note that the flight-crash events have been known for many years as ramp–cliff structures in the literature on turbulent mixing of passive tracers (Sreenivasan and Antonia, 1977; Sreenivasan, 2018). It is interesting to find out if the particles probing quantum turbulent flows have similar characteristics.

Due to the presence of quantized vortices in turbulent flows of He II, energy transport mechanisms are expected to be different from those occurring in similar flows of viscous fluids (diLeoni *et al.*, 2017). There is numerical evidence that in the zero-temperature limit the energy put into the system is dissipated by sound emission following vortex reconnections and excitation of Kelvin waves. However, above 1 K, in the two-fluid regime of He II, the liquid viscosity, carried by the normal component, cannot be neglected and should be expected to play a role in the mechanisms of energy dissipation.

Švančara and La Mantia (2019) showed experimentally that the mechanisms of energy transport in turbulent flows of He II are indeed strikingly different from those occurring in turbulent flows of viscous fluids. Using PTV visualization technique by employing small, micron-sized solid deuterium and deuterium hydride particles suspended in the liquid and illuminated by a planar laser sheet to determine the velocity increments along the particle trajectories. From these data, Léveque and Naso (2010) obtained the skewness of the corresponding statistical distributions. The skewness is known to be negative at all probed scales in homogeneous and isotropic turbulence, indicating that particle deceleration is on average more abrupt than acceleration, similar to the ramp model of Sreenivasan and Antonia (1977). Švančara and La Mantia (2019) observed that, in turbulent flows of He II, flight-crash events are less apparent than in classical turbulence, at least for scales of the order of quantum length scale ℓ. On the other hand, at larger, quasi-classical scales, the skewness of the particle velocity increment distributions is negative. In other words, the coupling action of the two fluids is important at scales appreciably larger than ℓ, whereas at smaller scales energy transport is ruled by the dynamics of the quantized vortex tangle.

10.7 Summary

Similar to studies of classical turbulent flows, a considerable amount of useful information can be gained by seeding turbulent quantum flows with small solid tracers and by following their motions.[1] Analysing the recorded movies, however, is generally more complex than in the classical case, primarily due to the two-fluid

[1] Ions and He_2^* excimer molecules as tracers are discussed separately, due to a different underlying physics.

behavior of quantum fluids at nonzero temperature, even though the application of classical flow visualization methods over last two decades, especially PTV, has successfully led to important results already. It is clearly desirable to apply visualization techniques in the zero-temperature limit, where there is no normal fluid, to study the behavior of pure superfluid turbulence. However, because of technical reasons, such experiments in both helium isotopes are highly challenging and are awaiting further experimental progress.

Although there was no doubt about the existence of quantized vortex lines in He II, it took long time to visualize them in situ in the form of a rectilinear lattice mimicking solid body rotation, the rotating bucket. It took an even longer time to visualize them in their natural state. It was shortly followed by in situ visualization of vortex reconnections and Kelvin waves. In low-velocity thermal counterflow, the particles clearly demonstrated the existence of two velocity fields. At higher counterflow velocities as well as in mechanically generated coflow of He II, the visualization can be tuned to probe various length scales, ℓ_{exp}, which could be of the order of the quantum scales (i.e., $\ell_{\text{exp}} \lesssim \ell$, characterized by quantum signatures of the flow) and quasi-classical scales (i.e., $\ell_{\text{exp}} > \ell$, where the quantum flow under study displays classical features), including the crossover between the two. Moreover, often the same visualization technique can be applied in complementary flows of He I, allowing direct comparison of classical and quantum flows.

Recorded movies of particle positions in quantum flows can be processed in various sophisticated ways, leading to useful statistical quantities such as PDFs of velocity, velocity increments and accelerations, skewness and flatness, Lagrangian pseudo-vorticity, and their dependence on the types of flow, as well their temperature, pressure, geometry, and length scale. The visualization of quantum flows, which is just at its infancy, will undoubtedly produce new and important results.

11

Decay of Quantum Turbulence

Decay of turbulence in the absence of sustained production, especially for the case of nearly homogeneous and isotropic turbulence, is one of the most important and extensively explored problems in fluid dynamics and a centerpiece in the study of turbulence. In this chapter, we review the basic facts and approaches for decaying classical turbulence in viscous fluids and then discuss the decay of various forms of quantum turbulence. We shall see that classical and quantum decays are fairly similar for certain conditions but that important differences exist.

11.1 Decay of Classical Turbulence in Viscous Fluids

We shall focus on the seminal problem of decaying classical turbulence whose steady state can be characterized as nearly homogeneous and isotropic, representing one of the most important and extensively explored problems in fluid dynamics. Numerous theoretical and experimental studies followed the influential work of Taylor (1935), von Kármán and Howarth (1938), and Kolmogorov (1941a,b). Most of the experimental work has been related to turbulence generated by a grid of bars in wind tunnels, where the turbulence is studied as it decays downstream. By using special arrangements such as a secondary contraction of the test section (Comte-Bellot and Corrsin, 1966) the grid-generated turbulence was rendered closely isotropic some distance downstream. In most experiments, however, the spanwise and streamwise components of the velocity variance differ typically by 10–30% and the turbulence can be thought of as only approximately homogeneous and isotropic, perhaps closer to being axisymmetric; see Sreenivasan and Narasimha (1978). Direct numerical simulations in periodic boxes, which reveal some important features, have their own limitations (John *et al.*, 2022). Even so, much is known about this particular problem of turbulence.

The comparison of experimental data with theory can be done directly by using an oscillating grid (De Silva and Fernando, 1994) or by towing a grid through a stationary sample of fluid (van Doorn *et al.*, 1999). In wind tunnels, one usually assumes the validity of the frozen turbulence hypothesis of Taylor (1935), i.e., $\partial/\partial t = U_0 \partial/\partial x$, where U_0 is the mean velocity and x denotes the downstream distance from the grid. Numerical simulations, which do not have some of the problems characteristic of wind tunnel flows, have depended on a periodic-box configuration with various forcing schemes at large scales. These artifacts influence the details of the decay process.

An early theoretical result of Kolmogorov (1941b) is that the kinetic energy of turbulence, given by $E = \boldsymbol{u}^2/2$, where $\boldsymbol{u} = \boldsymbol{u}(\boldsymbol{r}, t)$ is the velocity vector, decays as $E(t) \propto t^{-\gamma}$, with $\gamma = 10/7$. The experimental data display large variability in the decay exponent γ; as discussed, among others, by Skrbek and Stalp (2000) and John *et al.* (2022), at least some of this variation is due to the strong influence of the so-called virtual origin chosen for time. Numerical studies also yield substantially varying results for γ (John *et al.*, 2022), confirming that these differences could be the effects of finite computational domains or of specific initial conditions (Yakhot, 2004). On the whole, the problem of decaying classical turbulence in viscous fluids, even its simplest homogeneous and isotropic case, cannot be treated as satisfactorily understood, even though a consistent story seems to be emerging (John *et al.*, 2022).

The basic equation for turbulent energy decay in the classical case is

$$-\frac{dE}{dt} = \epsilon = \nu \overline{\omega^2}, \tag{11.1}$$

where E is the turbulence energy, ϵ is the energy dissipation rate, $\overline{\omega^2}$ is the mean-square vorticity, and ν is the kinematic viscosity. The energy E is the integral of the spectral density $\Phi(k)$ over the wavenumber k. This relation also gives the decay rate of mean-square vorticity.

Indeed, if the outer scale of classical homogeneous and isotropic turbulence is equal to the size of the system (more general cases will be discussed in Section 11.2), then Eq. (11.1) can be approximated as $\epsilon \propto E^{3/2}/D$, consistent with the familiar relation of dissipative anomaly (Sreenivasan, 1984),

$$\epsilon = \epsilon_0 \frac{E^{3/2}}{\ell_{\mathrm{e}}}, \tag{11.2}$$

where the energy containing length scale ℓ_{e} is assumed to be constant and proportional to the size of the system, D. It is then immediately evident that the kinetic energy decays as

$$E(\tau) \propto \frac{D^2}{(t + t^*)^2} = \frac{D^2}{\tau^2}, \tag{11.3}$$

where t^* is the virtual origin in time that depends on the starting value of energy as well as the transient processes that take place at the start of decay.

For decaying vorticity of classical homogeneous and isotropic turbulence for this same period of time, this corresponds to

$$\omega(\tau) \propto \frac{D}{(t+t^*)^{3/2}} = \frac{D^2}{\tau^{3/2}} \,. \tag{11.4}$$

Despite the inevitability of the above result, there are no experimental data on decaying classical homogeneous and isotropic turbulence that would confirm it convincingly. This is so because every grid-turbulence experiment has aimed to keep the length scale small compared with the size of the apparatus. Towed grid experiments that can be run for very long periods of time show a tendency to approach this state (van Doorn *et al.*, 1999). Numerical simulations (John *et al.*, 2022), especially those of Touil *et al.* (2002), appear closer. In particular, the latter authors investigated the decay of turbulence in a bounded domain without mean velocity using direct and large-eddy simulations, as well as the eddy-damped quasi-normal Markovian (EDQNM) closure. The effect of the finite geometry of the domain is accounted for by introducing a low wavenumber cut-off into the energy spectrum of isotropic turbulence. It is found that, once the turbulent energy-containing length scale saturates, the root-mean-square vorticity follows a power law with the $-3/2$ exponent, Eq. (11.4).

11.2 Spectral Decay Model of Classical Turbulence

In order to proceed further, it is useful to describe the spectral decay model for classical homogeneous and isotropic turbulence. It is based on the form of the energy spectrum, which was discussed in Chapter 9. Here, we have to take into account that the spectral density in decaying turbulence behaves differently in different scale ranges and obeys different forms of similarity, with appropriate crossovers between them. One can then use Eq. (11.1) to calculate the decay in terms of crossover scales and other spectral constants such as scaling exponents.

This procedure was first used by Comte-Bellot and Corrsin (1966) and revised by Saffman (1967a,b). It was extended by Skrbek and Stalp (2000) and Skrbek *et al.* (2000) by taking account of ideas of Eyink and Thomson (2000) to develop a spectral decay model for classical homogeneous and isotropic turbulence and relate it to the available data on decaying turbulent energy. Following the updated approach of Skrbek and Sreenivasan (2013), the model can be described briefly as follows:

$$\Phi(k) = 0, \qquad k < 2\pi/D; \tag{11.5}$$

$$\Phi(k) = Ak^m, \qquad 2\pi/D \le k \le k_e(t); \tag{11.6}$$

$$\Phi(k) = C\epsilon^{2/3}k^{-5/3}, \qquad k \ge k_e(t). \tag{11.7}$$

The form of Eq. (11.7) has already been discussed in earlier chapters. The form of Eq. (11.5) reflects the fact that scales larger than the size of the turbulence box, D, do not exist. Additionally, we neglect intermittency corrections and all details of the spectrum at high k and assume that the Kolmogorov scaling holds for all $k > 2\pi/\ell_e$. In the vicinity of the energy-containing scale $\ell_e = 2\pi/k_e(t)$, where $k_e(t) \geq 2\pi/D$, the spectral energy density displays a broad maximum whose shape does not have to be exactly specified.

We shall now consider three stages of the decay. Evaluating the total turbulent energy by integrating the 3D spectrum over all k leads to a differential equation for decaying turbulent energy; and applying $\epsilon = \nu\omega^2$ leads to a differential equation for decaying vorticity, the quantity of particular interest to us.

It is not difficult to show that the first decay regime is of the form

$$E(t + t_{\mathrm{vo1}}) = E(\tau) \propto \tau^{-2\frac{m+1}{m+3}}; \quad \omega(t + t_{\mathrm{vo1}}) = \omega(\tau) \propto \tau^{-\frac{3m+5}{2m+6}}, \tag{11.8}$$

where t_{vo1} is a virtual origin in time. Assuming the validity of the Birkhoff–Saffman invariant ($m = 2$), we obtain the first regime for decaying energy, $E \propto \tau^{-6/5}$ (i.e., we have recovered the original result of Birkhoff [1954] and Saffman ([1967a; 1967b]) and for vorticity, $\omega \propto \tau^{-11/10}$. On the other hand, assuming the validity of the Loitsianskii invariant ($m = 4$), we obtain the alternative expression for decaying energy, $E \propto \tau^{-10/7}$, and the vorticity, $\omega \propto \tau^{-17/14}$, i.e., we have recovered the original result of Kolmogorov (1941b).

As the turbulence decays further and ℓ_e grows, the lowest physically significant wave number becomes closer to the broad maximum around $2\pi/\ell_e$. The low wave number part of the spectrum can no longer be approximated as Ak^m. Instead, it can be characterized by an effective power m_{eff} that decreases as the turbulence decays, such that $0 < m_{\mathrm{eff}} < m$ and the decay rate slows down. As ℓ_e approaches D, m_{eff} essentially vanishes to zero and we arrive at the second regime of the decay, $\omega \propto \tau^{-5/6}$. At the saturation time, t_{sat}, the vorticity reaches its saturation value, ω_{sat}, and the growth of ℓ_e is completed.

The universal third regime is predicted as

$$\omega(t + t_{\mathrm{vo3}}) = \omega(\tau) = \frac{3\sqrt{3}D}{2\pi}\sqrt{\frac{C^3}{\nu}}\tau^{-3/2}, \tag{11.9}$$

with the appropriate (generally different) virtual origin time t_{vo3}. This regime is universal in the sense that a decaying system must sooner or later reach it, independently of the initial level of turbulence (in practice it must be high enough to eliminate the influence of viscosity that we have neglected so far). Transitions from one decay regime to the next are gradual, and so care must be taken while searching for the decay exponent from log-log plots of the decay data, as each of the decay regimes could have its own virtual origin – to the extent that the notion of virtual origin makes good sense.

In subsequent sections, we shall separately study the cases of coflows and counterflows of He II.

11.3 Basics on the Decay of Quantum Turbulence in He II

Solving the problem of decaying classical turbulence simultaneously with that in quantum turbulence, based especially on the experimental data of grid-generated turbulence in classical and quantum fluids, helps to resolve this cornerstone problem of turbulence in fluids. The impediment, however, is that the results of most experiments (though not all of them; see Section 11.8) on decaying quantum turbulence are given not in terms of decaying energy but vortex line density, $L(t)$. In any attempt to link this to energy decay, we should appreciate the role of quantized vortices.

The simplest system for that purpose is a rotating bucket of He II, which we discussed in Chapter 2. In its steady state, the normal fluid is in solid body rotation, with vorticity $\boldsymbol{\omega}_{\rm n} = \operatorname{curl} \boldsymbol{u}_{\rm n} = 2\boldsymbol{\Omega}$, and He II is threaded with a rectilinear array of quantized vortex lines parallel to the axis of rotation. Although the superfluid vorticity is restricted to vortex lines, their areal density evolves to match the vorticity of the fluid in solid body rotation, so that on length scales larger than the inter-vortex line distance, the superfluid is also in solid body rotation, with a velocity field equal to that of the normal fluid, i.e., $\langle \omega_{\rm s} \rangle = \omega_{\rm n}$. Even though quantum turbulence is far more complex to be fully understood by the paradigm of aligned vortices, it is helpful to bear the above observation in mind.

One can offer an alternative picture, starting with the view that turbulence in the superfluid component takes the form of an apparently random tangle of vortex lines, which interact with each other and, through a mutual friction force, also with the normal fluid. This type of quantum turbulence is often named Vinen or ultraquantum turbulence, where the turbulent energy in the superfluid component is confined to scales comparable to or less than the quantum length scale ℓ.

As we shall discuss in this chapter, these two pictures represent the two extreme cases. In practice, quantum turbulence is in most cases an interpolation of the two, and a smooth crossover between these types can be observed by changing experimental conditions, such as the energy input. This crossover can be thought of as partial polarization of the originally random tangle, resulting in the appearance of coherent structures, both numerically predicted (Baggaley *et al.*, 2012b), and directly observed (Rusaouen *et al.*, 2017b), as in classical turbulence.

Mechanical forcing of He II at some large scale results in Kolmogorov-like quantum turbulence: The turbulent superflow at length scales sufficiently exceeding the quantum length scale ℓ resembles classical flow possessing an effective kinematic viscosity, ν_{eff}. Then the relationship between turbulent energy dissipation rate per unit volume and the mean square vorticity should apply, and we should have

$$\epsilon = -\frac{dE}{dt} = \nu_{\text{eff}}(\kappa L)^2, \tag{11.10}$$

i.e., vorticity would be defined as κL in analogy with the rotating bucket. This is one of the key relationships for comparison of quantum turbulence with its classical counterpart. Its underlying physics were first discussed in detail by Vinen (2000); see also the review by Vinen and Niemela (2002). If this equation holds, it allows superfluids to be treated as a hypothetical classical fluid possessing an effective kinematic viscosity ν_{eff}.

Assuming the validity of Eq. (11.10) and the energy-containing length scale ℓ_e to be a constant and equal to the size of the system, D, for decaying vortex line density in quantum turbulence, we get

$$\omega(\tau) = \kappa L \propto \frac{D}{(t+t^*)^{3/2}} = \frac{D^2}{\tau^{3/2}}\,. \tag{11.11}$$

This fundamental result, obtained under rather rough assumptions that are yet to be justified in detail, has been confirmed for the decay, after an initial adjustment period, of vortex line density in many experiments on decaying quantum turbulence, both in He II and ^{3}He-B.

11.4 Application for Turbulent Coflows of He II

Application of the spectral decay model to decaying Kolmogorov-like quantum turbulence is straightforward, assuming that the spectral energy density is of the form of Eq. (11.7) and the relationship $\omega = \kappa L$ holds. It is natural to compare the predictions of the spectral decay model with experimental data on decaying quantum turbulence, generated by methods of classical turbulence, in both the zero-temperature limit and at finite temperatures, generated by methods that force both the superfluid and the normal fluid in the same way to generate the coflowing quantum turbulence.

Historically, there are many relevant sources of data of this kind, mostly in the form of decaying vortex line density $L(t)$ resulting from second sound experiments. They include the towed grid data He II from Oregon (Smith *et al.*, 1993; Stalp, 1998; Stalp *et al.*, 1999; Skrbek and Stalp, 2000; Skrbek *et al.*, 2000), obtained with the square grid of nonconventional design consisting of six parallel 0.5 mm tines that are 1.17 mm apart joined from corner to corner by another tine as shown in Fig. 1b in Smith *et al.* (1993). More recent Oregon data of Niemela *et al.* (2005) and Niemela and Sreenivasan (2006) was obtained with a more standard grid (20 × 20) tines

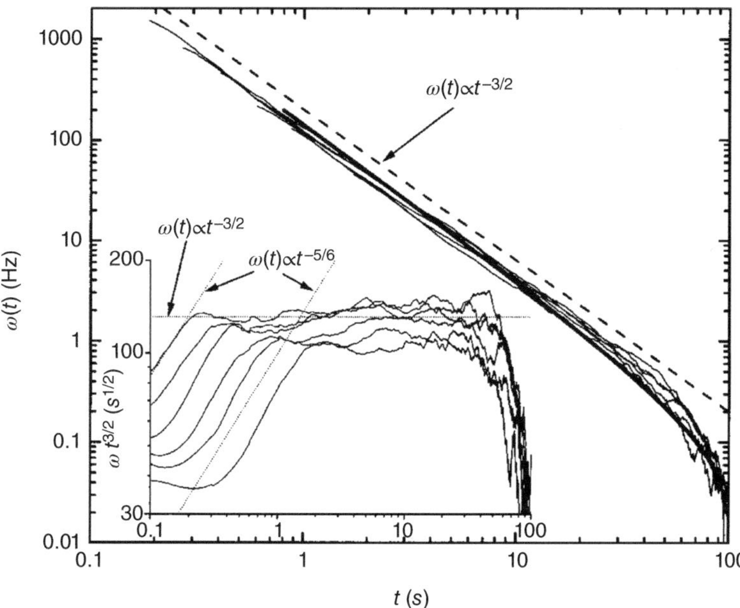

Figure 11.1 Vorticity decay data obtained in He II at $T = 1.75\,\mathrm{K}$ and mesh Reynolds numbers of 2×10^5, 1.5×10^5, 10^5, 6×10^4, 4×10^4, and 2×10^4. (Solid line) Theoretically predicted decay as explained in text. (Inset) Same vorticity decay data multiplied by $t^{3/2}$ where t is the experimental time. Dotted lines represent the $t^{-3/2}$ and $t^{-5/6}$ decay power laws. Reprinted figure with permission from Skrbek *et al.* (2000). Copyright 2000 by the American Physical Society.

of square cross section; even more recent data comes from Tallahassee (Gao *et al.*, 2016a; Varga *et al.*, 2018) and Gainsville (Yang and Ihas, 2018). Also relevant are the Lancaster oscillating grid data in ^{3}He-B in the zero-temperature limit of Bradley *et al.* (2006), measured by the Andreev scattering method (to be discussed in Section 11.7.2), shown together with the He II Oregon towed-grid data of Stalp *et al.* (1999); see Fig. 11.11.

The pronounced feature of the Oregon towed-grid data is that the different decay regimes switch from one to the other successively as the decay proceeds and the spectrum evolves (Skrbek *et al.*, 2000). Focusing now on the first regime, the Oregon group concluded that the decay exponent for decaying vortex line density (see Eq. (11.8)) is in agreement with the spectral decay model predicting the $-11/10$ decay exponent, based on the validity of the Birkhoff–Saffman invariant, which assumes that the total momentum of the flow is conserved. As a consequence, the spectral energy density $\Phi(k) \propto k^2$ at low k. If, however, one assumes the validity of Loitsianskii invariant, i.e., $\Phi(k) \propto k^4$ at low k, the decay exponent ought to be

$-17/14$. More recently, Skrbek and Sreenivasan (2013) critically reanalyzed the Oregon data. Given the fact that the two competing decay exponents in the first regime differ only slightly, and taking into account the experimental accuracy and that the temporal resolution of the second sound technique, a firm conclusion cannot be drawn about which scenario works better. This should not be surprising given the complexity of the classical decay itself (John *et al.*, 2022).

The second regime, which is more in the nature of a slow changeover, displays a slower decay $L(t) \propto t^{-5/6}$, as displayed in Fig. 11.1. It does not depend on the validity of the Loitsianskii or the Birkhoff–Saffman invariant, as long as there is a broad maximum in spectral energy density and therefore ought to occur in all experiments. Indeed, this feature does exist in all Oregon towed-grid data, obtained with the nonconventional grid design and the standard grid design of Niemela *et al.* (2005) and Niemela and Sreenivasan (2006), and is consistent with other data on decaying vortex line density of mechanically generated quantum turbulence in He II and ^{3}He-B (these cases will be discussed separately in Section 11.7.2).

The third universal decay regime is a direct consequence of Eq. (11.9), assuming the validity of Eq. (11.1). The expression for decaying vortex line density therefore reads as

$$L(t + t_{\mathrm{vo3}}) = \frac{3\sqrt{3}D}{2\pi\kappa}\sqrt{\frac{C^3}{\nu_{\mathrm{eff}}}}\tau^{-3/2}. \tag{11.12}$$

This form of decaying vortex line density for times larger than some initial adjustment time, $L(t) \propto t^{-3/2}$, is observed in a large number of He II as well as ^{3}He-B experiments, where quantum turbulence is generated by mechanical and other means. This is understood reasonably well, as its only requirement is that the entire energy spectrum, starting at large scale – the size of the turbulent box D in the superfluid component – should be characterized (neglecting intermittency) by the $-5/3$ roll-off exponent of K41. This can be the case in the zero-the temperature limit for pure superfluid turbulence both in He II and ^{3}He-B and in the two-fluid regime at finite temperature for He II, as long as there is negligibly small dissipative mutual friction between the normal and superfluid velocity fields. At finite temperature, this basically means that the normal and superfluid eddies (composed of bundles of quantized vortices) are locked together. The high-frequency end of the superfluid spectrum can then be thought to be terminated by effective kinematic viscosity, just as the kinematic viscosity terminates the high-k spectrum in classical turbulence.

Equation (11.12) allows the extraction of effective kinematic viscosity of turbulent superfluid and its temperature dependence from observed decays of vortex line density in various experiments, as was first done by Stalp (1998) and Stalp *et al.* (2002), and later by others, e.g., Walmsley *et al.* (2007), Walmsley and Golov

(2008, 2017), Chagovets *et al.* (2007), Babuin *et al.* (2015), Varga *et al.* (2015), and Gao *et al.* (2018) in He II and Bradley *et al.* (2006, 2011a) in ^{3}He-B.

The physical quantity that we have introduced to interpret the decay of the Kolmogorov-type turbulence of He II, the effective kinematic viscosity (of *turbulent* superfluid), $\nu_{\rm eff}$, was independently tested by Babuin *et al.* (2014a) in the steady-state case. In this work, $\nu_{\rm eff}$ was deduced from measurements of the vortex line density in a grid flow. The scaling of L with velocity confirms the validity of the heuristic relation defining Eq. (11.1). Within the $1.17 \leq T \leq 2.16\,$K range, the effective kinematic viscosity deduced in this way is consistent with that obtained from decay measurements, if due allowance is made for the uncertainties in flow parameters.

The concept of effective kinematic viscosity in Kolmogorov-like quantum turbulence thus appears rather robust within the achievable range of parameters in this experimental configuration. Later in this chapter, we shall return to a more general concept of effective kinematic viscosity defined for various forms of quantum turbulence, including those in the zero-temperature limit.

11.5 Decay of Vinen-Type (Ultraquantum) Turbulence

So far, we have discussed decaying Kolmogorov-like quantum turbulence and found it to be similar to decaying classical turbulence. We now outline a strikingly different decay of Vinen-type or ultraquantum turbulence.

In Chapter 6, we discussed the phenomenological, physically motivated Vinen equation, Eq. (6.5), which was introduced in order to describe the steady state and evolution of vortex line density L under the assumption that the vortex tangle is random and the quantum length scale ℓ is the only relevant length scale of the problem. Neglecting the (small) term $g(u_{\rm ns})$, for tangles of relatively low density and small deviations from the steady state, the Vinen equation was rewritten by Schwarz (1985) in the form of Eq. (6.7). Following Schwarz and Rozen (1991), if one is interested in the situation in which the vortex line density changes from the equilibrium value $L_{\rm i}$ to $L_{\rm f}$, triggered by a sudden change of the counterflow velocity, the temporal development (of both growth and decay) is given by the analytical solution

$$\tau = -l^{-1/2} - \ln\left|1 - l^{-1/2}\right| + \text{const.}, \tag{11.13}$$

where $l = L(t)/L_{\rm f}$ and $\tau = \alpha c_2^2 t L_{\rm f}/2$. In the particular case of $L_{\rm f} = 0$, we therefore expect the decay of the form $L(t) \sim 1/(t + t_{\rm vo})$, where $t_{\rm vo}$ is the virtual origin in time. Its physical meaning is that at $t_{\rm vo}$ the equilibrium vortex line density would have been infinite. In other words, at long times we expect decay of the simple form $L(t) \sim 1/t$.

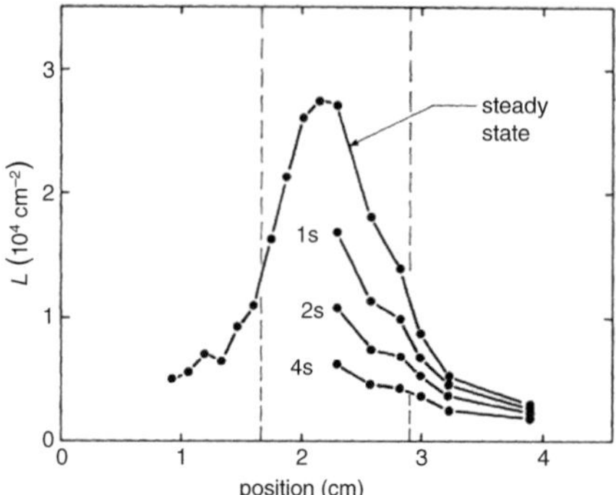

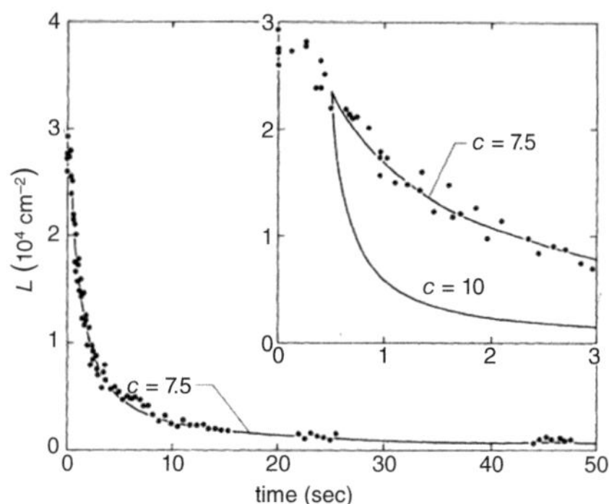

Figure 11.2 (Left) The top curve is the steady-state profile of vortex line density, L, obtained by sweeping the sample point across the helium container shown in Fig. 4.11. Lower curves are 1, 2, and 4 s after the ultrasound transducers have been turned off. (Right) Local characterization of the decay process, with measurements taken at 1.45 K in the middle of the cell. The curves are obtained using the equation $L(t) = L_0/(1 + \beta L_0 t/4c)$, where β is defined in Eq. (6.8). See the text for further details. Reprinted figure with permission from Milliken *et al.* (1982). Copyright 1982 by the American Physical Society.

At finite temperature, where He II displays the two-fluid behavior, this simple prediction rarely holds in practice. One significant example is shown in Fig. 11.2, whose main results come from the finite-temperature experiment of Milliken *et al.* (1982) on decaying inhomogeneous quantum turbulence, created by ultrasonic transducers in He II introduced in Fig. 4.11. At the temperature of 1.45 K, for rather low starting vortex line density about $10^4\,\mathrm{cm}^{-2}$, the authors were able to verify the inverse linear time dependence $L(t) \sim 1/t$ up to about 50 s.

Moreover, the left panel of Fig. 11.2 shows the persistence of structure as the vortex line density profile decays. It strongly suggests that the vortex line density decays locally and any diffusion of quantized vorticity is not appreciably important. Under this assumption, the nonlinear decay rate $\propto L^2$ would first lead to a rapid homogenization of the tangle, which would further decay as $L(t) \propto 1/t$. While this scenario is contrary to the claim that the spatial profile is maintained during decay, it is surprisingly consistent with the raw data in the left panel of Fig. 11.2, where, in the center of the helium apparatus, the initially high L drops by a factor of about 4.5 within the first 4 s, while in the border regions a significantly smaller drop of a factor of 2 is observed. This validates the scenario of predominantly local decay of nonhomogeneous quantum turbulence, which results in $L(t) \propto 1/t$ at late times.

We shall return to this aspect in Sections 11.5 and 11.7 of this chapter, when considering alternative mechanisms of decay.

11.6 Two Effective Kinematic Viscosities of Turbulent He II

Generally, the Kolmogorov and Vinen forms of quantum turbulence in He II display two distinctly different forms of decay. The Vinen turbulence represents an essentially random tangle of quantized vortex lines, above about 1 K existing in the presence of a nonturbulent normal fluid. Its decay follows the prediction of Vinen (1957), given in Eq. (6.5), which is

$$\frac{dL}{dt} = -\frac{\zeta_2 \kappa}{2\pi} L^2, \tag{11.14}$$

where ζ_2 is a dimensionless parameter of the order unity. Following Walmsley and Golov (2017) and Gao *et al.* (2018), noting that the energy per unit mass associated with a random tangle of vortex lines is given by $\left(\rho_s \kappa^2\right)/(4\pi\rho)\, L \ln(\ell/\xi_0)$, where ξ_0 is the vortex core parameter, we see that the turbulent energy per unit mass would then decay as $\epsilon = -dE/dt = \nu' \kappa^2 L^2$. Using Eq. (11.10), we obtain

$$\nu' = \frac{\zeta_2 \rho_s}{8\pi^2 \rho} \kappa \ln(\ell/\xi_0) \tag{11.15}$$

as the effective kinematic viscosity for the Vinen-type, or ultraquantum, turbulence. At low temperatures its value is of order 0.1 κ (Zmeev *et al.*, 2015).

The other form of turbulence (i.e., Kolmogorov or quasi-classical) represents a coupled turbulent motion of the two fluids in the two-fluid regime of He II, often exhibiting quasi-classical characteristics on scales larger than the quantum length scale ℓ. We have already explained that Eq. (11.12) allows the extraction of kinematic viscosity of turbulent superfluid, ν_{eff}, and its temperature dependence from observed late decays of vortex line density is of the form $L(t) \propto t^{-3/2}$ in various experiments, related to the decay of energy given by Eq. (11.10), $\epsilon = -dE/dt = \nu_{\text{eff}}(\kappa L)^2$. Recently, Gao *et al.* (2016a) argued that, for more reliable determination of the effective kinematic viscosity ν_{eff} of quasi-classical turbulence in He II, it is useful to simultaneously measure both the decaying energy, $E(t)$, using a recently developed flow visualization technique based on excimer molecules, as well as the vortex line density, $L(t)$, via second sound attenuation. The values of Gao *et al.* (2016a) are somewhat higher, as seen in Fig. 11.3, but this difference is within the experimental spread of the data given for this temperature range by other authors; see e.g., Walmsley *et al.* (2007); Walmsley and Golov (2008); Babuin *et al.* (2014a). With dropping temperature, the values of ν_{eff} decrease from being constant at about 0.5 κ for $T > 1.5$ K, to about 0.1 κ for lower temperatures.

We therefore have two forms of effective kinematic viscosity, which can be extracted from decaying turbulent He II depending on whether the decaying He II turbulence was of Vinen or Kolmogorov type. The existence of these two forms is not a consequence of the two-fluid behavior, since they readily exist, as we shall now show, in the zero-temperature limit as well. Indeed, while ν_{eff} is a property of

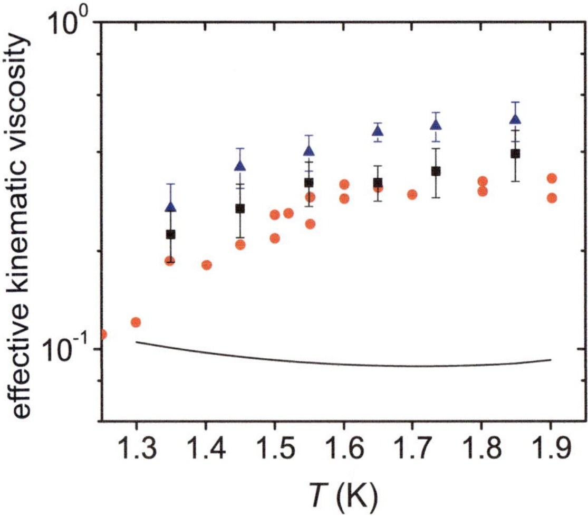

Figure 11.3 The effective kinematic viscosity ν_{eff} of He II in units of κ, obtained from the third universal late decay regime, as deduced from the Oregon towed grid data (Stalp *et al.*, 2002) and later corrected by Chagovets *et al.* (2007) (circles), shown together with the data (squares) obtained using the same method (i.e., from the decaying vortex line density only) by Gao *et al.* (2016a) (triangles). The black solid curve is the kinematic viscosity of the normal-fluid of He II alone (Donnelly and Barenghi, 1998). Reprinted figure with permission from Gao *et al.* (2016a). Copyright 2016 by the American Physical Society.

turbulent flow of the entire He II, i.e., of both fluids whose turbulent motions are coupled at large enough scales, ν' is the property of a single fluid flow, namely of the turbulent superfluid component, coupled by the mutual friction force to the normal fluid. This difference – coupling to the normal fluid as well as the factor ρ_s/ρ – ought to disappear with dropping temperature, as ρ_s approaches ρ and the mutual friction gradually ceases to operate. For reasons such as ill-defined boundary conditions in He II below 1 K (see Section 11.7), the values of ν' and ν_{eff} are so far not determined with sufficient accuracy to unequivocally confirm that ν' equals ν_{eff} at low temperature; however, it appears that they are both of order 0.1κ (Zmeev *et al.*, 2015).

Although the interpretation of ν_{eff} and ν' is relatively straightforward, it is difficult to compare their numerical values with those of classical turbulence, where the effective viscosity with its definition closer to ν_{eff} increases essentially linearly with the large-scale Reynolds number. This difficulty is compounded by the fact that the Reynolds number in the classical and quantum cases cannot be compared unambiguously. Two features are noteworthy: (i) ν_{eff} and ν' are comparable to each other, which is somewhat astonishing given their different sources; and (ii) they are both of the order of the kinematic viscosity of He II (using the total density in defining the kinematic viscosity).

11.7 Decay of Quantum Turbulence in the $T \to 0$ Limit

Let us now consider the experimentally challenging case, the decay of quantum turbulence in the $T \to 0$ limit. There is no normal fluid and we therefore have to deal with the intellectually simpler pure superfluid turbulence. In practice, these conditions are satisfied in purified He II (otherwise ^{3}He impurities would act effectively as the normal fluid and alter the physical properties of He II) at temperatures of order 100 mK and below. The practical limit for considering ^{3}He-B quantum turbulence as pure superfluid turbulence is up to about $0.2\,T/T_c$, i.e., below about 200 μK.

11.7.1 Mechanisms Operating in the Decay of Pure Superfluid Turbulence in He II

The first attempt to generate quantum turbulence in He II in the $T \to 0$ limit by an oscillating grid and to detect its temporal decay by means of ions was by the Lancaster group (Davis *et al.*, 2000), using the apparatus shown schematically in Fig. 4.9. Steady-state superfluid turbulence was produced by oscillating the tightly stretched grid near its resonant frequency, causing the decrease of ionic current because of the trapping of ions by vortex lines in the tangle. After stopping the grid drive, the recovery of the ionic current amplitude to its vortex-free level took several seconds, attributable to the time taken for the vortex tangle to decay. This illuminating experiment clearly demonstrated the production and decay of pure superfluid in the zero-temperature limit, although only qualitatively.

Quantitative data on decaying He II turbulence in the zero-temperature limit have been obtained in Manchester (Walmsley *et al.*, 2007, 2008; Walmsley and Golov, 2008, 2017; Zmeev *et al.*, 2015) using negative ions injected into the experimental cell by a sharp field-emission tip and manipulated by an applied electric field. This technique, introduced in Fig. 4.1, has been used for both the generation and detection of quantum turbulence.

We first focus on the impulsive spin-down technique, which proved suitable for temperatures down to at least 80 mK. Its success relied on rapidly bringing to rest a rotating cubic-shaped container of He II, 4.5 cm in size; the range of angular velocities of initial rotation was $0.05 \leq \Omega \leq 1.5$ rad/s. Before obtaining each experimental point, the cryostat was kept at steady rotation for at least 300 s before decelerating to a stop; data were taken after waiting for a time interval t. Then the probed tangle was discarded and a new one generated. Hence, different

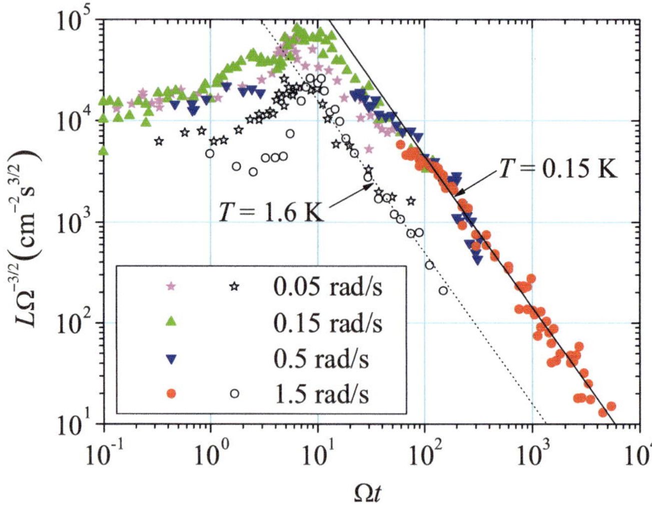

Figure 11.4 The deduced decaying vortex line density, normalized by $\Omega^{-3/2}$, plotted versus Ωt (note the log–log scale) for $T = 0.15$ K (filled symbols) and $T = 1.6$ K (open symbols). Dashed and solid lines have a slope of $-3/2$ to guide the eye through the late-time decay. Reprinted figure with permission from Walmsley *et al.* (2007). Copyright 2007 by the American Physical Society.

data points in Fig. 11.4 represent different realizations. The origin of time was chosen at the start of the deceleration. It is a common feature of the data (measured along both horizontal and vertical direction at any temperature; see Fig. 4.1) that after a certain transient the vortex line density decays as $L(t) \propto t^{-3/2}$. To stress the scaling of the characteristic times with the initial turnover time Ω^{-1}, the data shown in Fig. 11.4 for different Ω are rescaled accordingly. At all starting angular velocities of rotation, Ω, the observed transients are basically universal. The scaling of the transient times with the turnover time $1/\Omega$ indicates that the transient flows are similar at different initial velocities, which is expected for flow instabilities in classical liquids. Eventually, after $\Omega t \approx 100$, the decays observed at 0.15 K take the late-time form $L(t) \propto t^{-3/2}$, expected for vorticity for classical isotropic turbulence decaying in the bounded domain. We note that, in the temperature range of overlap, a fair agreement exists with the Oregon decay data using towed grid, discussed above. Over the low end of attainable temperature range, $0.08 < T < 0.5$ K, the measured $L(t)$ was independent of the temperature. This result means that the effective kinematic viscosity, ν_{eff}, is temperature independent here.

Additionally, Fig. 11.4 compares the decaying vortex line densities $L(t)$ observed at low (0.15 K) and high (1.6 K) temperatures. Note the appreciable difference in the prefactors (i.e., also in values of ν_{eff}) in the late-time dependence $L(t) \propto t^{-3/2}$ at 0.15 K and 1.6 K. It implies that, at low temperatures, the inertial cascade with a

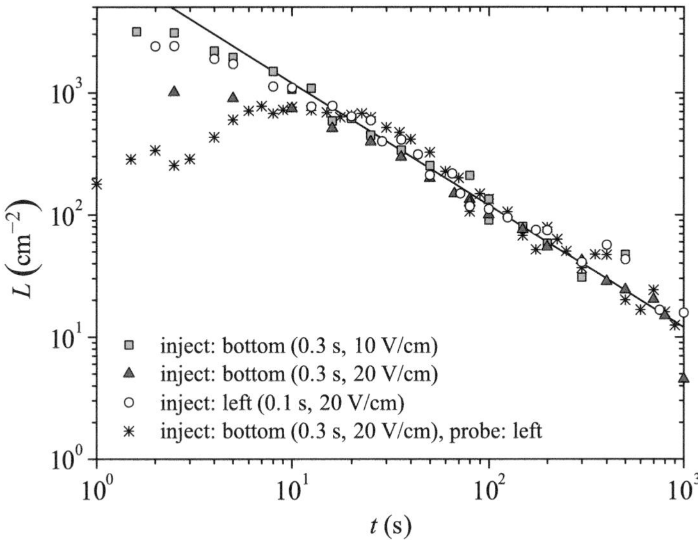

Figure 11.5 Free decay of a tangle at $T = 0.15\,\mathrm{K}$ generated by short pulses of charged vortex rings resulting from the injection of ions from a sharp tip. The injection direction and its duration, as well as the driving field are indicated. Probing with pulses of charged vortex rings of duration 0.1–0.3 s were done in the same electric field as the initial injection, and also in the same direction except in one case, marked (*) (for details, see Fig. 4.1). The line $L \propto 1/t$ corresponds to effective kinematic viscosity in Vinen (ultra-quantum) turbulence $\nu' = 0.1\kappa$. Reprinted figure with permission from Walmsley and Golov (2008). Copyright 2008 by the American Physical Society.

saturated energy-containing length and constant energy flux requires a greater total vortex line density. A crucial, perhaps surprising, observation is that even in the zero-temperature limit the impulsive spin down is followed by the temporal decay of vortex line density of classical character (after some transient period during which the vortex tangle becomes established).

Let us now consider the temporal decay of vortex tangles of relatively low starting vortex line density, generated at very low temperature $T = 0.15\,\mathrm{K}$ using a jet of charged vortex rings resulting from injection of negative ions. We emphasize that this technique, schematically illustrated in the lower part of Fig. 4.1, requires no moving parts in the cryostat. Walmsley *et al.* (2007, 2008), Walmsley and Golov (2008, 2017), and Golov *et al.* (2010) found that the properties of these tangles can be quite different. This difference is most clearly demonstrated while observing their temporal decay rates.

Representative examples of available experimental data are displayed in Figs. 11.5 and 11.6. They show that, at late times, two distinctly different power laws of decay can be observed, depending on experimental conditions and history of preparing the vortex tangle. Basically, instances of quantum turbulence produced

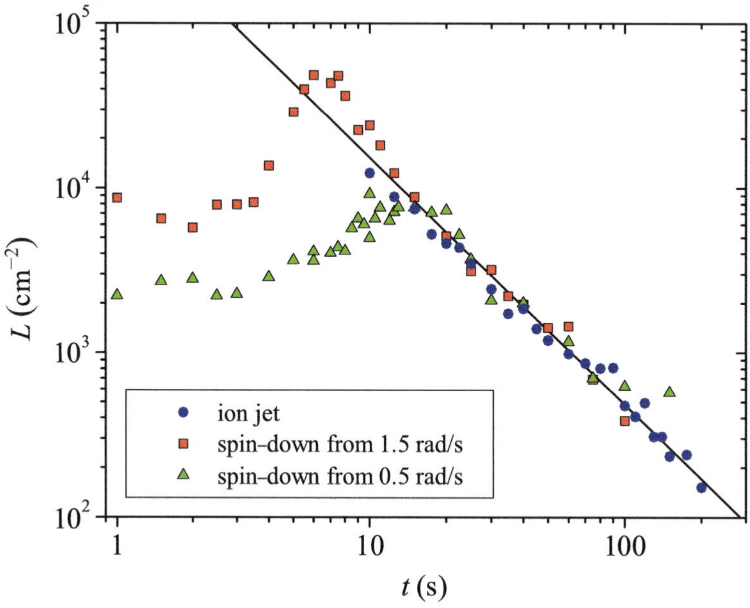

Figure 11.6 Free decay of a tangle produced by a jet of ions (blue circles) from the bottom injector into a 10 V/cm field for 150 s, as well as by an impulsive spin-down to rest (Walmsley *et al.*, 2007) from 1.5 rad/s and 0.5 rad/s, at $T = 1.6$ K. All tangles were probed by pulses of free ions in the horizontal direction (spin-down data: probe field 20 V/cm, pulse length 0.5 s; ion jet data: probe field 10 V/cm, pulse length 1.0 s). The line $L \propto t^{-3/2}$ corresponds to Eq. (11.12) with $\nu_{\text{eff}} = 0.2\kappa$. Reprinted figure with permission from Walmsley and Golov (2008). Copyright 2008 by the American Physical Society.

after brief injection of ions at low temperatures displays $L(t) \propto 1/t$ decay, while that generated after long injection at high temperatures display a $L(t) \propto 1/t^{3/2}$ decay law at late times, in agreement with Kolmogorov-like quantum turbulence generated by the rapid spin-down technique. For temperatures $0.7 \leq T \leq 1.6$ K, the temporal decays of vortex line density generated by these two different techniques were found to be identical to within experimental uncertainty.

Walmsley and Golov (2017) have investigated in more detail how the observed differences in the decay depend on the history of the original vortex tangle. They developed a technique of generating, in the $T \to 0$ limit, pure superfluid turbulence with the known amplitude of flow velocity at the integral length scale, the size of the cell D; see Fig. 11.7. We have to bear in mind that, generated this way, the tangle is charged and anisotropic, with large-scale flow imposed by the charge; the intensity of this large flow grows with the charging time. It is therefore hardly surprising that the authors found that, for the form of the decay the generation, history plays an essential role. Indeed, for the range of injection conditions, at

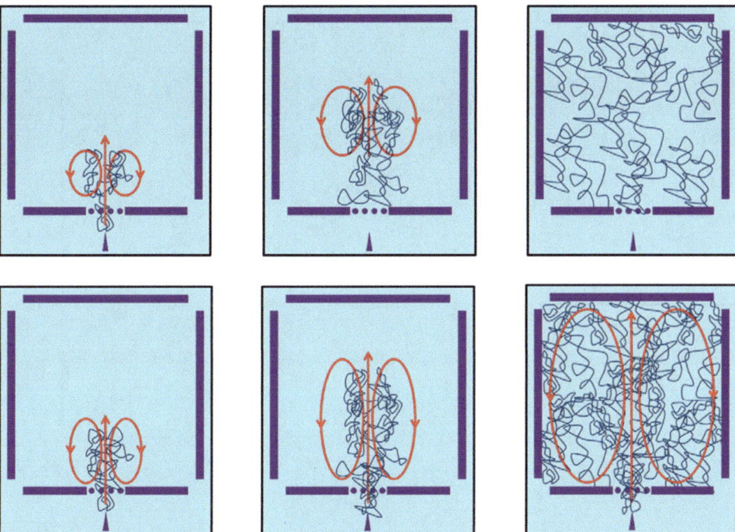

Figure 11.7 Cross section of the experimental cell with the injector tip and grid at the bottom. The top row, left to right, illustrates the development of the vortex tangle (blue) and large scale flow (red) after a brief injection. The bottom row shows the development during continuous injection. Reproduced from Walmsley and Golov (2017), used under CC BY 4.0 (https://creativecommons.org/licenses/by/4.0).

$T = 0.17\,\mathrm{K}$, they created tangles of initial vortex line density about $6 \times 10^3\,\mathrm{cm}^{-2}$ but variable amplitude of fluctuations of flow velocity U_0 at the largest length scale and measured their free decay. If U_0 is small, the excess random component of the vortex line length first decays as $L(t) \propto 1/t$, until it becomes comparable to the structured component responsible for the classical velocity field, and the decay changes to $L(t) \propto 1/t^{3/2}$ decay law. This behavior is nicely illustrated in Fig. 11.8. To quantitatively describe this behavior of this decay of charged, inhomogeneous and anisotropic superflow turbulence, Walmsley and Golov (2017) developed a model of coexisting cascades of quantum and classical energies.

Zmeev *et al.* (2015) compared the decay of turbulence in He II produced by a moving grid to the decay of turbulence created by either impulsive spin-down to rest or by intense ion injection. In all cases, the vortex line density L decayed at late time as $L(t) \propto 1/t^{3/2}$. It is worth mentioning that studies of the towed-grid turbulence in He II at temperatures below 1 K are technically demanding. However, a special towed-grid apparatus has been developed, as shown in Fig. 4.17. The interesting result of this comparison, shown in Fig. 11.9, is that, at temperatures above 0.8 K, all methods yielded the same decay rate. Below 0.8 K, however, the spin-down turbulence maintained the initial rotation and decayed more slowly than grid turbulence and ion-jet turbulence.

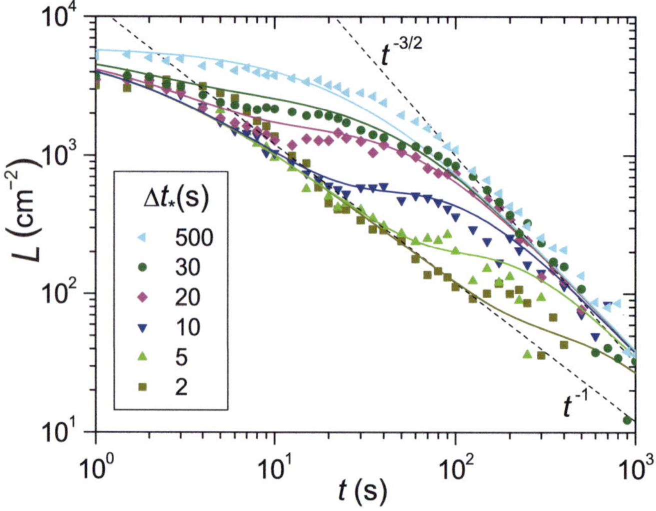

Figure 11.8 $L(t)$ for decaying tangles forced by the same ionic current of 466 pA at 380 V, but for different durations Δt. Dashed lines correspond to the model developed by Walmsley and Golov (2017) – see their Eqs. (4) and (6) – while solid lines are solutions of their Eq. (10) with $L_0 = 6 \times 10^3\,\mathrm{cm}^{-2}$. See text for details. Reproduced from Walmsley and Golov (2017), used under CC BY 4.0 (https://creativecommons.org/licenses/by/4.0).

Zmeev *et al.* (2015) therefore concluded that at $T > 0.8\,\mathrm{K}$ the late-time turbulence is the same whatever the initial flow, i.e., approximately isotropic and homogeneous. But at lower temperatures, the spin-down turbulence at all times was different from that generated by other methods. This might be explained by the observation that the memory of initial rotation is retained during the late-time decay, presumably in the form of a vortex tangle rotating at angular velocity about 0.1 rad/s near the vertical axis of the cell, preserving some of the initial angular momentum. The authors claimed that during the transient that follows the spin-down of a rectangular cell, much of the fluid's initial angular momentum is transferred to the walls through pressure fluctuations from large eddies, eventually creating turbulence with a broad distribution of length scales. At late times, when the remains of that angular momentum supposedly survive only near the axis, the pressure fluctuations at walls (pressure drag) become inefficient, and only the traction at the walls (frictional drag) exerts the torque. If this traction becomes too small for $T < 0.8\,\mathrm{K}$ to reduce the remaining angular momentum within the decay time, there ought to exist a changeover in effective boundary conditions: With falling temperature they will gradually change from the usual no-slip to the slip type, which is believed to describe the flow of the superfluid component free of vortices. The spin-down flow then maintains rotation and slows down the decay of turbulence.

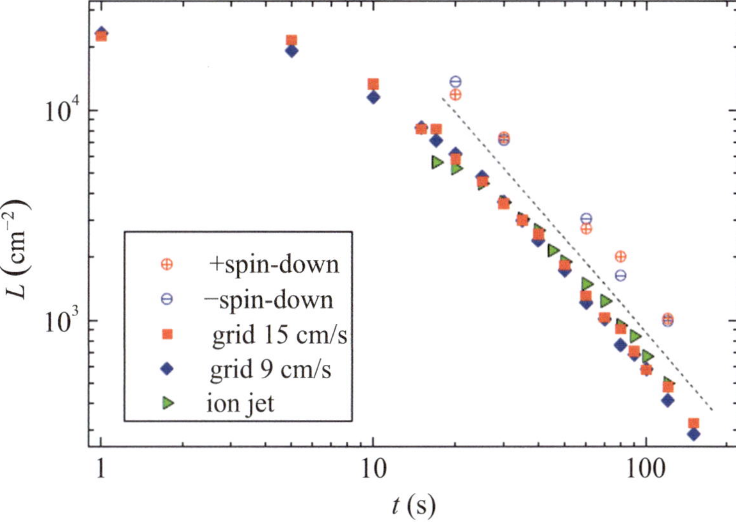

Figure 11.9 Decay of vortex line density $L(t)$ for turbulence generated by different means. "+/− spin-down": spin down from $\Omega = +/-1.5$ rad/s; "grid 15 cm/s (9 cm/s)": turbulence generated by a 3 mm mesh grid, towed with a velocity 15 cm/s (9 cm/s); "ion jet": turbulence generated by injection of negative ions at a current of 700 pA lasting for 100 s. The dashed line shows the $t^{-3/2}$ dependence. $T = 80$ mK. Reproduced from Zmeev *et al.* (2015), used under CC BY 3.0 (https://creativecommons.org/licenses/by/3.0).

11.7.2 Special Features of Decaying Pure Superfluid Turbulence in ^{3}He-B

In Chapter 8 we discussed spin-up of a cylindrical container with ^{3}He-B, leading to the propagating vortex front and peculiar twisted vortex state. Hosio *et al.* (2012) studied the dynamics of quantized vortices in superfluid ^{3}He-B, including the temporal decay of pure superfluid turbulence, after a rapid stop of rotation. The ^{3}He-B sample was contained in a smooth-walled fused quartz cylinder. The bottom part of it, a tube of 30 mm length and 3.6 mm inner diameter, was open to the (rough) heat exchanger, cooled by the nuclear demagnetization stage of the Helsinki rotating cryostat. We discuss here the main results of spin-down experiments of Hosio *et al.* (2012), performed using two slightly different configurations at 29 and 0.5 bar.

To create the vortex array, the cryostat was rotated at constant angular velocity Ω around its axis. The vortex line density in the equilibrium state was determined by minimizing the free energy in the rotating frame and was given by the solid-body rotation value $L = 2\Omega/\kappa$ (see discussion on rotating bucket, Chapter 4). After bringing the container to rest at $t = 0$, the vortex line density was inferred from the fraction of Andreev-reflected thermal excitations, trajectories of which are shown by arrows passing through the orifice, as shown in the inset of Fig. 11.10.

The main panel of Fig. 11.10 shows the key results of this spin-down experiment. There is an initial temperature increase during the deceleration caused mainly by

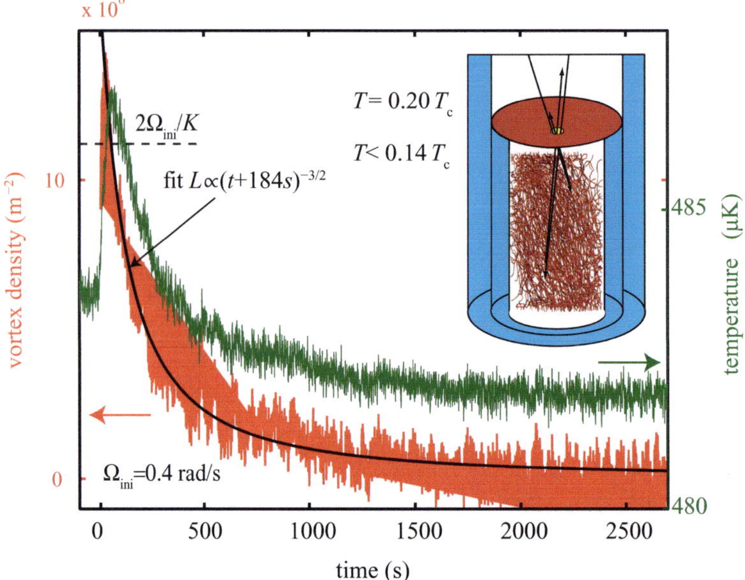

Figure 11.10 Temporal dependence of temperature and vortex line density after bringing the container to rest from $\Omega = 0.5$ rad/s at $t = 0$. The solid line is the best fit to $L(t) \propto (t + t_0)^{-3/2}$ as shown; the vortex line density was inferred from the fraction of Andreev-reflected thermal excitations. The inset shows the principle of the measurement: The vortices below the orifice Andreev reflect some fraction of the thermal excitations back to the bolometer. The fraction depends on the density and configuration of these vortices at the lower temperature $T < 0.14\,T_c$. The measurement is performed at 29 bar liquid ^{3}He pressure. Reprinted figure with permission from Hosio *et al.* (2012). Copyright 2012 by the American Physical Society.

the spin-down-induced turbulence; the temperature then relaxes to its equilibrium value for $\Omega = 0$. The result for $L(t)$ carries two characteristic signatures of turbulent spin-down: (i) an initial overshoot with the maximum at around 30 s, indicating that the kinetic energy of the rotating superfluid is converted to a turbulent tangle, and (ii) subsequent temporal decay of the classical form $L \propto t^{-3/2}$. Although the overshoot is appreciably smaller here than that observed in a cubic cell by Walmsley *et al.* (2007) (see Fig. 11.4) in the turbulent spin-down of He II, it is clear that both forms of the decay of pure superfluid turbulence observed in He II and ^{3}He-B are quasi-classical in character.

We note that, just as Zmeev *et al.* (2015) found in spin-down experiments of He II, Hosio *et al.* (2013) observed that the rotation of superfluid ^{3}He-B decouples from container walls when the temperature is lowered.

In Chapter 8 we also discussed how the Lancaster group generated pure superfluid turbulence in ^{3}He-B by an oscillating grid and investigated its generation and steady

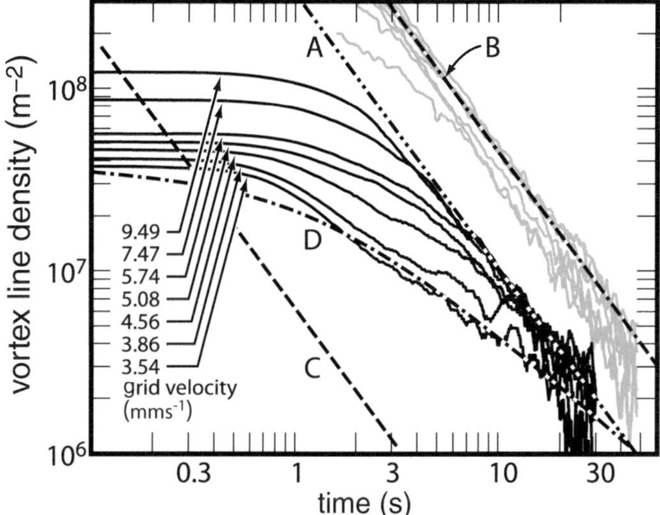

Figure 11.11 (Solid black curves) Inferred vortex line density as a function of time after the cessation of grid motion in ^{3}He-B in the zero-temperature limit for initial grid velocities indicated. Line A is the limiting behavior $L(t) \propto t^{-3/2}$ as discussed in the text. The halftone data are from the Oregon towed grid experiments of Stalp *et al.* (1999) in He II, with line B showing the same power law that is applicable to the late-time limiting behavior, while line C shows this behavior assuming the classical dissipation law in thick normal ^{3}He. Curve D illustrates the behavior for a random tangle in superfluid ^{3}He when the decay originates from the steady-state of lower vortex line density. Reprinted figure with permission from Bradley *et al.* (2006). Copyright 2006 by the American Physical Society.

state, using the Andreev scattering technique available for fermionic superfluids. In addition, Bradley *et al.* (2006) investigated the temporal decay of vortex line density in grid-generated pure superfluid turbulence; we present their representative results together with the Oregon grid data in Fig. 11.11 and point out, again, that the decaying vortex line density displays the quasi-classical character.

This fact is further strengthened in the subsequent Lancaster experiment of Bradley *et al.* (2011a), schematically shown in Fig. 11.12. The energy dissipation was measured directly, using a "black-body radiator." It is a thin-walled box with a small orifice in one side, immersed in superfluid ^{3}He-B, and cooled by the powerful Lancaster-style nuclear refrigerator. Inside the radiator is a thermometer wire resonator, a heater wire resonator to inject power, and a very low-amplitude resonating grid, which generates turbulence by the initial production of microscopic vortex rings, as discussed in Section 6.6.3. The grid resonator consists of a 5 × 5 mm goalpost shaped Ta wire carrying a 5 × 3.5 mm Cu grid mesh of about 200 lines per cm. The experiments were performed in the low-temperature ballistic

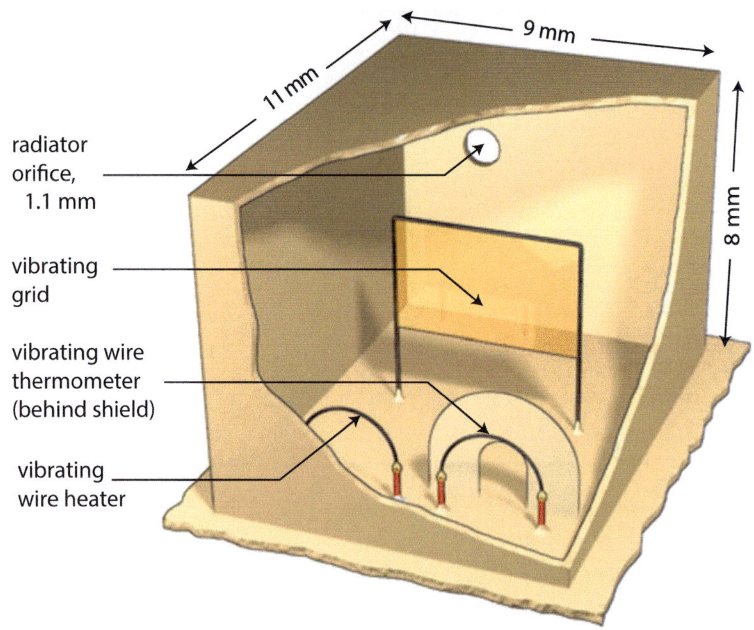

Figure 11.12 The quasiparticle black-body radiator for quasiparticle excitations (rather than conventional ones for photons). Inside the box, one vibrating-wire resonator acts as a thermometer by probing the quasiparticle excitation flux. A second resonator acts as a heater by injecting power directly into the superfluid when accelerated beyond the critical velocity for pair breaking. The grid oscillates at ≈ 1 kHz, but with negligible amplitude (of the order of hundreds of nanometers), to create turbulence. The box acts as a bolometer measuring the energy released by the decaying turbulence inside the enclosure. Reproduced with permission from Bradley *et al.* (2011a).

regime below 200 μK; here the thermal mean free path of quasiparticle excitation greatly exceeds the container dimension. Heat entering the radiator from any source produces ballistic quasiparticles, which thermalize by scattering off the walls, finally emerging as a beam of excitations from the orifice. At steady state, the power emitted in the beam balances the power entering the radiator. The damping on the thermometer wire is dominated by quasiparticle scattering, which provides very sensitive thermometry. The thermal flux of excitations can therefore be inferred from the measured damping on the thermometer wire, and from this quantity the authors obtained the power leaving the black-body radiator. When quantum (pure superfluid) turbulence was present, the excess power leaving the box provided a direct measure of the energy released by the freely decaying tangle.

Bradley *et al.* (2011a) observed that at late times the power leaving the black-body radiator decays as $\epsilon \propto t^{-3}$, so that the turbulent energy decays as $E \propto t^{-2}$, in agreement with the late decay of turbulent energy in classical turbulence in a bounded domain, once the saturation of the energy-containing length scale occurs.

This experiment provides the first direct measurement of the energy released by freely decaying quantum turbulence. It is remarkable that, leaving aside important caveats such as homogeneity and isotropy of the turbulence inside the black box radiator, the decay of pure superfluid turbulence – a tangle of quantized vortex lines – was found to be surprisingly similar to the known decay of classical isotropic and homogeneous turbulence. The results also confirm that the key phenomenological relationship, Eq. (11.10), $\epsilon = -dE/dt = \nu_{\mathrm{eff}}(\kappa L)^2$, introduced earlier in this chapter, is meaningful for quasi-classical or Kolmogorov-like pure superfluid turbulence. However, the situation is more complicated in the two-fluid regime, as we shall discuss in Section 11.8.

11.8 Decay of Counterflow Turbulence in He II

Experiments have suggested that the simple prediction $L(t) \propto 1/t$ does not hold for many cases for long times, as one would have expected for the ultraquantum turbulence from the Vinen equation. The reason is that the decaying counterflow turbulence (assuming that it was generated thermally in a long channel of the characteristic width D) involves additional physical processes, more closely connected to classical hydrodynamic turbulence. This situation is true in the case of a zero-temperature regime and even more so at finite temperature, where one has to take into account possible turbulence in the normal fluid. Still, as we discussed in some detail in Chapter 6, equilibrium vortex line density in the steady state of counterflow turbulence is quantitatively well described by the Vinen Eq. (6.5) (or Eq. (6.7) due to Schwarz). It was therefore a long-standing puzzle as to why the temporal decay of $L(t)$ did not follow the simple prediction $L(t) \propto 1/t$ at late times. Only recently, thanks to improved experimental resolution, has this inverse time decay been observed for low values of L of the order 10^3 cm^{-2}; see Fig. 11.13.

In thermal counterflow of He II, turbulence can also occur in the normal fluid, where it is similar to that in a classical viscous fluid, except for the modification due to mutual friction between the two fluids. We already know that counterflow turbulence in relatively wide channels can exist in two states, T I and T II. Although it has been long expected that in the T I state the normal fluid flow is laminar while it is turbulent in the T II state, experimental confirmation came only recently via visualization techniques based on neutral He_2^* triplet molecules (McKinsey *et al.*, 2005), introduced in Chapter 4. Being effectively a part of the normal fluid with no vortex line trapping above 1 K, these neutral He_2^* triplet molecules are ideal tracers of the motion of the viscous normal fluid, providing information about normal fluid velocity fluctuations, leading to the deduction of structure functions and therefore

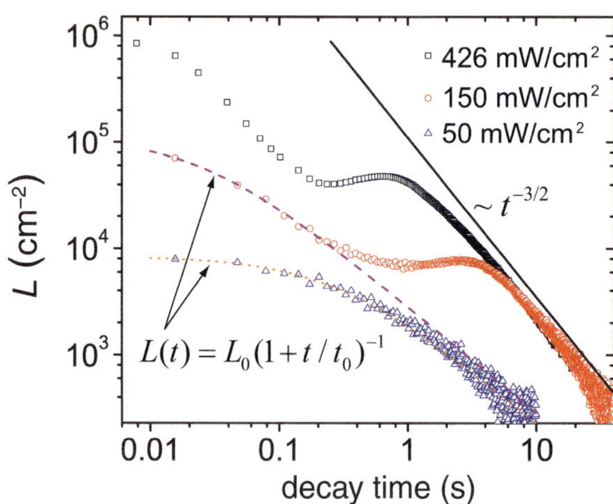

Figure 11.13 Observed decay of the vortex line density for different values of the heat flux in the steady state at 1.65 K; the lines show the two theoretical predictions. Reproduced with permission from Gao *et al.* (2016b).

also the spectra, as was described in Chapter 9; see examples shown in Fig. 11.14 (Gao *et al.*, 2016b).

It is well known in classical turbulence that the energy spectrum rolls off with the exponent $-5/3$, according to K41. Indeed, this is the backbone consideration in the decay of coflowing He II. In the T II state of counterflow turbulence, due to the action of a dissipative mutual friction force acting at all length scales, the turbulent energy of the normal fluid is dissipated at all scales, resulting in a steeper roll-off. The inertial range of scales, strictly speaking, does not exist. The implementation of novel visualization techniques based on neutral He_2^* triplet molecules has shown that the normal fluid motion in the T II state of thermal counterflow is turbulent but has not provided direct information on the turbulent energy spectrum in the superfluid component. We can, however, guess its form to depend additionally on the quantum of circulation (see Section 9.6) to arrive at the spectral density of the general form

$$\Phi(k) = C\epsilon^{2/3}k^{-5/3}f\left(\frac{\epsilon}{k^4\kappa^3}\right). \tag{11.16}$$

Strictly speaking, the arguments leading to Eq. (11.16) are valid if dissipative mechanisms, such as mutual friction or phonon irradiation, are neglected and it is assumed that the turbulent energy is supplied at large scale as in Kolmogorov turbulence. This condition is not satisfied in steady-state counterflow turbulence, where energy is always injected on the quantum scale ℓ, by the mechanism identified by Schwarz (1988). In the T II steady state, energy must be injected also at a large scale D, the size of the channel, since this is the only large scale in the problem.

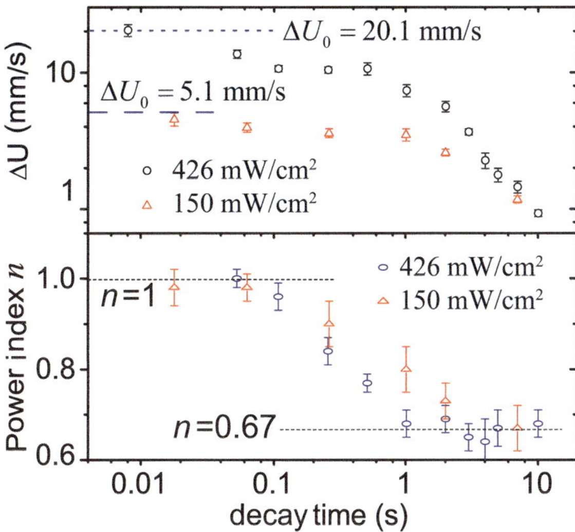

Figure 11.14 Decays in thermal counterflow at $T = 1.65\,\mathrm{K}$ observed by visualization (Gao *et al.*, 2016b). (Upper graph) The temporal decay of the rms velocity fluctuations in the turbulent normal fluid. At late times (i.e., after the saturation of the energy containing length scale by the size of the channel), the observed decay is consistent with the classical decay of the turbulent energy $\propto t^{-2}$. (Lower graph) Time dependence of the index n in the second-order transverse structure function, $S_2(r) \propto r^n$. The corresponding energy spectra have the form $\Phi(k) \propto k^{-\zeta}, -\zeta = n + 1$ (Frisch, 1995). Reproduced with permission from Gao *et al.* (2016b).

Energy injected at the quantum scale ℓ is most probably confined to the superfluid component and is dissipated on a similar scale by mutual friction. It gives rise to a quantum peak in the energy spectrum around $k_Q = (\epsilon/\kappa^3)^{1/4}$. Energy injected at scale D must involve both fluids to a large degree because: (i) visualization technique based on neutral He_2^* triplet molecules has identified a (steeper that K41) pseudo-inertial range in the normal fluid, and (ii) mutual friction coupling the normal and superfluid velocity fields will assure the existence of a similar part of energy spectrum in the superfluid component, with the superimposition of the quantum peak around k_Q.

When switching off the heater, in the absence of a steady counterflow energy injected at the scale D will be divided between the two fluids. Their motion will become coupled on scales significantly larger than ℓ, and the dissipation ceases on those scales. The turbulent energy will cascade down the inertial range; without a mean counterflow, there will hardly be any dissipation due to mutual friction, thus we have a true inertial range. The inertial range is terminated either by viscous dissipation in the normal component in the standard way or by the dominance of mutual friction on scales of the order ℓ, where coupling is no longer possible.

There is a gradual evolution toward a K41 spectrum in a timescale of the order the turnover time of the largest eddies, as Gao *et al.* (2016b) verified by studying the evolution using the differential equation due to Leith (1967) and the Sabra shell model due to Boue *et al.* (2015). The observed evolution of the scaling exponent shown in Fig. 11.14 confirms this expectation. As long as the scaling exponent is steeper than $-5/3$, the flux of energy from large scales to the quantum scales remains small because of the dissipation occurring at large scales due to mutual friction. As the Kolmogorov spectrum becomes established at large scales, the flux of energy into quantum scales rises and causes an increase of vortex line density, needed to accommodate this incoming energy at small scales.

The physical origin of the bump lies in the fact that, after relatively quick decay of the quantum peak that exists in the steady-state counterflow, energy from the decaying large-scale eddies reaches the dissipation scales of order ℓ with some delay – roughly the turnover time of the large-scale eddies, which decreases with increasing steady-state heat flux, as observed.

11.9 Comparison of Decay in Coflow, Pure Superflow, and dc/ac Counterflow in He II

We stressed in Chapter 6 that in the temperature range above about 1 K up to T_λ, where He II displays the two-fluid behavior, several types of turbulent flows of He II can be generated in a pipe or channel. The particular type of flow depends on the mean (net) flow of the normal and superfluid components in steady state relative to each other and/or to the channel walls.

For completeness, we should also mention the possibility of generating "unbounded" quantum turbulence in He II, where interaction with walls is absent or minimal. Such possibilities indeed exist; we already discussed the experiment of Milliken *et al.* (1982) on decaying unbounded inhomogeneous quantum turbulence created by ultrasonic transducers in He II, schematically shown in Fig. 4.11, displaying the $L(t) \propto 1/t$ type of decay, predicted by the Vinen equation.

Another possibility of generating unbounded quantum turbulence is spherical or cylindrical counterflow, as discussed in Section 6.6. Yet another possibility is provided by the Kibble–Zurek mechanism (Zurek, 1985), experimentally verified in superfluid ^{3}He-B by Bauerle *et al.* (1996) and independently by Ruutu *et al.* (1996). This cosmic-ray-like mechanism creates a tangle of vortex lines as a result of point-like heating by cosmic rays and subsequent cooling, quickly crossing the second-order phase transition. Here, the exact form of the decay is unfortunately not known.

We have already discussed experimental results on the temporal decays originating from various types of steady-state two-fluid flows. In addition, Fig. 11.15

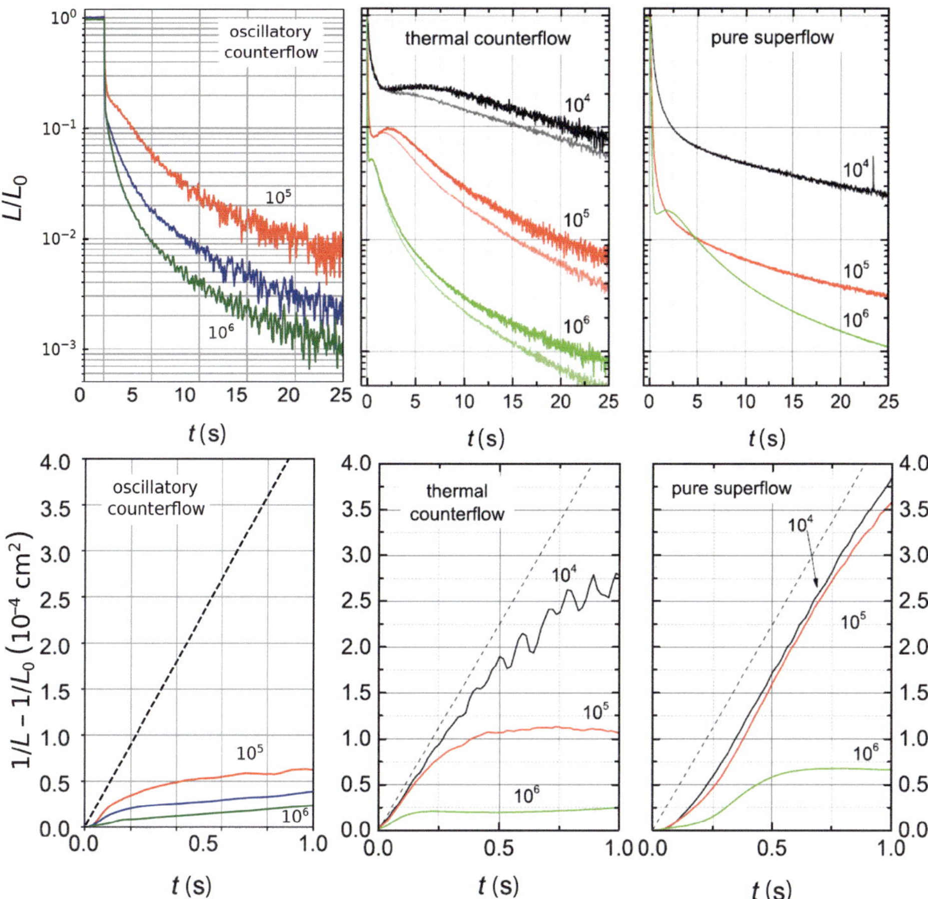

Figure 11.15 (Upper panels) Direct comparison of decays and (Lower panels) comparison of the ranges of validity of the initial $L(t) \sim 1/t$ scaling for (Left) ac counterflow (Midlik, 2019), (Middle) dc counterflow, and (Right) pure superflow (Babuin *et al.*, 2015). All flows were created in the same channel with square $1 \times 1\,\mathrm{cm}^2$ cross section at 1.65 K. The initial steady-state densities for the decays are 10^4, 10^5, 5×10^5, and $10^6\,\mathrm{cm}^{-2}$, as indicated. The vertical axis in the upper panels displays values of vortex line density, L, relative to the initial vortex line density L_0. The black dotted lines in the lower panels indicate the theoretical prediction for the early time decay given by the Vinen equation, Eq. (6.5).

illustrates the differences in the temporal decay of three types of these He II flows: oscillatory ac counterflow, conventional dc thermal counterflow, and pure superflow, studied in Prague in the same channel of the same cross section, $1 \times 1\,\mathrm{cm}^2$. First, no evident bumpy increase of $L(t)$ was revealed for ac counterflow except at the highest temperature of 1.83 K investigated,[1] while it was always seen for dc

[1] Here, we need to bear in mind that not only oscillatory counterflow, but also, as explained above, an

counterflow for the same experimental conditions and for pure superflow with its decay originating from the highest initial vortex line density.

The Prague group (Babuin *et al.*, 2015; Midlik, 2019) also analyzed the time ranges for which the initial $L(t) \propto 1/t$ scaling holds, as shown in the lower panels of Fig. 11.15. It is evident that in the case of ac counterflow the decay according to Vinen's prediction holds for very short times of the order of tens of milliseconds. The overall qualitative behavior of the decay from different initial vortex line densities is however the same – in the sense that the decay form $L(t) \propto 1/t$ occurs for longer times for decays starting from a lower initial vortex line density, followed by a crossover.

In order to understand these complex decay features, we have to discuss the possible forms of both the steady-state and decaying energy spectra. Unfortunately, detailed measurements over the entire form of the energy spectra in highly turbulent superfluid component are not yet available. The best that one can offer is a qualitative arguments leading to the spectrum shown in Fig. 11.16 (Babuin *et al.*, 2016). The suggested form of the spectra reflects complementary experimental, numerical, and theoretical studies of turbulent coflow, counterflow, and pure superflow of He II in the same channel. They result in a physically transparent and relatively simple model of decaying quantum turbulence that accounts for interactions of coexisting quantum and classical components of turbulent He II.

Following Babuin *et al.* (2016), we surmise the shape of the superfluid energy spectrum in coflow (Fig. 11.16(a)) as follows. The mean velocity profiles of the normal and superfluid velocity are practically similar almost everywhere in the channel, except perhaps in the boundary layers on channel walls, because the pressure drop is the same for both fluid components. The relatively large mutual friction tries to lock the mean normal and superfluid velocities. Thus the energy spectra of the normal and superfluid components practically coincide in the entire energy containing and inertial intervals of scales. Ignoring intermittency effects, we have classical K41 energy spectra for both components of He II. At the same time, the experimentally observed steady-state vortex line density is significantly larger than the K41 estimate of the classically generated value. One should therefore find some additional mechanism of vortex generation aside from the classical flow instabilities in the channel flow. Such a mechanism may be provided by the difference between the normal and the superfluid velocities (i.e., counterflow). The normal and superfluid energy spectra depart from each other on scales of the order of the quantum length scale. Moreover, the velocity coupling is also violated in the narrow

unspecified dc counterflow is involved, and may influence the results. Furthermore, the bump observed at 1.83 K behaves similarly to the one typically occurring in dc counterflow, in the sense that it gets broader for smaller initial L values. For more precise characterization of turbulent decay in the pure ac counterflow, one has to devise a dedicated experiment where the dc counterflow is completely suppressed.

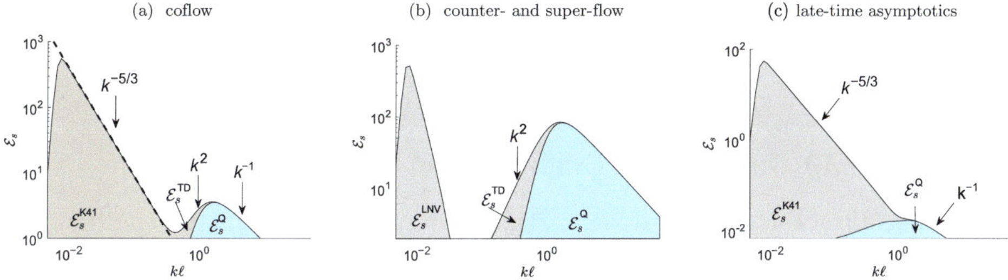

Figure 11.16 Sketch of the steady-state superfluid turbulent energy spectra in log–log coordinates. The spectrum consists of a quasi-classical and a quantum part, colored in gray and cyan, respectively. (a) In the coflow, there is a classical K41 part, denoted as $\varepsilon_s^{K41} \propto k^{-5/3}$, and a thermodynamic equilibrium part $\varepsilon_s^{TD} \propto k^2$. (b) In the counterflow and pure superflow, the quantum contribution and the classical thermal bath part look similar to that in coflow, while the cascade part terminates at some k due to mutual friction and does not provide energy to the quantum vortex tangle in the steady-state regime. After switching off the counterflow or pure superflow, the spectrum shown on the top right of (b) evolves as in (c), switching on the energy flux toward quantum vortex tangle after some delay. Reprinted figure with permission from Babuin *et al.* (2016). Copyright 2016 by the American Physical Society.

region near the walls, where normal and superfluid components satisfy different boundary conditions: $v_n \neq v_s = 0$. In these areas, the velocities of both components differ and excite a random vortex tangle (via the mechanism identified long ago by Schwarz), leading to a peak in the energy spectrum near the crossover scale, shown in Fig. 11.16(a). This peak has large-k asymptote $\propto 1/k$ originating from the velocity field of a single vortex line. The low-k side of it is adjoined by the classical region with the thermodynamic equilibrium spectrum $\propto k^2$, describing equipartition of energy among degrees of freedom (Connaughton & Nazarenko, 2004).

In the steady-state regimes of the dc counterflow and pure superflow, the energy spectrum of the superfluid component has a qualitatively different form, as sketched in Fig. 11.16(b). The physical reason for that is the decoupling of the normal and superfluid turbulent velocity fluctuations, caused by the nonzero counterflow velocity in the dc counterflow and pure superflow.

Babuin *et al.* (2016) then developed a basic and improved model of decaying quantum turbulence in He II (for mathematical details and further references see the original paper). In particular, the authors explained the energy-flux delay in the decaying counterflow turbulence, resulting in the experimentally observed bump in decaying vortex line density. The level of agreement between the experimental observations and the analytical predictions for the time evolution of the vortex-line density in decaying quantum turbulence allows one to conclude that the improved model adequately reflects the underlying physical processes responsible for the decay of quantum turbulence, originating from various types of steady superfluid He II flows.

11.10 Diffusion and Evaporation of Quantized Vorticity

Following Skrbek and Sreenivasan (2012), we now discuss another problem of turbulence decay that applies to many experimental situations in which the turbulent vortex tangle does not fill the experimental cell, but is initially concentrated in a small part of a larger volume. The typical case is quantum turbulence generated by an oscillating structure. If the forcing and oscillations of the structure cease to operate, we may expect that the quantum turbulence will spread out and decay in analogy to classical fluid dynamics. In a classical incompressible viscous fluid, in fact, this process is governed by the vorticity transport equation. Taking the curl of Eq. (5.1), we obtain

$$\frac{D\boldsymbol{\omega}}{Dt} = \nu \nabla^2 \boldsymbol{\omega}. \tag{11.17}$$

On the other hand, in pure ($T = 0$) quantum turbulence the kinematic viscosity ν is zero. What happens then?

Early studies of the problem by Barenghi and Samuels (2002) showed that, in superfluid helium, the quantization of the circulation introduces a new decay mechanism, which was named *evaporation of quantized vorticity*.[2] The physical principle is that, within a vortex tangle, reconnections that will result in some loops escaping at a rate that depends on the average intervortex distance ℓ. Consider the situation when the vortex tangle (for simplicity at $T = 0$) exists only in a localized region of space, and the liquid helium is vortex-free outside this region. If a vortex loop, smaller than ℓ, is located near the surface of the tangle, and points in the outward direction, it will escape from the tangle, carrying energy and momentum away to infinity (or to the walls of the container). The average intervortex distance within the tangle will become slightly larger, increasing the probability of another vortex loop escaping. Slowly but surely, the vortex tangle will disintegrate in a gas of evaporating vortex loops. We can describe this effect as a peculiar form of diffusion in an inviscid fluid. Numerical simulations confirm this scenario: Fig. 11.17 shows small vortex loops evaporating from a vortex tangle at $T = 0$ computed using the vortex filament method. The evaporation of quantized vorticity occurs also in 2D, taking the form of vortex-antivortex pairs evaporating from a turbulent region consisting of random positive and negative quantized vortices, and has been described using classical point vortices as well as the Gross–Pitaevskii equation (Rickinson *et al.*, 2018).

[2] Not to be confused with *quantum evaporation*, the process in which an excitation is annihilated at the free surface of helium by ejecting an atom (Wyatt, 1991).

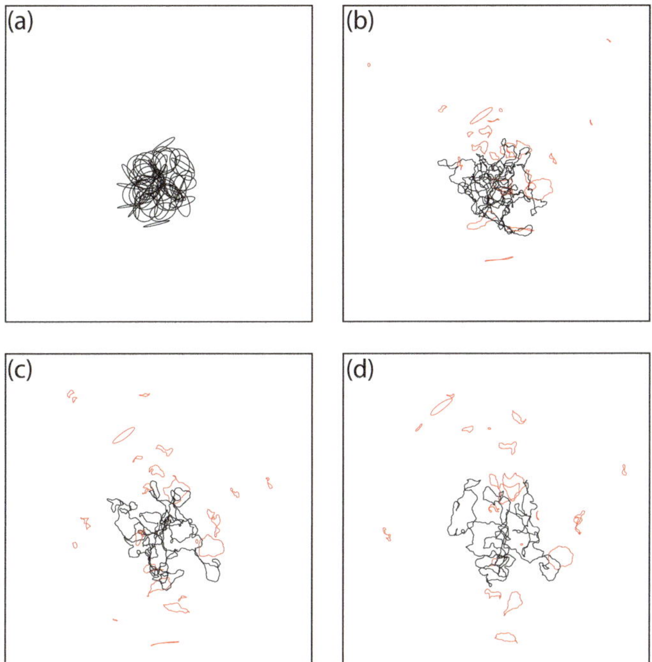

Figure 11.17 Evaporation of a small vortex tangle at $T = 0$ computed using the vortex filament method. Snapshot of the vortex configurations at times (a) $t = 0\,\mathrm{s}$, (b) $t = 100\,\mathrm{s}$, (c) $t = 200\,\mathrm{s}$, and (d) $t = 300\,\mathrm{s}$. All panels show the region $-2.5\,\mathrm{cm} \leq x, y \leq 2.5\,\mathrm{cm}$ projected onto the $z = 0$ plane. The computational domain is infinite (no boundaries). Relatively smaller vortex loops containing fewer than 200 points are shown in red, with the remaining vortices shown in black. Reprinted figure with permission from Rickinson *et al.* (2019). Copyright 2019 by the American Physical Society.

If one looks carefully at Fig. 11.17, one notices that the evaporating vortex loops leave behind a larger, less dense vortex tangle. The question is whether this spatial decay or diffusion of the main vortex tangle is similar to the classical diffusion described by Eq. (11.17).

Tsubota *et al.* (2003b) numerically generated a dense tangle of quantized vortex in a cubic domain by applying a thermal counterflow of normal and superfluid component, removed one half of the vortex tangle from the right-hand side of the experimental cube, set the temperature to zero (i.e., removed the normal fluid and the counterflow that was used to create the tangle), and computed the further evolution of the vortex tangle in time and space. They found that the quantized vorticity expands into the empty space according to the 1D modified Vinen equation

$$\frac{dL}{dt} = -\chi_2 \frac{\kappa}{2\pi} L^2 + D_{\mathrm{d}} \nabla^2 L, \tag{11.18}$$

where χ_2 is Vinen's decay parameter and D_d is an effective diffusion coefficient. We know that for $D_\mathrm{d} = 0$ this equation predicts the vortex line density decay to be $L(t) \propto 1/t$ for large t. The presence of an added diffusion term in Eq. (11.18) generalizes the temporal decay by taking into account the inhomogeneity of the initial vortex tangle. One can argue that the only relevant length scale is the quantum length scale ℓ and the only relevant velocity scale is of order of κ/ℓ, hence D_d must be of the order of κ. By numerically fitting the temporal and spatial evolution of the vortex tangle, Tsubota *et al.* (2003b) found $D_\mathrm{d}/\kappa \approx 0.1$. However, their procedure depends on the parameter χ_2, which must be determined independently, and does not distinguish the evaporating vortex loops (which move ballistically and independently) from the vortex tangle left behind (which obeys different physics, consisting of interacting vortices).

Rickinson *et al.* (2019) considered the problem starting from random vortex loops as in Fig. 11.17(a). They removed the evaporating vortex loops from the analysis, and determined D_d using two procedures. The first was the same procedure as Tsubota *et al.* (2003b), and consisted of fitting in space and time the solution to Eq. (11.18). The second procedure considered the deviation in the trajectories of diffusing fluid tracers (Sikora *et al.*, 2017), and consisted of computing the rms deviation $d_\mathrm{rms}(t)$ of N points along the vortex lines from their initial positions:

$$d_\mathrm{rms}(t) = \sqrt{\frac{1}{N(t)} \sum_{i=1}^{N(t)} \left(\Delta x_i^2(t) + \Delta y_i^2(t) + \Delta z_i^2(t)\right)}, \tag{11.19}$$

where $\Delta x_i(t) = x_i(t) - x_i(0)$ and similarly for $\Delta y_i(t)$ and $\Delta z_i(t)$; note that the number of points, N, depends on time, because points belonging to evaporating loops are not used in this analysis. The diffusion constant is then determined by

$$D_\mathrm{d} = \frac{d_\mathrm{rms}^2}{4t}. \tag{11.20}$$

These authors found that the first procedure yields values of D_d/κ that depend on the initial vortex line density, giving $D_\mathrm{d}/\kappa \approx 0.3$; if the evaporating loops are not removed from the analysis, $D_\mathrm{d}/\kappa \approx 0.1$ to 0.2, in agreement with Tsubota *et al.* (2003b). Using the second procedure based on the rms deviations, they found a more consistent value $D_\mathrm{d}/\kappa \approx 0.5$ independently of the initial condition, more in agreement with the order of magnitude estimate $D_\mathrm{D}/\kappa \approx 1$, and with values in the range from $D_\mathrm{d}/\kappa \approx 0.4$ to 0.5 obtained in 2D (Rickinson *et al.*, 2018).

It is interesting to estimate the possible role of the quantum evaporation mechanism in the observed decay in the zero-temperature limit of the Lancaster experiment by Bradley *et al.* (2006), in which the tangle was created by an oscillating grid in ^{3}He-B. It seems to be a conservative estimate to assume that the vortex line density in the inhomogeneous case does not vanish faster than that over the energy containing length scale; thus the diffusive term is of order $\kappa\nabla^2 L \approx \kappa L/\ell^2$. The observed classical decay of the form $L(t) \propto t^{-3/2}$ indicates that the energy-containing length scale is saturated and equals the distance of the sensor (vibrating wire) from the grid, i.e., of the order of 1 mm. This estimate may not seem to be correct if we are unaware of the Kolmogorov type of quantum turbulence, but it seems to provide a plausible estimate for the gradient based on the distance between the sensors or their distance from the grid, or the size of the grid, whichever is smaller. For the initial vortex line density of order 10^8 m^{-2} (see Fig. 11.11), the contribution of the evaporative decay mechanism thus ought to be small.

Nevertheless, Nemirovskii (2010) attempted to apply evaporating decay mechanism to the Lancaster experiments of Bradley *et al.* (2006). Although the simulated decay curves, shown in Fig. 7 of Nemirovskii (2010), look similar to the experimental data shown in Fig. 19, they lack the pronounced feature of the experimental decay data curves merging subsequently at the third universal classical regime of the form $L(t) \propto t^{-3/2}$. Even though this observation seems to justify the approach described in this chapter, based on the decay of homogeneous and isotropic turbulence, we conclude that the evaporative decay mechanism is interesting in its own right and should remain a subject of further research.

11.11 Summary

As in the case of classical turbulence, decay of quantum turbulence in the absence of sustained production is one of the most important and extensively explored problems in quantum turbulence. In comparison with classical turbulence, the problem here is more complex for the following reasons.

Intellectually simplest but experimentally the most difficult to investigate is the decay of pure superfluid turbulence in the zero-temperature limit. Still, there are two types of 3D turbulence here, the ultraquantum or Vinen type that consists of a tangle of apparently random vortex lines, and the quasi-classical or Kolmogorov turbulence where partial polarization or vortex lines organized in bundles results in the existence of large eddies at scales exceeding the quantum length scale ℓ. These vortex structures behave similarly to classical eddies, undergoing advection and stretching as in the Richardson cascade. In the pure superfluid, however, there is no viscosity and therefore no mechanism for the classical viscous decay. The Richardson cascade continues in the form of a Kelvin wave cascade along vortex lines until dissipation via phonon irradiation (He II) or heating the Caroli–Matricon states in vortex cores of ^{3}He-B. After thermalization, the final destiny of the turbulent energy is the creation of the normal component via heating.

At nonzero temperatures, where quantum fluids display the two-fluid behavior, both the superfluid component and the normal component might be turbulent (as in He II), or they affect each other by the action of the mutual friction force, attempting to couple the normal and superfluid velocity fields at all length scales larger than the quantum length scale ℓ. The decaying velocity fields then affect each other and the dissipative mechanism includes mutual friction and viscosity of the normal fluid.

In the case of unbounded or inhomogeneous quantum turbulence, an alternative dissipation mechanism is quantum evaporation: During reconnections in the tangle, a gas of small vortex loops is created and those of suitable orientation fly away from the tangle carrying energy and momentum with them, before they annihilate at the container walls.

Despite these complex features, there are many turbulent quantum flows (both at zero and nonzero temperatures) displaying quasi-classical decay with very similar features as decaying classical turbulence. The universal decay regime of Kolmogorov quantum turbulence then allows for the definition of effective kinematic viscosity ν_{eff}, deduced from the observed decays of turbulent energy $E(t) \propto t^{-2}$ and vortex line density $L(t) \propto t^{-3/2}$. Decaying vortex line density in Vinen's type of quantum turbulence of the form $L(t) \propto t^{-1}$ allows for an alternative definition of effective kinematic viscosity, ν'. In He II, both ν_{eff} and ν' are closely similar in magnitude, and both of them gradually decrease with dropping temperature, being of order $0.5\,\kappa$ close to T_λ and $0.1\,\kappa$ at very low temperature.

The exact form of the temporal decay of quantum turbulence depends on various factors such as the fermionic or bosonic nature, temperature, boundary conditions (open or bound domain), and the mean flow of both components of quantum fluid. Many open questions still await further scientific enquiry; some of them have been highlighted in this chapter.

12
Regimes of Quantum Turbulence

In previous chapters we have described a variety of flows that exhibit quantum turbulence, its generation, measurement, and decay. The aim of this chapter is to make a synthesis of the resulting understanding. In particular, we wish to address the following natural questions. Can different regimes of quantum turbulence be identified without ambiguity? If so, how do these regimes compare to classical turbulence? What is their adequate phenomenological description on the basis of the development of the last several chapters? These questions are easier to answer in the zero-temperature limit where we have the luxury of dealing with only pure superfluid turbulence. In a sense, this chapter pulls together much of our understanding already presented.

12.1 Quantum Turbulence at Zero Temperature

We stated in earlier chapters that, in the zero-temperature limit, there are two limiting types of quantum turbulence: (i) ultraquantum (or Vinen-type) turbulence and (ii) quasi-classical (or Kolmogorov-type) turbulence. As a summary, Vinen turbulence consists of a tangle of apparently random vortex lines and its decay is characterized by an inverse time law for the vortex line density of the form $L(t) \propto 1/t$ at late times, as predicted by the Vinen equation. In Kolmogorov turbulence, partial polarization (vortex lines organized in bundles) results in the existence of large eddies at scales exceeding the quantum length scale. These large eddies behave as classical eddies, undergoing advection and stretching, resulting in the Richardson cascade. In a pure superfluid, however, there is no viscosity and therefore the classical viscous decay mechanism is absent. The Richardson cascade of eddies therefore extends to larger wavenumbers in the form of a Kelvin wave cascade along individual vortex lines, until dissipation occurs, either via phonon radiation (in He II) or via heating the Caroli–Matricon states in the vortex cores (in ^{3}He-B). If the energy input is halted, Kolmogorov turbulence in bounded domains is characterized by the vortex line density decaying as $L(t) \propto 1/t^{3/2}$ for late times (after saturation of the energy-containing length scale). Effective values of the kinematic viscosity, determined in both Kolmogorov and Vinen regimes, are of the order of 0.1κ to 0.5κ in He II.

We stress that the two distinct regimes of the Kolmogorov and Vinen types are limiting cases: Although there are turbulent flows that clearly belong to one or the other type, there appear to be intermediate cases, either because they are truly so, or because the available incomplete information about them suggests so. For example, the 3D numerical simulations are limited in k-space (typically three orders of magnitude that, for statistically steady state, must include forcing and dissipation). They also do not cover very long times (compared to the turnover time of the large eddies) because of the total CPU time involved. Similarly, experimental measurements may depend on the probes themselves. In both experiments and simulations, it is often the case that not all relevant quantities can be measured or computed, and researchers have to make sense of the flow of interest despite some missing pieces of the puzzle.

We now present a simple phenomenological description of these two regimes of pure superfluid turbulence and a comparison with ordinary turbulence in classical viscous fluids. For the sake of simplicity, we restrict our analysis to homogeneous isotropic turbulence (HIT) and neglect intermittency properties. This neglect is justified as long as our interest is in low-order statistical properties.

12.1.1 Phenomenology of Pure Superfluid Turbulence

A phenomenological description of classical turbulence requires physical explanations of a few length scales. The largest scale, the outer scale D, is the characteristic size of the turbulent region, pipe, or channel that limits the size of turbulent eddies. There are also other related scales such as the correlation length scale (also called the integral scale), the scale around which most of the energy resides (called the energy containing scale), the Taylor microscale, etc. The scaling relations among these scales are universal but the precise numerical factors depend on the particular flow. We shall not be concerned with all of them here but, for classical HIT, we assume that the turbulent energy is supplied at some characteristic scale $M \leq D$, which can be the mesh size of a grid. Here, the turbulent energy, E, is supplied at the rate $\epsilon = -dE/dt$. In the steady state, the supplied energy is transferred by the Richardson cascade through the inertial range of scales and is dissipated essentially at the Kolmogorov scale $\eta = \left(\nu^3/\epsilon\right)^{1/4}$, where ν is the kinematic viscosity; the corresponding Kolmogorov wavenumber is $k_\eta = 2\pi/\eta$. No scales smaller than η are thought to be relevant to turbulent motion – though intermittency modifies this result (Yakhot and Sreenivasan, 2004). Using ν and ϵ, we can define a velocity scale $u = (\nu\epsilon)^{1/4}$. Notice that the Reynolds number based on u and η is unity: $Re_\eta = u\eta/\nu \equiv 1$. At the small length scale η, viscous forces become as large as inertial forces, thus damping the energy cascade.

In pure superfluid turbulence there is no viscosity, therefore the dissipative Kolmogorov scale η does not exist. In principle, turbulent flow exists down to the smallest length scale, which is the size of the vortex core $\xi \approx 0.1$ nm in He II and ≈ 10–60 nm in ^{3}He-B. One can, however, define a superfluid Reynolds number by replacing the kinematic viscosity ν with the quantum of circulation κ, and ask at what scale ℓ this newly defined Reynolds number becomes unity: $Re_\kappa = u\ell/\kappa \equiv 1$. We remark that this definition of superfluid Reynolds number does not arise formally from the governing equations of motion (which we do not know; see Chapter 5), but guides our argument physically. Following Skrbek *et al.* (2021), by analogy with classical turbulence we postulate that this quantum length scale is $\ell = (\kappa^3/\epsilon)^{1/4}$, and define the corresponding wavenumber $k_Q = 2\pi/\ell$. The physical meaning of ℓ is that it distinguishes the quasi-classical scales for which quantization of circulation does not play any significant role (i.e., the "granularity" of quantum turbulence does not matter) from the purely quantum scales, for which quantum restrictions are essential. The very existence of ℓ is the quantum effect that distinguishes quantum turbulence from classical turbulence; in fact, in the quasi-classical limit of vanishing Planck constant we recover $\ell \to 0$. The Kolmogorov cascade cannot proceed beyond k_Q simply because individual vortex lines are quantized and thus cannot be stretched. There is no dissipation at these scales (except due to reconnections as discussed in earlier chapters) and the energy cascade continues transferring the energy down the scales, mediated by the Kelvin waves cascade on individual vortex lines. Dissipation occurs at much higher $k \approx k^*$ thanks to phonon emission by Kelvin waves in He II or by exciting Caroli–Matricon states in vortex cores in ^{3}He-B, as described in earlier chapters, smoothing the vortex lines.

Let us return to the evolution of the steady-state energy spectrum for a viscous fluid in a box of size D, assuming an energy input at the scale M (e.g., the mesh size M). If the energy input at the scale M is too small so that the Kolmogorov dissipation length $\eta = \left(\nu^3/\epsilon\right)^{1/4}$ is larger than or equal to M itself, no turbulence is possible and the flow is laminar. Upon increasing the driving, which manifests in increased ϵ, η becomes smaller and gradually a steady-state energy spectrum develops between M and η, ultimately developing an inertial range of scales. The small-k part of the spectrum most likely acquires a k^2-slope, in agreement with equipartition or the existence of the Birkhoff–Saffman invariant. The spectral energy density peaks around M (the energy-containing length scale), and the temporal decay of the grid turbulence is described by the classical spectral decay model discussed in detail in Chapter 11.

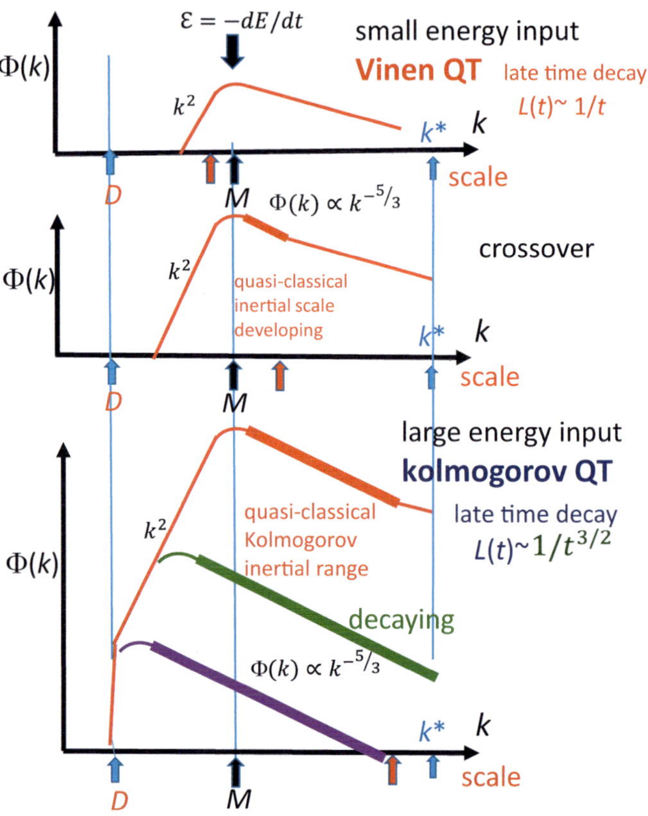

Figure 12.1 Schematic view of general shapes of the three-dimensional energy spectra $\Phi(k)$ plotted vs. k of pure superfluid turbulence, assuming that the turbulence is forced at a scale M (QT: quantum turbulence). (Top) Vinen turbulence, where quantum length scale ℓ (red arrow) is larger than M. (Middle) Crossover from Vinen to Kolmogorov turbulence. With increasing ϵ, ℓ crosses M. Energy cascade akin to the classical Richardson cascade starts to operate and an inertial range of scales (highlighted) gradually develops. (Bottom) For large enough ϵ, an inertial range of scales, similar to that in classical 3D HIT, is developed in steady state (red). When forcing stops, the decaying spectrum is shown for two subsequent times in green and violet. The energy-containing length scale grows, eventually becomes saturated by D, and classical-like decay of the form $L(t) \propto 1/t^{3/2}$ follows. For further details, see text.

Now we examine the same problem for a pure superfluid in the same box of size D with the same energy input at the scale M, with the constraint that $2\pi/k^* \ll M \ll D$; see Fig. 12.1. The motion of the grid will be dissipationless until, upon exceeding a critical velocity, vortices are nucleated, interact, and a vortex tangle forms. Over some transient time, a steady-state energy spectrum develops, the energy input ϵ cascading down the scales (via Kelvin wave cascade along individual vortex lines) being dissipated around k^* by the mechanisms described above. The actual form of the energy spectrum and the temporal decay that follows a sudden halt to the energy input depends on the interplay of two important scales, M and ℓ. For small enough ϵ, $\ell > M$, Kolmogorov cascade cannot operate and energy is transferred down the scales via a Kelvin waves cascade and dissipated at scales of order k^*. Due to the equipartition theorem, on the left side of M the spectrum acquires the k^2 form (see Fig. 12.1, top panel) and decays as k^{-1} for larger wavenumbers. This is Vinen (ultra-quantum) turbulence, where the vortex tangle is essentially random. The temporal decay of vortex line density in Vinen turbulence is (at late times) inversely proportional to t, as we discussed in detail in Chapter 11.

Let us now increase ϵ so that the quantum length scale ℓ crosses M and becomes appreciably smaller. On the left side of M the spectrum will still acquire the k^2 form, but the situation on the right-hand side changes: Between M and ℓ the inertial range of scale develops (see Fig. 12.1, middle panel), over which the energy becomes transferred quasi-classically by a Richardson cascade. With increasing ϵ, the quasi-classical inertial range of scales gradually increases and contains most of the energy content of the superflow. This is the Kolmogorov (quasi-classical or structured) turbulence, containing large vortex structures that can be thought of as composed of bundles of vortex lines, achieved by partial polarization of the tangle. The temporal decay of vortex line density in Kolmogorov turbulence follows the prediction of the spectral decay model of decaying classical HIT. It is applicable because the quasi-classical relation $\epsilon = \nu_{\text{eff}}(\kappa L)^2$ holds.

12.2 Kolmogorov and Vinen Turbulence Compared

Let us reemphasize that both Vinen-type and Kolmogorov-type regimes of quantum turbulence exist in pure forms for superfluid turbulence in the zero-temperature limit. Their existence is a direct consequence of classical vortex dynamics and the quantization of circulation in the superfluid. The normal fluid makes the issue more complex. We also emphasize that both regimes have been identified as transients in decaying turbulence and as statistically steady states in driven turbulence. In this section we compare the properties of Vinen and Kolmogorov turbulence. Additionally, we shall show that Vinen turbulence has been seen in other quantum fluids besides He II and ^{3}He-B.

The prototype finite-temperature steady-state Vinen turbulence is a vortex tangle driven by a uniform normal fluid – the subject of the pioneering numerical studies of Schwarz (1988) for modeling the counterflow experiments of Vinen. Figure 12.2 shows the resulting energy spectrum $\Phi(k)$. Note the lack of $k^{-5/3}$ Kolmogorov scaling, particularly the lack of energy in the large eddies. The spectral energy density $\Phi(k)$ has a broad peak at the mesoscales ($k \approx 300\,\text{cm}^{-1}$) not far from the wavenumber corresponding to the intervortex spacing $k_\ell = 2\pi/\ell \approx 890\,\text{cm}^{-1}$. At

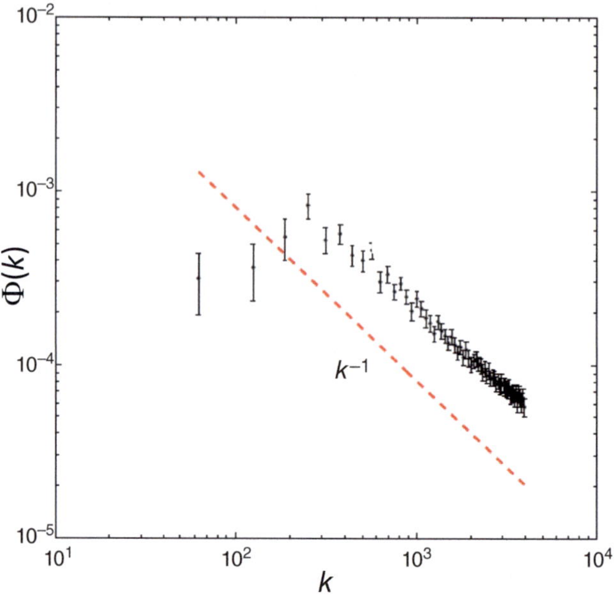

Figure 12.2 Energy spectrum $\Phi(k)$ of Vinen turbulence plotted vs. wavenumber k. The turbulence is generated numerically using the vortex filament method in a periodic domain at $T = 1.9\,\text{K}$ by imposing a uniform normal fluid at velocity $u_n = 0.75\,\text{cm/s}$; the resulting vortex line density is $L \approx 2 \times 10^4\,\text{cm}^{-2}$. The dashed line has a slope of -1. Reprinted figure with permission from Sherwin-Robson *et al.* (2015). Copyright 2015 by the American Physical Society.

larger k, $\Phi(k) \sim k^{-1}$. Since this is also the spectrum of a straight isolated vortex line, the observation of $\Phi(k) \sim k^{-1}$ in a vortex tangle suggests that the contributions of far-away randomly oriented vortex lines is weak, as verified in the simulations (Sherwin-Robson *et al.*, 2015). Figure 12.2 should be compared to the Kolmogorov spectrum shown in Fig. 9.5 that arises in a finite-temperature steady-state regime if the normal fluid is not uniform but (synthetic) turbulent. The key difference is the injection of energy, which occurs at large length scales for Kolmogorov turbulence and at the mesoscales (via the Donnelly–Glaberson instability) for Vinen turbulence. In physical space, the difference between Vinen and Kolmogorov turbulence becomes apparent if one computes the coarse-grained superfluid vorticity field ω_s. Figure 12.3 shows that in Vinen turbulence ω_s is weak and featureless, whereas in Kolmogorov turbulence there are regions of strong vorticity. Another difference between Vinen and Kolmogorov turbulence, already apparent in Fig. 12.3, is the statistical distribution of curvature along the vortex lines, shown in Fig. 12.4: Vinen turbulence contains more closed vortex loops and larger curvatures (defined as $C = |s''|$); that is, smaller radii of curvature $R = 1/C$; Kolmogorov turbulence contains a larger number of long vortex lines that extend across the periodic domain

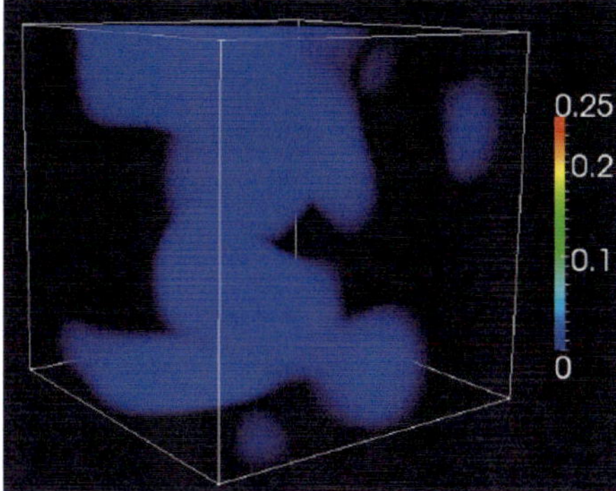

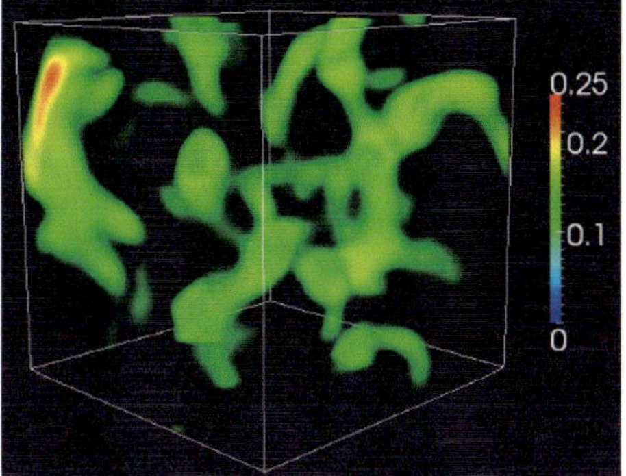

Figure 12.3 Coarse-grained superfluid vorticity for (Left) Vinen turbulence and (Right) Kolmogorov turbulence at the same temperature ($T = 1.9\,\mathrm{K}$) and vortex line density plotted on the same scale. Note that the former is featureless, whereas the latter contains intense vortical regions. Reprinted figure with permission from Baggaley *et al.* (2012d). Copyright 2012 by the American Physical Society.

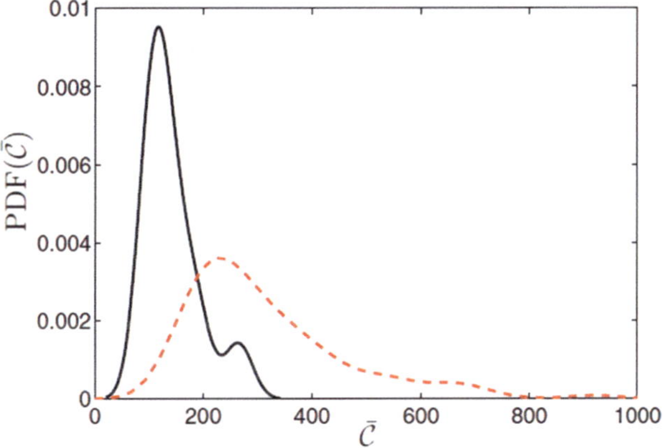

Figure 12.4 Probability density function (PDF) of the mean curvature per vortex loop for Kolmogorov turbulence (solid black line) and Vinen turbulence (dashed red line). Reprinted figure with permission from Baggaley *et al.* (2012d). Copyright 2012 by the American Physical Society.

with smaller curvatures (larger radii of curvature). This visually inspired remark is consistent with the energy spectrum. It is well known (Leonard, 1985) that the energy spectrum of a single vortex ring of radius R rises as k^2, peaks at $kR \approx 1$, and, at larger k, decays as k^{-1} with characteristic small-amplitude oscillations. Yurkina and Nemirovskii (2021) found that a gas of random vortex rings has a similar spectrum, the precise crossover between k^2 and k^{-1} depending on the distribution of R. An example is shown in Fig. 12.5. Since this spectrum is similar to Fig. 12.2, we can qualitatively think of Vinen turbulence as consisting of random vortex loops.

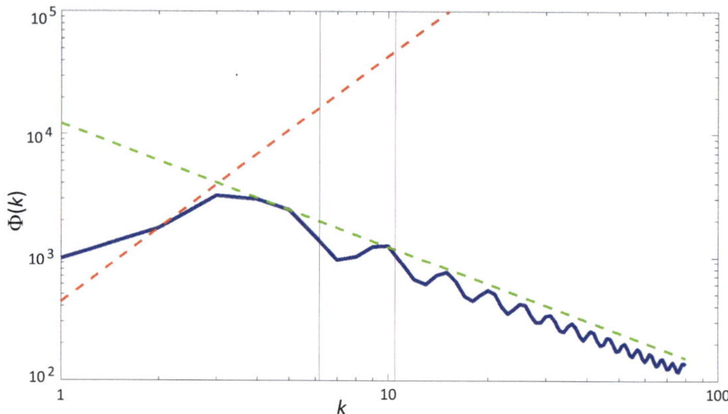

Figure 12.5 Energy spectrum of a gas of random vortex rings of radius R. The k^2 and k^{-1} scalings are shown as dashed lines. The vertical lines correspond (left to right) to $k_\ell = 2\pi/\ell$ and $k_R = 2\pi/R$, respectively. If the value of R is decreased, the spectrum takes a better defined k^2 form at small wavenumbers. Courtesy of L. Galantucci.

12.2.1 A Note on Circumstances in Which Vinen Turbulence Appears

Vinen turbulence appears in other quantum fluids beyond the specific contexts of He II and ^{3}He-B. The first example is the 3D atomic Bose–Einstein condensates. The relatively small size of these systems at the present state of research (the typical condensate is only 10 and 100 times larger than the vortex core size) prevents the formation of an inertial range in k-space, and hence Kolmogorov turbulence. A numerical study by Cidrim *et al.* (2017) has found that 3D turbulence in a typical atomic condensate, despite the very limited range of length scales and the small length of vortex lines compared to helium experiments, shows the characteristic features of Vinen turbulence: the lack of energy at the largest length scales, the energy peak at scales of the order of ℓ (rather than the pile-up at the lowest wave numbers typical of Kolmogorov scaling), the k^{-1} decay at larger k, and the typical $L \sim t^{-1}$ temporal decay. The interpretation of Vinen turbulence as an almost random distribution of vortex lines (in the sense that the interscale energy transfer is weak) is supported by the analysis of (normalized) velocity correlation function, defined as

$$f_i(r) = \frac{\langle u_i(\boldsymbol{x} + r\hat{\mathbf{e}}_i)u_i(\boldsymbol{x})\rangle}{\langle u_i(\boldsymbol{x})^2\rangle}, \qquad i = x, y, z, \tag{12.1}$$

where the symbol $\langle\cdots\rangle$ denotes the average over the position vector $\boldsymbol{x} = (x, y, z)$ and $\hat{\mathbf{e}}_i$ is the unit vector in the Cartesian direction $i = x, y, z$. Cidrim *et al.* (2017) found that this correlation function drops to $\approx 10\%$ at distances of the order of the average vortex distance, confirming the random nature of this flow.

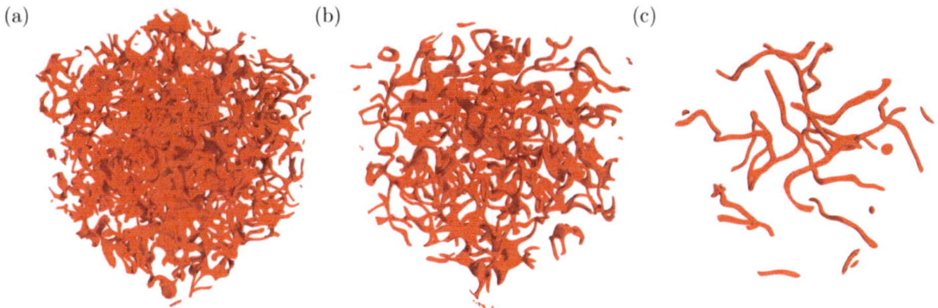

Figure 12.6 Thermal quench of a Bose gas: At the initial time $t = 0$ all states are equally occupied and the phase is random. During the evolution, the low-k states become macroscopically occupied, forming the condensate, and the high-k states, occupying less space, forming the thermal part. The initially random phase field forms an intense turbulent tangle of vortex lines that decays. Reproduced from Stagg *et al.* (2016), used under CC BY 3.0 (https://creativecommons.org/licenses/by/3.0).

The second example of Vinen turbulence is the thermal quench of a Bose gas. This is the fundamental process through which a coherent superfluid is formed from a non-equilibrium initial condition (Kagan and Svistunov, 1994). During the quench, phase defects in the superfluid component become vortex lines and form a turbulent tangle. Over time, this state of turbulence decays to a vortex-free equilibrium in which the condensate coexists with a finite-temperature gas. The gas is modeled as a classical condensed-matter field $\psi(\boldsymbol{r}, t)$ described by the Gross–Pitaevskii equation (GPE). The initial condition $\psi(\boldsymbol{r}, 0) = \sum_{\boldsymbol{k}} a_{\boldsymbol{k}} e^{i\boldsymbol{k}\cdot\boldsymbol{r}}$ represents uniform occupation numbers and random phase. Berloff and Svistunov (2002) found that the interactions lead to macroscopic occupation of the low-k modes (characteristic of the condensate) whereas the high-k modes have low occupation (and represent thermal excitations). The very intense vortex tangle that emerges and then decays (see Fig. 12.6) was studied by Stagg *et al.* (2016), who reported the absence of Kolmogorov scaling (the incompressible energy spectrum peaks at $k \approx k_\ell$ instead) and the typical $L \sim t^{-1}$ decay of the vortex length, as shown in Fig. 12.7.

The third example of Vinen turbulence appears in the study of Bose–Einstein condensed dark matter haloes. Here, the governing equation is the self-gravitating GPE. Mocz *et al.* (2017) demonstrated the formation of a turbulent tangle of reconnecting vortex lines and found a Vinen energy spectrum that peaks at the mesoscales and decays as $k^{-1.1}$.

12.3 Quantum Turbulence at Finite Temperature: General Considerations

We now return to quantum turbulence at finite temperatures. Both He II and ^{3}He-B are governed by two-fluid equations at finite temperatures, governing separately

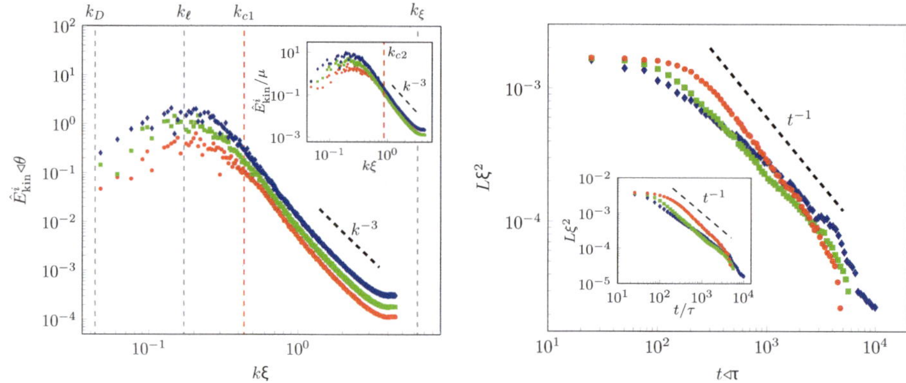

Figure 12.7 Thermal quench of a Bose gas, as in Fig. 12.6. (Left) Incompressible energy spectrum $\Phi^i_{\text{kin}}(k)$ vs. wavenumber k. The symbols refer to different condensate fractions (red circles: $\rho_0/\rho = 0.22$; green squares: $\rho_0/\rho = 0.48$; blue diamonds: $\rho_0\rho = 0.77$). The dashed line represents the k^{-3} spectrum associated with the vortex cores. Note that the spectrum peaks at $k \approx k_\ell$. (Right) Decay of the vortex length L vs. time t; note the $L \sim t^{-1}$ behavior. Symbols as in the left panel. The insets plot the same properties but use a different cutoff to separate the condensate from the thermal states, and demonstrate that they are essentially independent of the cutoff. Reproduced from Stagg *et al.* (2016), used under CC BY 3.0 (https://creativecommons.org/licenses/by/3.0).

the inviscid superfluid and the viscous normal component carrying all the entropy content of the liquid. The two velocity fields are generally coupled by the action of the mutual friction force. The relative fraction of normal fluid, ρ_n/ρ, depends strongly on temperature in both He II and ^{3}He-B. One distinction to be made among different regimes of turbulence is simply based on whichever fluid component is turbulent. Another is whether we are dealing with He II and ^{3}He-B.

12.3.1 Behavior of Quantum Turbulence in ^{3}He-B

We consider ^{3}He-B first for the following reason. In the hydrodynamic regime, i.e., above about $0.3\,T_c$, the viscosity of the normal fluid of ^{3}He-B is relatively high. As we have pointed out already, this means that the normal fluid remains effectively at rest in the frame of reference of the experimental container – e.g., in experiments performed inside a container in solid body rotation. The case of superfluid turbulence in stationary normal fluid can be phenomenologically understood as a next step in complexity after pure superfluid turbulence, with an added mutual friction term acting at all length scales. Although we formally deal (as in classical and pure superfluid turbulence) with two velocity fields, the velocity field of the normal fluid only determines the frame of reference of the container in which the normal fluid remains quiescent.[1] In this situation, vortex lines in the superfluid component of ^{3}He-B will not move freely but are subject to a friction force when they move in the background normal fluid; this friction is temperature dependent and becomes larger as T increases to T_c. We should therefore expect that pure superfluid turbulence is modified by the presence of damping. For example, the vortex lines will appear smoother (Tsubota *et al.*, 2000) than at lower temperatures, as short Kelvin waves and vortex cusps resulting from reconnections will be quickly damped.

In Chapter 8 we discussed the physical properties of quantum turbulence in ^{3}He-B at finite temperatures. In particular, we have shown that quantum turbulence cannot exist above a certain temperature due to strong damping of Kelvin waves by mutual friction. Temperature thus plays a similar role to (inverse) Reynolds number in classical 3D turbulence. Below the (pressure-dependent) critical temperature, the energy spectrum of the turbulent superfluid possesses special features:

(i) there is a maximum size of the turbulent eddy, limited by mutual friction;
(ii) dissipation due to mutual friction occurs at all length scales;
(iii) modified roll-off exponent to -3 at largest scales, which, according to predictions of continuum approximation, displays a crossover to the classical $-5/3$.

The temporal decay of this type of quantum turbulence, which can be loosely described as a dynamical vortex tangle in a thick "soup" of stationary normal fluid, is yet to be experimentally investigated.

12.3.2 Coflowing Cases of He II

We now consider superfluid ^{4}He at temperatures between about 1 K and T_λ where it displays the two-fluid behavior. Unlike ^{3}He-B, the normal fluid has very small viscosity and so moves easily, and hence becomes turbulent easily. We shall use the term *double turbulence* for the turbulent regime where both the normal and superfluid components of He II are turbulent.

The direction and the length scale of the forcing (which is necessary to sustain steady-state turbulence against losses) becomes crucial here. As already stated, we can either drive the normal fluid and the superfluid in the same direction (*coflows*) or in opposite directions (*counterflows*). Moreover, we can force the two fluid components at large length scales (e.g., by means of propellers) or at small length scales (via cavitation of small bubbles, or instabilities or reconnections occurring on the scale of the intervortex distance). In the case of coflows (away from boundaries, for simplicity) we expect that, at large scales, the normal fluid and superfluid velocity

[1] The existence of this special frame of reference is physically significant, for example for understanding the Kelvin–Helmholtz instability observed under rotation at the interface between ^{3}He-A and ^{3}He-B (Blaauwgeers *et al.*, 2002).

fields will be similar because they are driven together in the same direction (for example by a moving object). Because of mutual friction, the two fluids will tend to lock to each other behaving like a single fluid. To estimate the effect of the mutual friction it is useful to decompose normal fluid and superfluid velocity fields into large-scale flows and fluctuations: $\boldsymbol{u}_{\rm n} = \overline{\boldsymbol{u}_{\rm n}} + \boldsymbol{u}'_{\rm n}$ and $\boldsymbol{u}_{\rm s} = \overline{\boldsymbol{u}_{\rm s}} + \boldsymbol{u}'_{\rm s}$. Note that for coflows ($\overline{\boldsymbol{u}_{\rm n}} = \overline{\boldsymbol{u}_{\rm s}}$) the friction force (Eq. (5.47)) will be small, proportional to the differences between the two fluctuations, $\boldsymbol{F}_{\rm ns} \sim (\boldsymbol{u}'_{\rm n} - \boldsymbol{u}'_{\rm s})$; therefore the resulting turbulence will not be very different from the turbulence described by the classical Navier–Stokes equations. In contrast, in the case of counterflows (e.g., purely thermal counterflow, for which $\rho_{\rm n}\boldsymbol{u}_{\rm n} + \rho_{\rm s}\boldsymbol{u}_{\rm s} = \boldsymbol{0}$) the friction force will be large, $\boldsymbol{F}_{\rm ns} \sim (\rho/\rho_{\rm s})(\overline{\boldsymbol{u}_{\rm n}} + \boldsymbol{u}'_{\rm n})$, creating a dynamic that is unlike that of the Navier–Stokes equations.

We now consider coflows and counterflows separately keeping the focus on He II because ^{3}He-B has been discussed separately (Chapter 8).

12.4 Coflows of He II

Consider turbulent He II flow that is driven, e.g., by a grid at some (large) mesh scale M. To simplify things, let us artificially switch off the mutual friction and assume that the temperature is just below 1.6 K, where the kinematic viscosity of the normal fluid (based on the density of the normal fluid only), is $\nu_{\rm n} = O(\kappa)$. It follows that the quantum length scale ℓ in the superfluid component is approximately equal to the Kolmogorov dissipation scale η in the normal fluid, and that the turbulent spectra in the normal fluid and in the superfluid are exactly matched at length scales considerably larger than ℓ or η (except close to the wall, due to the difference in boundary conditions). The difference arises when ℓ and η are different. While in the normal fluid the Richardson cascade is nominally terminated at η, in the superfluid it continues in the form of a Kelvin wave cascade, and dissipation occurs only around k^* introduced above due to phonon radiation. Upon increasing/decreasing the temperature, due to steep temperature dependence $\nu_{\rm n}(T)$, η becomes smaller/larger than ℓ.

Turning on the mutual friction does not make much difference to scales significantly larger than η or ℓ, as appropriate, as they are matched in the two fluids. The matching cannot be complete when mutual friction starts to operate because one component starts to act as a source/drain for the other, resulting in the roll-off exponent that becomes gradually steeper. This results in an increase of intermittency corrections, as predicted by Boue *et al.* (2013) and experimentally confirmed by Varga *et al.* (2018).

Both the superfluid and normal energy spectra in steady state as well as the decaying case contain an inertial range of scales. This has been observed in a number

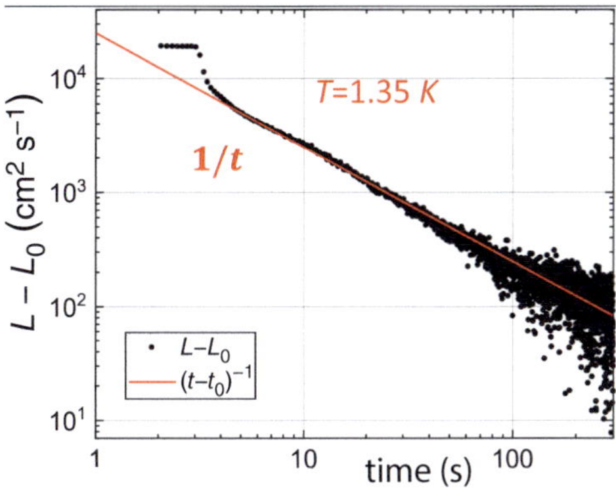

Figure 12.8 Temporal decay of vortex line density generated mechanically, by a vibrating fork. Due to small cross section of its prongs (75 μm × 90 μm) and relatively low steady-state vortex line density resulting in quantum length scale of the same order as the energy-containing length scale at which the flow is driven, the observed decay is of Vinen type, despite the fact that the flow was driven mechanically. Courtesy of S. Midlik and D. Schmoranzer.

of coflow experiments, first by Maurer and Tabeling (1998) in von Kármán flows, and, more recently, by Roche and collaborators (Salort *et al.*, 2010) in von Kármán flows as well as in superfluid wind tunnels (Salort *et al.*, 2012a) (see Fig. 9.10). In these experiments, velocity fluctuations were determined using a pitot tube, and the characteristic $f^{-5/3}$ frequency dependence of the spectrum was observed; this scaling (which, by Taylor's frozen field hypothesis, corresponds to $\Phi(k) \sim k^{-5/3}$) suggests the presence of a Kolmogorov energy cascade as in classical turbulence. As we saw geometrically in Chapters 3 and 11, it means that the turbulence contains within a tangle of random vortex lines, partially polarized bundles of vortex lines with relatively large coarse-grained superfluid vorticity.

Quantum turbulence in the Kolmogorov regime therefore appears similar to the ordinary turbulence in classical fluids, containing organized vortex structure embedded in a sea of random vorticity.

Although as a general rule, mechanically driven He II (in the two-fluid regime) results in coflow and represents Kolmogorov turbulence; however, an exception was recently found in Prague. In this experiment, quantum turbulence was driven mechanically by a small quartz tuning fork with prongs of small $75 \times 90\,\mu\text{m}^2$ cross section. Figure 12.8 shows that if the flow is driven at the scale smaller or comparable with the quantum length scale in the superfluid component, as well as Kolmogorov dissipation scale in the normal component, the temporal decay of vortex line density is of the Vinen form $L(t) \propto 1/t$, in accordance with Fig. 12.1 and subsequent discussion in the Section 12.3.

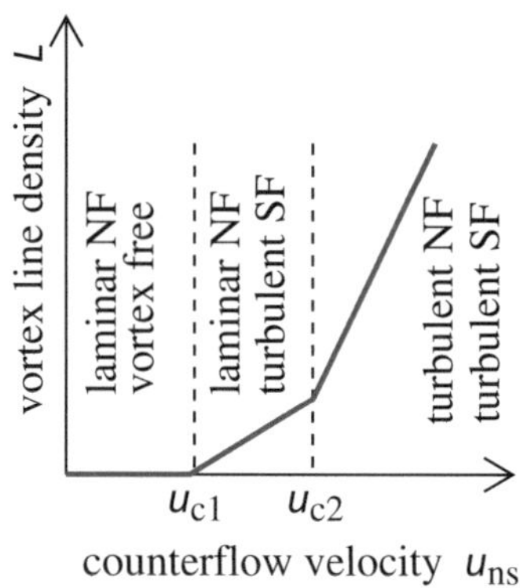

Figure 12.9 Regimes of thermal counterflow of He II in a channel: schematic flow diagram in terms of vortex line density and counterflow velocity. Above the second critical velocity u_{c2} in the double turbulence regime, both normal (NF) and superfluid (SF) velocity fields are turbulent.

12.5 Thermal Counterflow

Thermal counterflow is historically the first flow where quantum turbulence was studied by Vinen (1957) in his pioneering experiments. With hindsight, it is one of the most difficult quantum flows to understand and some of its aspects have been discovered only recently. We must consider it separately because, when driven in a conventional way, i.e., in a channel or pipe (1D counterflow) sufficiently hard, it belongs to the double turbulence category but fits neither the Kolmogorov regime nor the Vinen regime.

Experimental investigations as well as theoretical descriptions of thermal counterflow in He II were discussed in detail in Chapters 6 and 11. Here, we shall add a phenomenological description of this interesting quantum flow, based on the shape of the relevant energy spectra and simple dimensional arguments, in view of the earlier discussion of pure superfluid turbulence at zero temperature (Section 11.2).

Let us consider again the basic counterflow channel; see Fig. 2.9. If the flow is driven by a small heat flux, one can observe (though not in every geometry) transitions between regimes represented in Fig. 12.9. Sufficiently far away from the ends of the channel, at very small counterflow velocities u_{ns}, the normal fluid is laminar, possessing a classical Poiseuille parabolic profile, and the superfluid remains vortex-free (in the sense that any remanent vortex lines do not proliferate) with uniform velocity profile. Above the first critical velocity u_{c1}, a vortex tangle is created in the superfluid but the normal fluid remains laminar, although its profile is altered by the mutual friction, probably acquiring a flattened shape according to

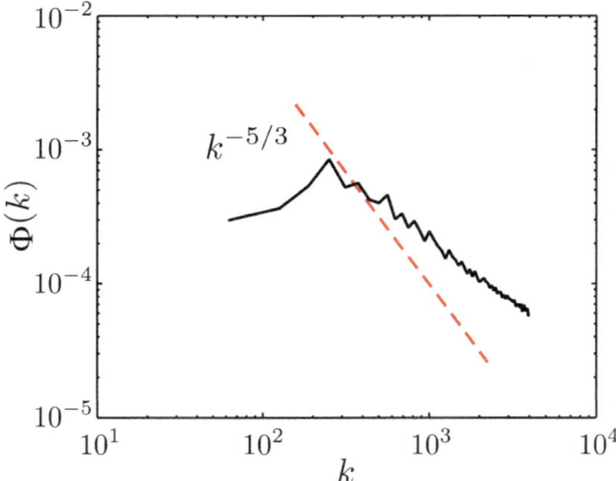

Figure 12.10 Vinen turbulence numerically generated by a small uniform heat flux. Energy spectrum $\Phi(k)$ vs. wavenumber k. Note that the energy density peaks at intermediate scales, and is quite different from the classical Kolmogorov $k^{-5/3}$ scaling, represented by the dashed line. Reproduced from (Barenghi *et al.*, 2016), used under CC BY 4.0 (https://creativecommons.org/licenses/by/4.0).

recent work (Marakov *et al.*, 2015; Yui and Tsubota , 2015); precisely what happens near the channel boundaries is under current investigation. It is likely, though, that the normal fluid profile in a wide channel is almost uniform far away from the boundary, and the turbulence in the superfluid component is approximately of the Vinen kind. This conclusion is supported by numerical simulations of thermal counterflow driven by a small uniform heat flux, which have demonstrated the spectral nature of the Vinen turbulence: The energy distribution peaks at intermediate wavenumbers $k \sim 1/\ell$, as shown in Fig. 12.10.

If we increase u_{ns}, we reach a second critical velocity u_{c2} past which the turbulence appears stronger. As suggested by Melotte and Barenghi (1998) and confirmed by flow visualization (Guo *et al.*, 2010), this transition marks the onset of turbulence in the normal fluid. For $u_{\mathrm{ns}} > u_{\mathrm{c2}}$, helium is in a remarkable double turbulent regime in which the large-scale velocity fields $\overline{\boldsymbol{u}_{\mathrm{n}}}$ and $\overline{\boldsymbol{u}_{\mathrm{s}}}$ flow in opposite directions. This strongly suggests that the main source of energy into the vortex tangle is the Donnelly–Glaberson instability, taking place at small length scales of the order of the quantum length scale, ℓ. There are no classical flows similar to this double turbulence, the closest candidate possibly being the turbulent magneto-hydrodynamic flow with interacting velocity and magnetic fields, if the latter is interpreted as a second fluid.

Let us now construct a schematic superfluid energy spectrum that reflects these basic facts characterizing thermal counterflow. The top panel of Fig. 12.11 shows the shape of three-dimensional superfluid energy spectrum in the T I state of thermal counterflow in He II. Flow of the viscous normal component is laminar. The energy input to the superfluid component, identified by Schwarz and discussed in Chapter 6, occurs at quantum length scale ℓ (indicated by the red arrow), which also plays a role in the energy-containing length scale. Likely because of the equipartition theorem, on the left side of ℓ the superfluid spectrum must acquire the k^2 form. The Kolmogorov cascade cannot operate at scales smaller than ℓ and energy is transferred down the scales via a Kelvin wave cascade and dissipated via mutual friction and phonon emission at scales of order k^*. As mutual friction increases with temperature, the shape of the superfluid energy spectrum between ℓ and k^* is temperature dependent. The state T I of counterflow turbulence therefore represents Vinen (ultra-quantum) turbulence, where the vortex tangle is approximately random. The temporal decay of vortex line density in T I is (at late times) inversely proportional to time; see Fig. 11.13.

Still in the T I state, the vortex line density increases with increasing heat flux and the quantum length scale becomes smaller. The energy input therefore shifts to smaller scales and dissipation due to mutual friction intensifies; the local slope of the superfluid energy spectrum between ℓ and k^* therefore becomes steeper. With increasing heat input, transition to turbulence occurs in the normal fluid upon reaching a second critical velocity v_{c2}. In analogy with classical channel flows, this happens at some critical Reynolds number $Re_{cr} = Du_{c2}\rho_n/\mu$; here, D is the characteristic size of the channel where the energy input in the normal fluid is assumed. The existence of large eddies in the normal fluid subsequently causes, via mutual friction, the creation of large superfluid eddies in the superfluid component, as shown in the middle panel of Fig. 12.11.

The bottom panels of Fig. 12.11 show the steady-state and decaying superfluid energy spectra in the T II state of counterflow turbulence. In the steady state, there are two energy inputs. Besides the one at the quantum length scale creating a quantum peak, there is now an essentially classical energy input at large scale. An inertial range of the Kolmogorov type, characterized by the $-5/3$ exponent, cannot develop, and the scaling exponent is steeper because mutual friction acts on all scales. Upon stopping the heat flux, the quantum energy peak quickly decays and the energy content at large scales gradually cascades down the scales, forming an inertial range that acquires classical Kolmogorov form, as discussed in Chapter 11. It results in a classical decay of the form $L(t) \propto 1/t^{3/2}$, with growing quantum length scale ℓ. The bottom right panel shows the shape of the decaying superfluid energy spectrum, after some transient period.

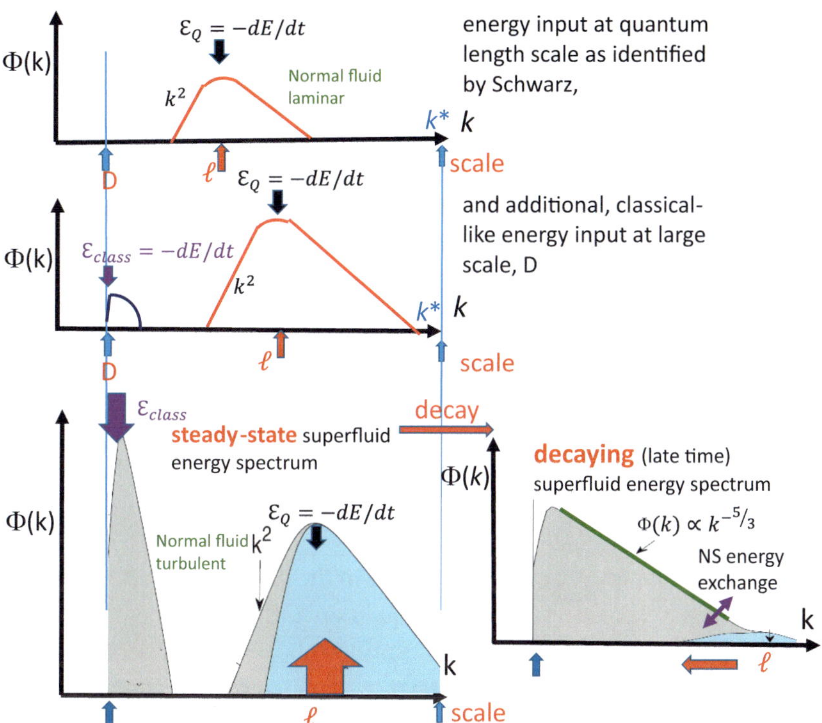

Figure 12.11 Schematic view of the general shapes of the three-dimensional superfluid energy spectra of thermal counterflow turbulence in He II. (Top) Vinen turbulence, where energy input identified by Schwarz occurs at quantum length scale ℓ (red arrow). The temporal decay is of the form $L(t) \propto 1/t$. (Middle) Upon increasing heat flux, there is additional classical-like energy input of energy in the normal component at large scale that, because of the mutual friction, occurs in the superfluid component too. (Bottom left) With increasing heat input, there is classical-like energy input at large scale, but inertial range of the Kolmogrov-type characterized by the -5/3 roll-off exponent cannot develop, as mutual friction acts on all scales, making this range much steeper. There is still a peak thanks to the energy input at quantum length scale ℓ, which itself shifts to the right with increasing heat input. (Bottom right) Upon stopping the heat flux, the quantum energy peak quickly decays and the energy content at large scales gradually cascades (Richardson cascade) down the scales, forming an inertial range that acquires classical Kolmogorov form. It results in a classical-like decay of the form $L(t) \propto 1/t^{3/2}$, with growing quantum length scale ℓ. See text for further details.

12.6 Summary

In this chapter, we have attempted to identify various regimes of quantum turbulence that occur in helium superfluids, namely in He II and ^{3}He-B.

We started with pure superfluid turbulence, where a single superfluid velocity field makes such identification much easier. The key parameter is the quantum length scale ℓ. Providing that there is a sink of energy at very small scales (by phonon-radiation in He II and by the excitation of Caroli–Matricon states in ^{3}He-B), one can identify two distinguished regimes called Vinen (or ultraquantum) and Kolmogorov (or quasi-classical) superfluid turbulence, depending on whether the turbulent flow is driven at scales smaller (Vinen) or larger (Kolmogorov) scales than ℓ. Vinen turbulence consists of a tangle of apparently random vortex lines and its decay is characterized by an inverse time law for the vortex line density. In Kolmogorov turbulence, partial polarization of vortex lines results in the existence of large eddies at scales exceeding ℓ. These large eddies behave as classical eddies, undergoing advection and stretching, resulting in the Richardson cascade. The cascade continues beyond ℓ in the form of a Kelvin wave cascade. In bounded domains, Kolmogorov quantum turbulence at late times is characterized by a classical decay of vortex line density, $L(t) \propto 1/t^{3/2}$. Effective values of the kinematic viscosity of order 0.1κ have been determined in both regimes.

A key point is that for classical turbulence in the unbounded case, there are only two length scales to consider: M and η. In quantum turbulence, there is an additional important scale, ℓ, which intervenes as described above. Pure superfluid turbulence therefore cannot be considered, in contrast to statements in the literature, a simple "prototype" of turbulence; despite the similarities, it is different and more complex than the classical case.

At finite temperatures helium superfluids display the two-fluid behavior. In ^{3}He-B, quantum turbulence exists only below a certain temperature, which plays a role similar to the inverse Reynolds number in classical turbulence. In this regime, quantum turbulence consists of a vortex tangle whose dynamics are affected by mutual friction with the normal fluid, which remains quiescent in the reference frame of the container.

Both Vinen and Kolmogorov regimes of quantum turbulence exist in He II, which serves as a playground for generating and studying a number of particular turbulent regimes, depending on how strongly and at which temperature and scale (or scales) the flow is driven. Both constituent flows may become turbulent, having their own velocity fields coupled by the mutual friction force. We speak about the double turbulence regime when both normal and superfluid velocity fields are turbulent. Each of these scenarios is considered separately in Chapter 13.

13
Outlook

Quantum turbulence comes in many forms. In this book we have focused mostly on three-dimensional quantum turbulence occurring in two quantum fluids, He II and ^{3}He-B, making selected references to quantum turbulence in atomic Bose–Einstein condensates. At this point, the range of accessible scales in Bose–Einstein condensates is comparatively narrow, but this limitation will likely disappear with time when experiments with larger condensates become possible. We have also substantially ignored other typical candidates for the application of quantum turbulence, such as neutron stars (Haskell *et al.*, 2020), containing both neutral and charged superfluids.

There are two excellent reasons for our focus on He II and ^{3}He-B. First, laboratory samples of these two quantum fluids are in most cases macroscopically large, allowing the excitation of a large number of degrees of freedom, similar to classical turbulence. This feature has enabled fruitful comparisons between turbulence in classical fluids and in quantum fluids. Second, quantum turbulence in these two helium superfluids has been intensely studied by many investigators over several decades and the numerous, independently verified, results can be generalized. It has been important that some of the same methods of study can be applied to both classical and quantum cases, allowing direct comparisons and bridging the knowledge of classical turbulence with facts obtained using methods from low-temperature physics research.

Because classical turbulence has served as the basis for our description of quantum turbulence, we should note the oft-repeated remark in the literature, attributed to Richard Feynman (without any attributable basis), that turbulence in fluids is the last unsolved problem of classical physics. Though it may be true at some intrinsic level, a large number of practical problems of turbulence in classical viscous fluids can be solved by a combination of theory, dimensional arguments, and high-precision numerical solutions. Quantum turbulence brings three exciting new ingredients into play: superfluidity, quantization of circulation, and a two-fluid behavior. Each of these ingredients, both singly and together, produce subtle and substantial effects, which are rich in their diversity. The phenomena change rapidly within small temperature ranges (a few K in He II and a few millikelvin in ^{3}He), and so great precision and ingenuity are called for in measurements as well as simulations. The phenomena require new tools, new ideas, and new effort; all together, they enrich the broad field of turbulence.

To take the case of superfluidity in the zero-temperature limit, where the two-fluid behavior is excluded, one can study the consequence of quantized vorticity in superfluid, which has no viscosity. This feature may appear to open up possibilities for comparisons with classical turbulence at infinitely large Reynolds numbers. However, several new elements, such as the new quantum length scale, have emerged, reflecting the severe quantum mechanical restriction that a superflow is potential everywhere except inside of vortex cores. A new form of turbulence, Vinen-type, has been identified, alongside the quasi-classical Kolmogorov-type, which resembles classical fluid turbulence. In the absence of viscous dissipation, new sinks of turbulent energy have been discovered, namely mutual friction, phonon radiation, and the excitation of Caroli–Matricon bound states in the vortex cores. These findings are robust and ought to be applicable to quantum turbulence occurring in other systems as well.

These new phenomena call to question the view that pure superfluid turbulence in He II and ^{3}He-B can be considered as a prototype of very high Reynolds number turbulence, consisting of a tangle of identical, singly quantized vortices moving in potential flow. A key point is that for classical turbulence in the unbounded case, there are essentially only two length scales to consider: the outer scale at which the turbulence is driven and the dissipation scale where the turbulent energy after passing the Richardson cascade becomes converted to heat. In pure superfluid turbulence, there is an additional important scale, namely the quantum length scale, which intervenes as we described in detail in Chapter 12. Despite the similarities to classical turbulence, pure superfluid turbulence is different and more complex than the classical case. However, the quasi-classical part of the picture is the same as that of classical 3D turbulence in viscous fluids (Skrbek *et al.*, 2021).

At finite temperatures the properties of quantum turbulence are even more complex because of the two-fluid behavior. In He II and ^{3}He-B the inviscid superfluid component coexists with viscous normal fluid. Moreover, at finite temperatures one expects even richer behavior thanks to the existence of the corresponding normal fluids. In the theoretically predicted but as yet undiscovered superfluid ^{3}He–^{4}He mixtures or in two-component atomic Bose–Einstein condensates, the two-fluid behavior takes place even in the zero-temperature limit. Investigations of these systems have already started and will very likely produce interesting results. In this instance, one of the two fluids can be turbulent or both, and their interactions can produce a rich variety of dynamical phenomena, as outlined in the book.

In this book we have discussed neither wave quantum turbulence nor two-dimensional quantum turbulence. Possibilities for investigating wave turbulence supported by quantized vortices, e.g., by forming and agitating a rectilinear vortex lattice in rotation with no or few reconnection events, certainly exist, in both He II and ^{3}He-B, and will surely become subjects of future investigations. And it will be very interesting to follow studies on the crossover from 2D to 3D quantum turbulence, which ought to be experimentally possible, e.g., by changing the confinement of atomic Bose–Einstein condensates from nearly flat to nearly spherical.

There are also practical reasons to study quantum turbulence. The superfluid phase of the common ^{4}He isotope, He II, is used as a cooling agent for superconducting magnets, both individual and those composing large research facilities like CERN. In order to run them safely and economically, one has to know the flow properties of the coolant, He II. Last but not least, He II has been seriously considered as a working fluid with known properties for various testing facilities with applications in aeronautics and ship hydrodynamics. The great advantage here is that, simply by changing the temperature and the pressure of He II, a model of an aircraft or ship can match several characteristic numbers (such as Reynolds and Froude numbers) simultaneously (Donnelly and Sreenivasan, 1998; Skrbek *et al.*, 1999).

From the scientific point of view, understanding various forms of quantum turbulence represents a formidable challenge that involves fluid dynamics, condensed-matter physics, atomic physics, low-temperature physics, and nuclear physics, as well as cosmology and astrophysics. We have attempted to explain most of the principal results, available in laboratory experiments as well as simulations, by a self-consistent combination of our knowledge of classical fluid dynamics and hydrodynamic turbulence, intuition based on the two-fluid model, and basic physics of quantized superfluids. Our focus is the physical understanding gained through a combination of experiment, theory, and simulations. We have used the available information to the full extent allowed within the scope of the book, and have attempted to point out outstanding problems. We have sometimes repeated a few key statements to provide immediacy to their importance. Putting together diverse information in one place, while integrating that information in a logical fashion, is the best way to point out existing gaps. It is our hope in doing so that this book will bring about a better understanding of this outstanding intellectual challenge.

References

Abo-Shaeer, J. R., C. Raman, J. M. Vogels, and W. Ketterle, Observation of vortex lattices in Bose–Einstein condensates, *Science* **292**, 476 (2001).

Achenbach, E., The effects of surface roughness and tunnel blockage on the flow past spheres, *J. Fluid Mech.* **65**, 113 (1974).

Adachi, H., S. Fujiyama, and M. Tsubota, Steady state counterflow quantum turbulence: Simulation of vortex filaments using the full Biot–Savart law, *Phys. Rev. B* **81**, 104511 (2010).

Ahlstrom, S. L., D. I. Bradley, S. N. Fisher, A. M. Guénault, E. A. Guise, R. P. Haley, S. Holt, O. Kolosov, P. V. E. McClintock, G. R. Pickett, M. Poole, R. Schanen, V. Tsepelin, and A. J. Woods, A quasiparticle detector for imaging quantum turbulence in superfluid ^{3}He-B, *J. Low Temp. Phys.* **175**, 725 (2014).

Allen, J. F., and M. Jones, New phenomena connected with heat flow in helium II, *Nature* **141** , 243 (1938).

Allen, J. F., and A. D. Misener, Flow of liquid He II, *Nature* **424** , 1022 (1938).

Allen, J. F., and J. Reekie, Momentum transfer and heat flow in liquid helium II, *Proc. Cambr. Phil. Soc.* **35**, 114 (1939).

Allum, D. R., P. V. E. McClintock, A. Phillips, and R. M. Bowley, The breakdown of superfluidity in liquid ^{4}He: An experimental test of Landau's theory. *Phil. Trans. Roy. Soc. A* **284**, 179 (1976).

Van Alphen, W. M., G. J. Van Haasteren, R. De Bruyn Ouboter, and K. W. Taconis, The dependence of the critical velocity of the superfluid on channel diameter and film thickness, *Phys. Lett.* **20**, 474 (1966).

Amelio, I., D. E. Galli, and L. Reatto, Probing quantum turbulence in ^{4}He by quantum evaporation measurements, *Phys. Rev. Lett.* **121**, 015302 (2018).

Andronikashvili, E., A direct observation of two kinds of motion in helium II, *J. Phys. USSR* **10**, 201 (1946).

Andronikashvili, E., Temperaturnaya zavisimost normalnoi plotnosti Geliya-II, *Zh. Eksp. Teor. Fiz.* **18**, 424 (1948).

Annett, J. F., *Superconductivity, Superfluids and Condensates*, Oxford University Press (2004).

Araki, T., M. Tsubota, and S. K. Nemirovskii, Energy spectrum of vortex tangle, *J. Low Temp. Phys.* **126**, 203 (2002a).

Araki, T., M. Tsubota, and S. K. Nemirovskii, Energy spectrum of superfluid turbulence with no normal-fluid component, *Phys. Rev. Lett.* **89**, 145301 (2002b).

Arp, V. D., and R. D. McCarty, *The Properties of Critical Helium Gas*, Tech. Rep., U. of Oregon (1998).

Ashton, A., L. B. Opatowsky, and J. T. Tough, Turbulence in pure superfluid flow, *Phys. Rev. Lett.* **46**, 658 (1981).

Autti, S., S. L. Ahlstrom, R. P. Haley, A. Jennings, G. R. Pickett, M. Poole, R. Schanen, A. A. Soldatov, V. Tsepelin, J. Vonka, T. Wilcox, A. J. Woods, and D. E. Zmeev, Fundamental dissipation due to bound fermions in the zero-temperature limit, *Nature Comm.* **11**, 4742 (2020).

Awschalom, D. D., and K. W. Schwarz, Observation of a remanent vortex-line density in superfluid helium, *Phys. Rev. Lett.* **52**, 49 (1984).

Awschalom, D. D., F. P. Milliken, and K. W. Schwarz, Properties of superfluid turbulence in a large channel, *Phys. Rev. Lett.* **53**, 1372 (1984).

Babuin, S., M. Stammeier, E. Varga, M. Rotter, and L. Skrbek, Quantum turbulence of bellows-driven ^{4}He superflow: Steady state, *Phys. Rev. B* **86**, 134515 (2012).

Babuin, S., E. Varga, and L. Skrbek, Effective viscosity in quantum turbulence: A steady state approach, *Europhys. Lett.* **106**, 24006 (2014a).

Babuin, S., E. Varga, and L. Skrbek, The decay of forced turbulent coflow of He II past a grid, *J. Low Temp. Phys.* **175**, 324 (2014b).

Babuin, S., E. Varga, W. F. Vinen, and L. Skrbek, Quantum turbulence of bellows-driven ^{4}He superflow: Decay, *Phys. Rev. B* **92**, 184503 (2015).

Babuin, S., V. S. L'vov, A. Pomyalov, L. Skrbek, and E. Varga, Coexistence and interplay of quantum and classical turbulence in superfluid ^{4}He: Decay, velocity decoupling, and counterflow energy spectra. *Phys. Rev. B* **94**, 174504 (2016).

Baehr, M. L., L. B. Opatowsky, and J. T. Tough, Transition from dissipationless superflow to homogeneous superfluid turbulence, *Phys. Rev. Lett.* **51**, 2295 (1983).

Baehr, M. L., L. B. Opatowsky, and J. T. Tough, Critical velocity in two-fluid flow of He II, *Phys. Rev. Lett.* **53**, 1669 (1984).

Baggaley, A. W., The sensitivity of the vortex filament method to different reconnections models, *J. Low Temp. Phys.* **68**, 18 (2012).

Baggaley, A. W., and C. F. Barenghi, Vortex density fluctuations in quantum turbulence, *Phys. Rev. B* **84**, 020504(R) (2011a).

Baggaley, A. W., and C. F. Barenghi, Quantum turbulent velocity statistics and quasiclassical limit, *Phys. Rev. E* **84**, 067301 (2011b).

Baggaley, A. W., and C. F. Barenghi, Tree method for quantum vortex dynamics, *J. Low Temp. Phys.* **166**, 3 (2012).

Baggaley, A. W., and C. F. Barenghi, Acceleration statistics in thermally driven superfluid turbulence, *Phys. Rev. E* **89**, 033006 (2014).

Baggaley, A. W., and R. Hänninen, Vortex filament method as a tool for computational visualization of quantum turbulence, *Proc. Nat. Acad. Sci. USA* **111** suppl. 1, 4667 (2014).

Baggaley, A. W., and J. Laurie, Kelvin-wave cascade in the vortex filament model, *Phys. Rev. B* **89**, 014504 (2014).

Baggaley, A. W., and J. Laurie, Thermal counterflow in a periodic channel with solid boundaries, *J. Low Temp. Phys.* **178**, 35 (2015).

Baggaley, A. W., C. F. Barenghi, and Y. A. Sergeev, Quasiclassical and ultraquantum decay of superfluid turbulence, *Phys. Rev. B* **85**, 060501(R) (2012a).

Baggaley, A. W., C. F. Barenghi, A. Shukurov, and Y. A. Sergeev, Coherent vortex structures in quantum turbulence, *Europhys. Lett.* **98**, 26002 (2012b).

Baggaley, A. W., J. Laurie, and C. F. Barenghi, Vortex-density fluctuations, energy spectra, and vortical regions in superfluid turbulence, *Phys. Rev. Lett.* **109**, 205304 (2012c).

Baggaley, A. W., L. K. Sherwin, C. F. Barenghi, and Y. A. Sergeev, Thermally and mechanically driven quantum turbulence in helium II, *Phys. Rev. B* **86**, 104501 (2012d).

Baggaley, A. W., C. F. Barenghi, and Y. A. Sergeev, Three-dimensional inverse energy transfer induced by vortex reconnections, *Phys. Rev. E* **89**, 013002 (2014).

Baggaley, A. W., V. Tsepelin, C. F. Barenghi, S. N. Fisher, G. R. Pickett, Y. A. Sergeev, and N. Suramlishvili, Visualizing pure quantum turbulence in superfluid ^{3}He: Andreev reflection and its spectral properties, *Phys. Rev. Lett.* **115**, 015302 (2015).

Balibar, S., The discovery of superfluidity, *J. Low Temp. Phys.* **146**, 441 (2007).

Barenghi, C. F., *Quantum Turbulence*. Ph.D. thesis, University of Oregon (1982).

Barenghi, C. F., Vortices and the Couette flow of helium II, *Phys. Rev. B* **45**, 2290 (1992).

Barenghi, C. F., and R. J. Donnelly, Vortex rings in classical and quantum systems, *Fluid Dyn. Res.* **41**, 051401 (2009).

Barenghi, C. F., and N. G. Parker, *A Primer on Quantum Fluids*, Springer (2016).

Barenghi, C. F., and D. C. Samuels, Evaporation of a packet of quantized vorticity, *Phys. Rev. Lett.* **89**, 155302 (2002).

Barenghi, C. F., C. E. Swanson, and R. J. Donnelly, Induced vorticity fluctuations in counterflowing He II, *Phys. Rev. Lett.* **48**, 1187 (1982).

Barenghi, C. F., R. J. Donnelly, and W. F Vinen, Friction on quantized vortices in helium II. A review, *J. Low Temp. Phys.* **52** 189 (1983).

Barenghi, C. F., D. C. Samuels, G. H. Bauer, and R. J. Donnelly. Superfluid vortex lines in a model of turbulent flow, *Phys. Fluids* **9**, 2631 (1997).

Barenghi, C. F., D. Kivotides, O. Idowu, and D. C. Samuels, Can superfluid vortex lines excite normal fluid turbulence in ^{4}He ? *J. Low Temp. Phys.* **121**, 377 (2000).

Barenghi, C. F., R. Hänninen, and M. Tsubota, Anomalous translational velocity of vortex ring with finite amplitude Kelvin waves, *Phys. Rev. E* **74**, 046303 (2006).

Barenghi, C. F., V. S. L'vov, and P.-E. Roche, Experimental, numerical and analytical velocity spectra in turbulent quantum fluid, *Proc. Nat. Acad. Sci. USA* **111** Suppl. 1, 4683 (2014a).

Barenghi, C. F., L. Skrbek, and K. R. Sreenivasan, Introduction to quantum turbulence, *Proc. Nat. Acad. Sci. USA* **111** Suppl. 1, 4647 (2014b).

Barenghi, C. F., Y. A. Sergeev, and A. W. Baggaley, Regimes of turbulence without an energy cascade, *Sci. Rep.* **6**, 35701 (2016).

Barenghi, C. F., P. McClintock, L. Skrbek, and K. R. Sreenivasan, W. F. (Joe) Vinen: An obituary, *Phys. Today* **75**, 60 (2022).

Barnes, J., and P. Hut, A hierarchical $O(N \log N)$ force calculation algorithm, *Nature* **324**, 446 (1986).

Barquist, C. S., W. G. Jiang, K. Gunther, N. Eng, and Y. Lee, Damping of a microelectromechanical oscillator in turbulent superfuid ^{4}He: A novel probe of quantized vorticity in the ultra-low temperature regime, *Phys. Rev. B* **101**, 174513 (2020).

Bauerle, C., Y. M. Bunkov, S. N. Fisher, H. Godfrin, and G. R. Pickett, Laboratory simulation of cosmic string formation in the early Universe using superfluid ^{3}He, *Nature* **382**, 332 (1996).

Bennett, J. C., and S. Corrsin, Small Reynolds number nearly isotropic turbulence in a straight duct and a contraction, *Phys. Fluids* **21**, 2129 (1978).

Benson, C. B., and A. C. Hollis-Hallett, The oscillating sphere at large amplitudes in liquid helium, *Canadian J. Phys.*, **34**, 668 (1956).

Benzi, R., L. Biferale, G. Paladin, A. Vulpiani, and M. Vergassola, Multifractality in the statistics of the velocity gradients in turbulence, *Phys. Rev. Lett.* **67**, 2299 (1991).

Berloff, N. G., Padé approximation of solitary wave solutions of the Gross–Pitaevskii equation, *J. Phys. A Math. Gen.* **37**, 11729 (2004).

Berloff, N. G., and C. F. Barenghi, Vortex nucleation by collapsing bubbles in Bose–Einstein condensates, *Phys. Rev. Lett.* **93**, 090401 (2004).

Berloff, N. G., and P. H. Roberts, Motions in a bose condensate: VI. Vortices in a nonlocal model, *J. Phys. A. Math Gen.* **32**, 5611 (1999).

Berloff, N. G., and P. H. Roberts, Motion in a Bose condensate: VIII. The electron bubble, *J. Phys. A: Math. Gen.* **34**, 81 (2001).

Berloff, N. G., and B. V. Svistunov, Scenario of strongly nonequilibrated Bose–Einstein condensation, *Phys. Rev. A* **66**, 013603 (2002).

Berloff, N. G., M. Brachet, and N. P. Proukakis, Modelling quantum turbulence of nonlinear classical matter fields, *Proc. Nat. Acad. Sci. USA* **111** Suppl. 1, 4675 (2014).

Bertolaccini, J., E. Leveque, and P.-E. Roche, Disproportionate entrance length in superfluid flows and the puzzle of counterflow instabilities, *Phys. Rev. Fluids* **2**, 123902 (2017).

Bevan, T. D. C., A. J. Manninen, J. B. Cook, H. Alles, J. R. Hook, and H. E. Hall, Vortex mutual friction in superfluid ^{3}He, *J. Low Temp. Phys.* **109**, 423 (1997).

Bewley, G. P., *Using Frozen Hydrogen Particles to Observe Rotating and Quantized Flows in Liquid Helium*. Ph.D. thesis, Yale University (2006).

Bewley, G. P., and K. R. Sreenivasan, The decay of a quantized vortex ring and the influence of tracer particles, *J. Low Temp. Phys.***156**, 84, (2009).

Bewley, G. P., D. P. Lathrop, and K. R. Sreenivasan, Superfluid helium. Visualization of quantized vortices, *Nature* **441**, 588 (2006).

Bewley, G. P., D. P. Lathrop, and K. R. Sreenivasan, Particles for tracing turbulent liquid helium, *Exp. Fluids* **44**, 887 (2008a).

Bewley, G. P., M. S. Paoletti, K. R. Sreenivasan, and D. P. Lathrop, Characterization of reconnecting vortices in superfluid helium, *Proc. Natl. Acad. Sci. USA* **105**, 13707 (2008b).

Biferale, L., Shell models of energy cascade in turbulence, *Ann. Rev. Fluid Mech.* **35**, 441 (2003).

Biferale, L., S. Musacchio, and F. Toschi, Inverse energy cascade in three-dimensional isotropic turbulence, *Phys. Rev. Lett.* **108**, 164501 (2012).

Biferale, L., D. Khomenko, V. S. L'vov, A. Pomyalov, I. Procaccia, and G. Sahoo, Turbulent statistics and intermittency enhancement in coflowing superfluid ^{4}He, *Phys. Rev. Fluids* **3**, 024605 (2018).

Birkhoff, G., Fourier synthesis of homogeneous turbulence, *Commun. Pure Appl. Math.* **7**, 19 (1954).

Blaauwgeers, R., V. B. Eltsov, G. Eska, A. P. Finne, R. P. Haley, M. Krusius, J. J. Ruohio, L. Skrbek, and G. E. Volovik, Shear flow and Kelvin–Helmholtz instability in superfluids, *Phys. Rev. Lett.* **89**, 155301 (2002).

Blaauwgeers, R., M. Blažková, M. Človečko, V. B. Eltsov, R. de Graaf, J. J. Hosio, M. Krusius, D. Schmoranzer, W. Schoepe, L. Skrbek, P. Skyba, R. E. Solntsev, D. E. Zmeev, Quartz tuning fork: Thermometer, pressure- and viscometer for helium liquids, *J. Low Temp. Phys.* **146**, 537 (2007).

Blažková, M., D. Schmoranzer, and L. Skrbek, Transition from laminar to turbulent drag in flow due to a vibrating quartz fork, *Phys. Rev. E* **75**, 025302 (2007).

Blažková, M., T. V. Chagovets, M. Rotter, D. Schmoranzer, and L. Skrbek, Cavitation in liquid helium observed in a flow due to a vibrating quartz fork, *J. Low Temp. Phys.* **150**, 194 (2008a).

Blažková, M., M. Človeçko, V. B. Eltsov, E. Gažo, R. de Graaf, J. J. Hosio, M. Krusius, D. Schmoranzer, W. Schoepe, L. Skrbek, P. Skyba, R. E. Solncev, and W. F. Vinen, Vibrating quartz fork: A tool for cryogenic helium research, *J. Low Temp. Phys.* **150**, 525 (2008b).

Blažková, M., D. Schmoranzer, L. Skrbek, and W. F. Vinen, Generation of turbulence by vibrating forks and other structures in superfluid ^{4}He, *Phys. Rev. B* **79**, 054522 (2009).

Bose, S. N. Plancks Gesetz und Lichtquantenhypothese, *Z. Phys.* **26**, 178 (1924).

Boue, L., R. Dasgupta, J. Laurie, V. L'vov, S. Nazarenko, and I. Procaccia, Exact solution for the energy spectrum of Kelvin wave turbulence in superfluids, *Phys. Rev. B* **84**, 064516 (2011).

Boue, L., V. S. L'vov, and I. Procaccia, Temperature suppression of Kelvin wave turbulence in superfluids, *Europhys. Lett.* **99**, 46003 (2012).

Boue, L., V. S. L'vov, Y. Nagar, S. V. Nazarenko, A. Pomyalov, and I. Procaccia, Enhancement of intermittency in superfluid turbulence, *Phys. Rev. Lett.* **110**, 014502 (2013).

Boue, L., V. S. L'vov, Y. Nagar, S. V. Nazarenko, A. Pomyalov, and I. Procaccia, Energy and vorticity spectra in turbulent superfluid ^{4}He from $T = 0$ to T_λ, *Phys. Rev. B* **91**, 144501 (2015).

Bradley, D. I., D. O. Clubb, S. N. Fisher, A. M. Guénault, C. J. Matthews, and G. R. Pickett, Vortex generation in superfluid ^{3}He by a vibrating grid, *J. Low Temp. Phys.* **134**, 381 (2004).

Bradley, D. I., D. O. Clubb, S. N. Fisher, A. M. Guénault, R. P. Haley, C. J. Matthews, G. R. Pickett, and K. L. Zaki, Turbulence generated by vibrating wire resonantors in superfluid ^{4}He at low temperatures, *J. Low Temp. Phys.* **138**, 493 (2005a).

Bradley, D. I., D. O. Clubb, S. N. Fisher, A. M. Guénault, R. P. Haley, C. J. Matthews, G. R. Pickett, V. Tsepelin, and K. Zaki, Emission of discrete vortex rings by a vibrating grid in superfluid ^{3}He-B: A precursor to quantum turbulence, *Phys. Rev. Lett.* **95**, 035302 (2005b).

Bradley, D. I., D. O. Clubb, S. N. Fisher, A. M. Guénault, R. P. Haley, C. J. Matthews, G. R. Pickett, V. Tsepelin, and K. Zaki, Decay of pure quantum turbulence in superfluid ^{3}He-B, *Phys. Rev. Lett.* **96**, 035301 (2006).

Bradley, D. I., S. N. Fisher, A. M. Guénault, R. P. Haley, V. Tsepelin, G. R. Pickett, and K. L. Zaki, The transition to turbulent drag for a cylinder oscillating in superfluid ^{4}He: A comparison of quantum and classical behavior, *J. Low Temp. Phys.* **154**, 97 (2009a).

Bradley, D. I., M. J. Fear, S. N. Fisher, A. M. Guénault, R. P. Haley, C. R. Lawson, P. V. E. McClintock, G. R. Pickett, R. Schanen, V. Tsepelin, and L. A. Wheatland, Transition to turbulence for a quartz tuning fork in superfluid ^{4}He, *J. Low Temp. Phys.* **156**, 116 (2009b).

Bradley, D. I., S. N. Fisher, A. M. Guénault, R. P. Haley, G. R. Pickett, D. Potts, and V. Tsepelin, Direct measurement of the energy dissipated by quantum turbulence, *Nat. Phys.* **7**, 473 (2011a).

Bradley, D. I., A. M. Guénault, S. N. Fisher, R. P. Haley, M. J. Jackson, D. Nye, K. O. Shea, G. R. Pickett, and V. Tsepelin, History dependence of turbulence generated by a vibrating wire in superfluid ^{4}He at 1.5K, *J. Low Temp. Phys.* **162**, 375 (2011b).

Bradley, D. I., M. Clovecko, S. N. Fisher, D. Garg, E. Guise, R. P. Haley, O. Kolosov, G. R. Pickett, V. Tsepelin, D. Schmoranzer, and L. Skrbek, Crossover from hydrodynamic to acoustic drag on quartz tuning forks in normal and superfluid ^{4}He, *Phys. Rev. B* **85**, 014501 (2012).

Bradley, D. I., S. N. Fisher, A. Ganshin, A. M. Guénault, R. P. Haley, M. J. Jackson, G. R. Pickett, and V. Tsepelin, The onset of vortex production by a vibrating wire in superfluid ^{3}He-B, *J. Low Temp. Phys.* **171**, 582 (2013).

Bradley, D. I., M. J. Fear, S. N. Fisher, A. M. Guénault, R. P. Haley, C. R. Lawson, G. R. Pickett, R. Schanen, V. Tsepelin, and L. A. Wheatland, Hysteresis, switching and anomalous behaviour of a quartz tuning fork in superfluid ^{4}He, *J. Low Temp. Phys.* **175**, 379 (2014).

Brewer, D. F., and D. O. Edwards, Heat conduction by liquid helium in capillary tubes. 2. Measurements of the pressure gradient, *Phil. Mag.* **6**, 1173 (1961).

Brewer, D. F., and D. O. Edwards, Heat conduction by liquid helium in capillary tubes. IV. Mutual friction under pressure, *Phil. Mag.* **6**, 1173 (1961). *J. Low Temp. Phys.* **43**, 327 (1981).

Buaria, D., and K. R. Sreenivasan, Dissipation range of the energy spectrum in high Reynolds number turbulence, *Phys. Rev. Fluids.* **5**, 092601(R) (2020).

Buaria, D., and K. R. Sreenivasan, Scaling of acceleration statistics in high Reynolds number turbulence, *Phys. Rev. Lett.* **128**, 234502 (2022).

Cao, N., S. Chen, and K. R. Sreenivasan, Properties of velocity circulation in three-dimensional turbulence, *Phys. Rev. Lett.* **76**, 616 (1996).

Castaing, B., B. Chabaud, and B. Hebral, Hot wire anemometer operating at cryogenic temperatures, *Rev. Sci. Instrum.* **63**, 4167 (1992).

Castiglione, J., P. J. Murphy, J. T. Tough, F. Hayot, and Y. Pomeau, Propagating and stationary superfluid turbulent fronts, *J. Low Temp. Phys.* **100**, 576 (1995).

Chagovets, T. V., Electric response in superfluid helium, *Physica B* **488**, 62 (2016).

Chagovets, T. V., and S. V. Van Sciver, A study of thermal counterflow using particle tracking velocimetry, *Phys. Fluids* **23**, 107102 (2011).

Chagovets, T. V., A. V. Gordeev, and L. Skrbek, Effective kinematic viscosity of turbulent He II, *Phys. Rev. E* **76**, 027301 (2007).

Chapman, S. J., A mean field model of superconducting vortices in three dimensions, *SIAM J. Appl. Math.* **55**, 1259 (1995).

Charalambous, D., L. Skrbek, P. C. Hendry, P. V. E. McClintock, and W. F. Vinen, Experimental investigation of the dynamics of a vibrating grid in superfluid 4He over a range of temperatures and pressures, *Phys. Rev. E* **74**, 036307 (2006).

Chase, C. E., Thermal conduction in liquid He II. 1. Temperature dependence, *Phys. Rev.* **127**, 361 (1962); Thermal conduction in liquid He II. 2. Effects of channel geometry, *Phys. Rev.* **131**, 1898 (1963).

Cheng, D. K., M. W. Cromar, and R. J. Donnelly, Influence of an axial heat current on negative-ion trapping in rotating helium II, *Phys. Rev. Lett.* **31**, 433 (1973).

Childers, R. K., and J. T. Tough, Critical velocities as a test of the Vinen theory, *Phys. Rev. Lett.* **31**, 911 (1973).

Chillà, F., and J. Schumacher, New perspectives in turbulent Rayleigh–Bénard convection, *Eur. Phys. J. E* **35**, 58 (2012).

Cidrim, A., A. C. White, A. J. Allen, V. S. Bagnato, and C. F. Barenghi, Vinen turbulence via the decay of multicharged vortices in trapped atomic Bose–Einstein condensates, *Phys. Rev. A* **96**, 023617 (2017).

Cockburn, S. P., and N. P. Proukakis, The stochastic Gross–Pitaevskii equation and some applications, *Laser Physics* **19**, 558 (2009).

Coddington, I., P. C. Haljan, P. Engels, V. Schweikhard, S. Tung, and E. A. Cornell, Experimental studies of equilibrium vortex properties in a Bose-condensed gas, *Phys. Rev. A* **70**, 063607 (2004).

Collin, E., J. Kofler, J.-S. Heron, O. Bourgeois, Yu. M. Bunkov, and H. Godfrin, Novel vibrating wire like NEMS and MEMS structures for low temperature physics, *J. Low Temp. Phys.* **158**, 678 (2010).

Comte-Bellot, G., and S. Corrsin, The use of a contraction to improve the isotropy of grid-generated turbulence, *J. Fluid Mech.* **25**, 657 (1966).

Connaughton, C., and S. Nazarenko, Warm cascades and anomalous scaling in a diffusion model of turbulence, *Phys. Rev. Lett.* **92**, 044501 (2004).

Cooper, R. G., M. Mesgarnezhad, A. W. Baggaley, and C. F. Barenghi, Knot spectrum of turbulence, *Sci. Reports* **9**, 10545 (2019).

Craig, P. P., and J. R. Pellam, Observation of perfect potential flow in superfluid, *Phys. Rev.* **108**, 1109 (1957).

Davis, S. I., P. C. Hendry, and P. V. E. McClintock, Decay of quantized vorticity in superfluid ^{4}He at mK temperatures, *Physica* B **280**, 43 (2000).

De Waele, A. T. A. M., and R. G. K. M. Aarts, Route to vortex reconnection, *Phys. Rev. Lett.* **72**, 482 (1994).

di Leoni, P. C., P. Mininni, and M. E. Brachet, Dual cascade and dissipation mechanisms in helical quantum turbulence, *Phys. Rev. A* **95**, 053636 (2017).

Dimotakis, P. E., and J. E. Broadwell, Local temperature measurements in supercritical counterflow in liquid He II, *Phys. Fluids* **16**, 1787 (1973).

Diribarne, P., M. Bon Mardion, A. Girard, J.-P. Moro, B. Rousset, F. Chilla, J. Salort, A. Braslau, F. Daviaud, B. Dubrulle, B. Gallet, I. Moukharski, E.-W. Saw, C. Baudet, M. Gibert, P.-E. Roche, E. Rusaouen, A. Golov, V. L'vov, and S. Nazarenko, Investigation of properties of superfluid ^{4}He turbulence using a hot-wire signal, *Phys. Rev. Fluids* **6**, 094601 (2021).

Donnelly, R. J., *Quantized Vortices in Helium II*, Cambridge University Press (1991).

Donnelly, R. J., and C. F. Barenghi, The observed properties of liquid helium at saturated vapor pressure, *J. Phys. Chem. Ref. Data* **27**, 1217 (1998).

Donnelly, R. J., and A. C. Hollis-Hallett, Periodic boundary layer experiments in liquid helium. *Ann. Phys.* **3**, 320 (1958).

Donnelly, R. J., and O. Penrose, Oscillation of liquid helium in a U-tube, *Phys. Rev.* **103** 1137 (1956).

Donnelly R. J., and K. R. Sreenivasan (eds), *Flow at Ultra-High Reynolds and Rayleigh Numbers*, Springer-Verlag (1998).

Donnelly, R. J., and C. E. Swanson, Quantum turbulence, *J. Fluid Mech.* **173**, 387 (1986).

van Doorn, E., C. M. White, and K. R. Sreenivasan, The decay of grid turbulence in polymer and sufactant solutions, *Phys. Fluids* **11**, 2387 (1999).

Dormy, E., and A. M. Soward, *Mathematical Aspects of Natural Dynamos*, Grenoble Sciences, CRC Press, Chapman and Hall (2007).

Dubrulle, B., Intermittency in fully developed turbulence: Log-Poisson statistics and generalized scale covariancea, *Phys. Rev. Lett.* **73**, 959 (1994).

Duda, D., M. La Mantia, M. Rotter, and L. Skrbek, On the visualization of thermal counterflow of He II past a circular cylinder, *J. Low Temp. Phys.* **175**, 331 (2014).

Duda, D., P. Švančara, M. La Mantia, M. Rotter, and L. Skrbek, Visualization of viscous and quantum flows of liquid ^{4}He due to an oscillating cylinder of rectangular cross section, *Phys. Rev. B* **92**, 064519 (2015).

Duda, D., M. La Mantia, and L. Skrbek, Streaming flow due to a quartz tuning fork oscillating in normal and superfluid ^{4}He, *Phys. Rev. B* **96**, 024519 (2017).

Dudley, M. L., and R. W. James, Time-dependent kinematic dynamos with stationary flows, *Proc. R. Soc. Lond.* **A 425**, 407 (1989).

Duri, D., C. Baudet, P. Charvin, J. Virone, B. Rousset, J.-M. Poncet, and P. Diribarne, Liquid helium inertial jet for comparative study of classical and quantum turbulence, *Rev. Sci. Instrum.* **82**, 115109 (2011).

Duri, D., C. Baudet, J.-P Moro, P.-E. Roche, and P. Diribarne, Hot-wire anemometry for superfluid turbulent coflows, *Rev. Sci. Instrum.* **86**, 025007 (2015).

Eckert, M., *The Turbulence Problem. A Persistent Riddle in Historical Perspective.* Springer (2019).

Einstein, A., Quantentheorie des einatomigen idealen Gases. Zweite Abhandlung (Quantum theory of the monoatomic ideal gas. Second treatise), *Sitzungsberichte* **I**, 3 (1925).

Ekinci, K. L., V. Yakhot, S. Rajauria, C. Colosquiac, and D. M. Karabacak, High-frequency nanofluidics: A universal formulation of the fluid dynamics of MEMS and NEMS, *Lab Chip* **10**, 3013 (2010).

Eltsov, V. B., and V. S. L'vov, Amplitude of waves in the Kelvin-wave cascade, *JETP Letters* **111**, 389 (2020).

Eltsov, V. B., A. P. Finne, R. Hänninen, J. Kopu, M. Krusius, M. Tsubota, and E. V. Thuneberg, Twisted vortex state, *Phys. Rev. Lett.* **96**, 215302 (2006).

Eltsov, V. B., A. I. Golov, R. de Graaf, R. Hänninen, M. Krusius, V. S. L'vov, and R. E. Solntsev, Quantum turbulence in a propagating superfluid vortex front, *Phys. Rev. Lett.* **99**, 265301 (2007).

Eltsov, V. B., R. de Graaf, R. Hänninen, M. Krusisus, R. E. Solntsev, V. S. L'vov, A. I. Golov, and P. M. Walmsley, Turbulent dynamics in rotating helium superfluids. In *Progress in Low-Temperature Physics*, vol. 14, W. P. Halperin (ed.). Elsevier (2009).

Eltsov, V. B., R. Hänninen, and M. Krusisus, Quantum turbulence in superfluids with wall-clamped normal component, *Proc. Nat. Acad. Sci. USA* **111** Suppl. 1, 4711 (2014).

Eltsov, V. B., A. Gordeev, and M. Krusius, Kelvin–Helmholtz instability of AB interface in superfluid 3He, *Phys. Rev. B* **99**, 054104 (2019).

Eyink, G. L., and D. J. Thomson, Free decay of turbulence and breakdown of self-similarity, *Phys. Fluids* **12**, 477 (2000).

Farge, M., G. Pellegrino, and K. Schneider, Coherent vortex extraction in 3D turbulent flows using orthogonal wavelets, *Phys. Rev. Lett.* **87**, 054501 (2001).

Farge, M., K. Schneider, G. Pellegrino, A. A. Wray, and R. S. Rogallo, Coherent vortex extraction in three-dimensional homogeneous turbulence: Comparison between CVS-wavelets and POD-Fourier decomposition, *Phys. Fluids* **15**, 2886 (2003).

Feynman, R. P., Application of quantum mechanics to liquid helium. In *Progress in Low-Temperature Physics*, vol. 1, C. J. Gorter (ed.). North Holland (1955).

Feynman, R. P., *The Feynman Lectures on Physics, Vol. II: The New Millennium Edition: Mainly Electromagnetism and Matter*. Hachette Book Group, Inc. (2011).

Fijimoto, K., and M. Tsubota, Spin turbulence with small spin magnitude in spin-1 spinor Bose–Einstein condensates, *Phys. Rev. A* **88**, 063628 (2013).

Finne, A. P., T. Araki, R. Blaauwgeers, V. B. Eltsov, N. B. Kopnin, M. Krusius, L. Skrbek, M. Tsubota, and G. E. Volovik, Observation of an intrinsic velocity-independent criterion for superfluid turbulence, *Nature* **424**, 1022 (2003).

Finne, A. P., V. B. Eltsov, G. Eska, R. Hänninen, J. Kopu, M. Krusius, E. V. Thuneberg, and M. Tsubota, Vortex multiplication in applied flow: A precursor to superfluid turbulence, *Phys. Rev. Lett.* **96**, 085301 (2006).

Fisher, S. N., and G. R. Pickett, Quantum turbulence in superfluid ^{3}He at very low temperatures. In *Progress in Low-Temperature Physics*, vol. 16, M. Tsubota and W. P. Halperin (eds), Elsevier (2009).

Fisher, S. N., A. J. Hale, A. M. Guénault, and G. R. Pickett, Generation and detection of quantum turbulence in superfluid ^{3}He-B, *Phys. Rev. Lett.* **86**, 244 (2001).

Fonda, E., D. P. Meichle, N. T. Oullette, S. Hormoz, and D. P. Lathrop, Direct observation of Kelvin waves excited by quantized vortex reconnection, *Proc. Nat. Acad. Sci. USA* **111** Suppl. 1, 4707 (2014).

Fonda, E., K. R. Sreenivasan, and D. P. Lathrop, Sub-micron solid air tracers for quantum vortices and liquid helium flows, *Rev. Sci. Instrum.* **87**, 025106 (2016).

Fonda, E., K. R. Sreenivasan, and D. P. Lathrop, Reconnection scaling in quantum fluids, *Proc. Nat. Acad. Sci. USA* **116**, 1924 (2019).

Frisch, U., *Turbulence. The Legacy of A. N. Kolmogorov*, Cambridge University Press (1995).

Fujiyama, S., A. Mitani, M. Tsubota, D. I. Bradley, S. N. Fisher, A. M. Guénault, R. P. Haley, G. R. Pickett, and V. Tsepelin, Generation, evolution, and decay of pure quantum turbulence: A full Biot–Savart simulation, *Phys. Rev. B* **81**, 026309R (2010).

Galantucci, L., M. Sciacca, and C. F. Barenghi, Coupled normal fluid and superfluid profiles of turbulent helium in channels, *Phy. Rev. B* **92**, 174530 (2015).

Galantucci, L., M. Sciacca, and C. F. Barenghi, Large-scale normal fluid circulation in helium superflows, *Phy. Rev. B* **95**, 014509 (2017).

Galantucci, L., A. W. Baggaley, N. G. Parker, and C. F. Barenghi, Crossover from interaction to driven regimes in quantum vortex reconnections. *Proc. Nat. Acad. Sci. USA* **116**, 12204 (2019).

Galantucci, L., A. W. Baggaley, C. F. Barenghi, and G. Krstulovic, A new self-consistent approach of quantum turbulence in superfluid helium, *Euro. Phys. J. Plus* **135**, (2020).

Galantucci, L., C. F. Barenghi, N. G. Parker, and A. W. Baggaley, Mesoscale helicity distinguishes Vinen from Kolmogorov turbulence in helium II, *Phys. Rev. B* **103**, 144503 (2021).

Galli, D. E., L. Reatto, and M. Rossi, Quantum Monte Carlo study of a vortex in superfluid ^{4}He and search for a vortex state in the solid, *Phys. Rev. B* **89**, 224516 (2014).

Gao, J., A. Marakov, W. Guo, B. T. Pawlowski, S. W. Van Sciver, G. G. Ihas, D. N. McKinsey, and W. F. Vinen, Producing and imaging a thin line of He_2^* molecular tracers in helium-4, *Rev. Sci. Instrum.* **86**, 093904 (2015).

Gao, J., W. Guo, and W. F. Vinen, Determination of the effective kinematic viscosity for the decay of quasiclassical turbulence in superfluid ^{4}He, *Phys. Rev. B* **94**, 094502 (2016a).

Gao, J., W. Guo, V. S. L'vov, A. Pomyalov, L. Skrbek, E. Varga, and W. F. Vinen, Decay of counterflow turbulence in superfluid ^{4}He, *JETP Letters* **103**, 648 (2016b).

Gao, J., E. Varga, W. Guo, and W. F. Vinen, Energy spectrum of thermal counterflow turbulence in superfluid helium-4, *Phys. Rev. B* **96**, 094511 (2017).

Gao, J., W. Guo, and W. F. Vinen, Dissipation in quantum turbulence in superfluid ^{4}He above 1K, *Phys. Rev. B* **97**, 184518 (2018).

Garcia-Orozco, A. D., L. Madeira, L. Galantucci, C. F. Barenghi, and V. S. Bagnato, Intra-scales energy transfer during the evolution of turbulence in a trapped Bose–Einstein condensate, *Europhys. Lett.* **130**, 46001 (2020).

Garg, D., V. B. Efimov, M. Giltrow, P. V. E. McClintock, L. Skrbek, and W. F. Vinen, Behavior of quartz forks oscillating in isotopically pure ^{4}He in the $T \to 0$ limit, *Phys. Rev. B* **85**, 144518 (2012).

Gaunt, A. L., T. F. Schmidutz, I. Gotlibovych, R. P. Smith, and Z. Hadzibabic, Bose–Einstein condensation of atoms in a uniform potential, *Phys. Rev. Lett.* **110**, 200406 (2013).

Gibbs, M. R., K. H. Andersen, W. G. Stirling, and H. Schober, The collective excitations of normal and superfluid ^{4}He: The dependence on pressure and temperature, *J. Phys. C* **11**, 603 (1999).

Glaberson, W. I., W. W. Johnson, and R. M. Ostermeier, Instability of a vortex array in He II, *Phys. Rev. Lett.* **33**, 1197 (1974).

Gledzer, E., System of hydrodynamic type admitting two quadratic integrals of motion, *Phys. Dok.* **18**, 216 (1973).

Golov, A. I., P. M. Walmsley, and P. A. Tompsett, Charged tangles of quantized vortices in superfluid He-4, *J. Low Temp. Phys.* **161**, 509 (2010).

Gorter, C. J., and J. H. Mellink, On irreversible processes in liquid helium-II, *Physica* **15**, 285 (1949).

Goto, R., S. Fujiyama, H. Yano, Y. Nago, N. Hashimoto, K. Obara, O. Ishikawa, M. Tsubota, and T. Hata, Turbulence in boundary flow of superfluid ^{4}He triggered by free vortex rings, *Phys. Rev. Lett.* **100**, 045301 (2008).

Greenspan, H. P., *The Theory of Rotating Fluids*, Cambridge University Press (1968).

Groszek, A. J., T. P. Simula, D. M. Paganin, and K. Helmerson, Onsager vortex formation in Bose–Einstein condensates in two-dimensional power-law traps, *Phys. Rev. A* **93**, 043614 (2016).

Guo, W., S. B. Cahn, J. A. Nikkel, W. F. Vinen, and D. N. McKinsey, Visualization study of counterflow in superfluid ^{4}He using metastable helium molecules, *Phys. Rev. Lett.* **105**, 045301 (2010).

Guo, W., M. La Mantia, D. P. Lathrop, and S. W. Van Sciver, Visualization of two-fluid flows of superfluid helium-4, *Proc. Nat. Acad. Sci. USA* **111** Suppl. 1, 4653 (2014).

Guthrie, A., S. Kafanov, T. Noble, Y. Pashkin, G. Pickett, V. Tsepelin, A. Dorofeev, V. Krupenin, and D. Presnov, Nanoscale real-time detection of quantum vortices at millikelvin temperatures, *Nat. Commun.* **12**, 2645 (2021).

Hall, H. E., and W. F. Vinen, The rotation of liquid helium II. I. Experiments on the propagation of second sound in uniformly rotating helium II, *Proc. Roy. Soc.* **A238**, 204 (1957); II. The theory of mutual friction in uniformly rotating helium II, **A238**, 215 (1957).

Hammel, E. F., and W. E. Keller, The flow of liquid helium II through a narrow channel, *Cryogenics* **5**, 245 (1965).

Hammer, P. W., K. R. Sreenivasan, and J. J. Niemela, Russell James Donnelly: An obituary, *Phys. Today* **68**, 59 (2015).

Hänninen, R., and W. Schoepe, Universal critical velocity for the onset of turbulence of oscillatory superfluid flow, *J. Low Temp. Phys.* **153**, 189 (2008).

Hänninen, R., and W. Schoepe, Universal onset of quantum turbulence in oscillating flows and crossover to steady flows, *J. Low Temp. Phys.* **158**, 410 (2010).

Hänninen, R., N. Hietala, and H. Salman, Helicity within the vortex filament model, *Sci. Rep.* **6**, 37571 (2016).

Hashimoto, N., R. Goto, H. Yano, K. Obara, O. Ishikawa, and T. Hata, Control of turbulence in boundary layers of superfluid ^{4}He by filtering out remanent vortices, *Phys. Rev. B* **76**, 020504 (2007).

Haskell, B., D. Antonopoulou, and C. F. Barenghi, Turbulent, Pinned superfluids in neutron stars and pulsar glitch recoveries, *Month. Roy. Astronom. Soc.* 499, 161–170 (2020).

Henberger, J. D., and J. T. Tough, Uniformity of the temperature gradient in the turbulent counterflow of He II, *Phys. Rev. B* **25**, 3123 (1982).

Henderson, K. L., and C. F. Barenghi, Vortex waves in a rotating superfluid, *Europhys. Lett.* **67**, 56 (2004).

Henderson, K. L., C. F. Barenghi, and C. A. Jones, Nonlinear Taylor–Couette flow of helium II, *J. Fluid Mech.* **283**, 329 (1995).

Hendry, P. C., and P. V. E. McClintock, Continuous-flow apparatus for preparing isotopically pure ^{4}He, *Cryogenics* **27**, 131 (1987).

Henn, E. A. L., J. A. Seman, G. Roati, K. M. F. Magalhães, and V. S. Bagnato, Emergence of turbulence in an oscillating Bose–Einstein condensate, *Phys. Rev. Lett.* **103**, 045301 (2009).

Hills, R. N., and P. H. Roberts, Superfluid mechanics of a high density of vortex lines, *Arch. Rat. Mech. Anal.* **66**, 43 (1977).

Hollis-Hallett, A. C., Experiments with oscillating disk systems in liquid helium II, *Proc. Roy. Soc. A* **210**, 404 (1952).

Holt, S., and P. Skyba, Electrometric direct current I/V converter with wide bandwidth, *Rev. Sci. Instrum.* **83**, 064703 (2012).

Hosio, J. J., V. B. Eltsov, R. de Graaf, P. J. Heikkinen, R. Hänninen, M. Krusius, V. S. L'vov, and G. E. Volovik, Superfluid vortex front at $T \to 0$: Decoupling from the reference frame, *Phys. Rev. Lett.* **107**, 135302 (2011).

Hosio, J. J., V. B. Eltsov, M. Krusius, and J. T. Mäkinen, Quasiparticle-scattering measurements of laminar and turbulent vortex flow in the spin-down of superfluid 3He-B, *Phys. Rev. B* **85**, 224526 (2012).

Hosio, J. J., V. B. Eltsov, P. J. Heikkinen, R. Hänninen, M. Krusius, and V. S. L'vov, Energy and angular momentum balance in wall-bounded quantum turbulence at very low temperatures, *Nat. Commun.* **4**, 1614 (2013).

Howe, M. S., *Acoustics of Fluid-Structure Interaction*, Cambridge University Press 1998.

Hrubcová, P., P. Švančara, and M. La Mantia, Vorticity enhancement in thermal counterflow of superflow helium, *Phys. Rev. B* **97**, 064512 (2018).

Ishihara, T., T. Gotoh, and Y. Kaneda, Study of high-Reynolds number isotropic turbulence by direct numerical simulations, *Annu. Rev. Fluid Mech.* **41**, 165 (2009).

Iyer, K. P., K. R. Sreenivasan, and P. K. Yeung, Circulation in high Reynolds number isotropic turbulence is a bifractal, *Phys. Rev. X* **9**, 041006 (2019).

Jackson, B., N. P. Proukakis, C. F. Barenghi, and E. Zaremba, Finite-temperature vortex dynamics in Bose–Einstein condensates, *Phys. Rev. A* **79**, 053615 (2009).

Jager, J., B. Schuderer, and W. Schoepe, Turbulent and laminar drag of superfluid helium on an oscillating microsphere, *Phys. Rev. Lett.* **74**, 566 (1995).

John, J. P., D. A. Donzis, and K. R. Sreenivasan, Laws of turbulence decay from direct numerical simulations, *Phil. Trans. R. Soc.* **A 380**, 20210089 (2022).

Kafkalidis, J. F., and J. T. Tough, Thermal counterflow in a diverging channel: A study of radial heat transfer in He II, *Cryogenics* **31**, (1991) 705.

Kafkalidis, J. F., G. Klinich III, and J. T. Tough, Superfluid turbulence in a nonuniform rectangular channel, *Phys. Rev. B* **50**, 15909 (1994).

Kagan, Y., and B. V. Svistunov, Kinetics of the onset of long-range order during Bose condensation in an interacting gas, *JETP* **78**, 187 (1994).

Kamppinen, T., and V. B. Eltsov, Nanomechanical resonators for cryogenic research, *J. Low Temp. Phys.* **196**, 82 (2018).

Kapitza, P. L., Viscosity of liquid helium below the λ-point, *Nature* **424**, 1022 (1938).

von Kármán, T., and L. Howarth, On the statistical theory of isotropic turbulence, *Proc. Roy. Soc.* **A164**, 192 (1938).

Kerr, R. M., Vortex stretching as a mechanism for quantum kinetic energy decay, *Phys. Rev. Lett.* **106**, 224501 (2011).

Keulegan, G. H., and L. H. Carpenter, Forces on cylinders and plates in an oscillating fluid, *J. Res. Natl. Bur. Stand.* **60**, 423 (1958).

Khalatnikov, I. M., *Introduction to the Theory of Superfluidity*, Addison–Wesley (1965).

Khomenko, D., L. Kondaurova, V. S. L'vov, P. Mishra, A. Pomyalov, and I. Procaccia, Dynamics of the density of quantized vortex lines in superfluid turbulence, *Phys. Rev. B* **91**, 180504 (2015).

Khomenko, D., V. S. L'vov, A. Pomyalov, and I. Procaccia, Reply to 'Comment on dynamics of the density of quantized vortex lines in superfluid turbulence', *Phys. Rev. B* **94**, 146502 (2016).

Khurshid, S., D. A. Donzis, and K. R. Sreenivasan, Energy spectrum in the dissipation range, *Phys. Rev. Fluids* **3**, 082601 (2018).

Kida, S., A vortex filament moving without change of form, *J. Fluid Mech.* **112**, 397 (1981).

Kim, S. K., and A. W. Troesch, Streaming flows generated by high-frequency small-amplitude oscillations of arbitrarily shaped cylinders, *Phys. Fluids* A **1**, 975 (1989).

Kistler, A. L., and T. Vrebalovich, Grid turbulence at large Reynold numbers, *J. Fluid Mech.* **26**, 37 (1966).

Kivotides, D., Motion of a spherical solid particle in thermal counterflow turbulence, *Phys. Rev. B* **77**, 174508 (2008).

Kivotides, D., C. F. Barenghi, and D. C. Samuels, Triple vortex ring structure in superfluid helium II, *Science* **290**, 777 (2000).

Kivotides, D., J. C. Vassilicos, D. C. Samuels, and C. F. Barenghi, Kelvin waves cascade in superfluid turbulence, *Phys. Rev. Lett.* **86**, 3080 (2001).

Kivotides, D., C. F. Barenghi, and Y. A. Sergeev, Interactions between particles and quantized vortices in superfluid helium, *Phys. Rev. B* **77**, 014527 (2008a).

Kivotides, D., Y. A. Sergeev, and C. F. Barenghi, Dynamics of solid particles in a tangle of superfluid vortices at low temperatures, *Phys. Fluids* **20**, 055105 (2008b).

Kleckner, D., and W. T. M. Irvine, Creation and dynamics of knotted vortices, *Nature Phys.* **9**, 253 (2013).

Kleckner, D., L. H. Kauffman, and W. T. M. Irvine, How superfluid knots untie, *Nature Phys.* **12**, 650 (2016).

Kobayashi, M., and M. Tsubota, Kolmogorov spectrum of superfluid turbulence: Numerical analysis of the Gross–Pitaevskii equation with a small-scale dissipation, *Phys. Rev. Lett.* **94**, 065302 (2005).

Kobayashi, M., and M. Tsubota, Quantum turbulence in a trapped Bose–Einstein condensate, *Phys. Rev. A* **76**, 045603 (2007).

Kolmogorov, A. N., The local structure of turbulence in incompressible viscous fluid for very large Reynolds number, *Dokl. Akad. Nauk. SSSR* **30**, 9 (1941a).

Kolmogorov, A. N., Energy dissipation in locally isotropic turbulence, *Dokl. Akad. Nauk. SSSR* **32**, 19 (1941b).

Kondaurova, L., and S. K. Nemirovskii, Full Biot–Savart numerical simulation of vortices in He II, *J. Low Temp. Phys.* 138, 555 (2005).

Koplik, J., and H. Levine, Vortex reconnection in superfluid helium, *Phys. Rev. Lett.* **71**, 1375 (1993).

Korhonen, J. S., A. D. Gongadze, Z. Janu, Y. Kondo, M. Krusius, Y. M. Mukharsky, and E. V. Thuneberg, Order parameter textures and boundary conditions in rotating vortex-free ^3B, *Phys. Rev. Lett.* **65**, 1211 (1990).

Kotsubo, V., and G. W. Swift, Vortex turbulence generated by second sound in superfluid He, *Phys. Rev. Lett.* **62**, 2604 (1989).

Kotsubo, V., and G. W. Swift, Generation of superfluid vortex turbulence by high-amplitude second sound in ^{4}He, *J. Low Temp. Phys.* **78**, 351 (1990).

Kozik, E., and B. V. Svistunov, Kelvin-wave cascade and decay of superfluid turbulence, *Phys. Rev. Lett.* **92**, 035301 (2004).

Krstulovic, G., Kelvin-wave cascade and dissipation in low-temperature superfluid vortices, *Phys. Rev. E* **86**, 055301 (2012).

Kruglov, V. I., Critical velocities and two mechanisms of transition in superfluid liquid ^{4}He, *Phys. Lett. A* **375**, 4058 (2011).

Kubo, W., and Y. Tsuji, Lagrange trajectory of small particles in superfluid He II, *J. Low Temp. Phys.* **187**, 611 (2017).

Kursa, M., K. Bajer, and T. Lipniacki, Cascade of vortex loops initiated by a single reconnection of quantum vortices, *Phys. Rev. B* **83**, 014515 (2011).

Kwan, W. J., J. H. Kim, S. W. Seo, and Y. Shin, Observation of von Karman vortex street in a Bose–Einstein condensate, *Phys. Rev. Lett.* **117**, 245301 (2016).

Ladner, R., and J. T. Tough, Temperature and velocity dependence of superfluid turbulence, *Phys. Rev. B* **20**, 2690 (1979).

La Mantia, M., Particle trajectories in thermal counterflow of superfluid helium in a wide channel of square cross section, *Phys. Fluids* **28**, 024102 (2016).

La Mantia, M., and L. Skrbek, Quantum, or classical turbulence? *Europhys. Lett.* **102**, 46002 (2014a).

La Mantia, M., and L. Skrbek, Quantum turbulence visualized by particle dynamics, *Phys. Rev. B* **90**, 014519 (2014b).

La Mantia, M., T. V. Chagovets, M. Rotter, and L. Skrbek, Testing the performance of a cryogenic visualization system on thermal counterflow by using hydrogen and deuterium solid tracers, *Rev. Sci. Instrum.* **83**, 055109 (2012).

La Mantia, M., D. Duda, M. Rotter, and L. Skrbek, Lagrangian accelerations of particles in superfluid turbulence, *J. Fluid Mech.* **770**, R9 (2013).

La Mantia, M., P. Svancara, D. Duda, and L. Skrbek, Small-scale universality of particle dynamics in quantum turbulence, *Phys. Rev. B* **94**, 184512 (2016).

Landau, L. D., The theory of superfluidity of helium II, *J. Phys. USSR* **5**, 356 (1941).

Landau, L. D., On the theory of superfluidity of helium II, *J. Phys. USSR* **11**, 91 (1947).

Landau, L. D., and E. M. Lifshitz, *Fluid Mechanics*, 2nd ed., Pergamon Press (1987).

Leadbeater, M., T. Winiecki, D. C. Samuels, C. F. Barenghi, and C.S Adams, Sound emission due to superfluid vortex reconnections, *Phys. Rev. Lett.* **86**, 1410 (2001).

Leadbeater, M., D. C. Samuels, C. F. Barenghi, and C. S. Adams, Decay of superfluid vortices due to Kelvin wave radiation, *Phys. Rev. A* **67**, 015601 (2002).

Leith, C. E., Diffusion approximation to internal energy transfer in isotropic turbulence, *Phys. Fluids* **10**, 1409 (1967).

Leonard, A., Computing three-dimensional incompressible flows with vortex elements, *Ann. Rev. Fluid. Mech.* **17**, 523 (1985).

Leonardo da Vinci, *Studies of Water*, 1507. Web Gallery of Art, wga.hu/frames-e.html?/html/l/leonardo/12engine/2water2.html.

Léveque, E., and A. Naso, Introduction of longitudinal and transverse Lagrangian velocity increments in homogeneous and isotropic turbulence, *Europhys. Lett.* **108**, 54004 (2014).

Lipniacki, T., Dynamics of superfluid ^{4}He: Two-scale approach, *Eur. J. Mech. B/Fluids* **25**, 435 (2006).

London, F., The λ-phenomenon of liquid helium and the Bose–Einstein degeneracy, *Nature* **141**, 643 (1938).

Lu, S. T., and H. Kojima, Observation of second sound in superfluid He 3-B, *Phys. Rev. Lett.* **55**, 1677 (1985).

Luzuriaga, J., Measurements in the laminar and turbulent regime of superfluid ^{4}He by means of an oscillating sphere, *J. Low Temp. Phys.* **108**, 267 (1997).

L'vov, V. S., and S. Nazarenko, Spectrum of Kelvin-wave turbulence in superfluids, *JETP Lett.* **91**, 428 (2010).

L'vov, V. S., and A. Pomyalov, Statistics of quantum turbulence in superfluid He, *J. Low Temp. Phys.* **187**, 497 (2017).

L'vov, V. S., E. Podivilov, A. Pomyalov, I. Procaccia, and D. Vandembroucq, Improved shell model of turbulence, *Phys. Rev. E* **1811**, (1998).

L'vov, V. S., S. V. Nazarenko, and G. E. Volovik, Energy spectra of developed superfluid turbulence, *JETP Lett.* **80**, 479 (2004).

L'vov, V. S., S. V. Nazarenko, and O. Rudenko, Bottleneck crossover between classical and quantum superfluid turbulence, *Phys. Rev. B* **76**, 024520 (2007).

L'vov, V. S., L. Skrbek, and K. R. Sreenivasan, Viscosity of liquid ^{4}He and quantum of circulation: Are they related? *Phys. Fluids* **26**, 041703 (2014).

Mäkinen, J. T., S. Autti, P. J. Heikkinen, J. J. Hosio, R. Hänninen, V. S. L'vov, P. M. Walmsley, V. V. Zavjalov, and V. B. Eltsov, Rotating quantum wave turbulence, *Nat. Phys.* (2023). https://doi.org/10.1038/s41567-023-01966-z

Marakov, A., J. Gao, W. Guo, S. W. Van Sciver, G. G. Ihas, D. N. McKinsey, and W. F. Vinen, Visualization of the normal-fluid turbulence in counterflowing superfluid ^{4}He, *Phys. Rev. B* **91**, 094503 (2015).

Martin, K. P., and J. T. Tough, Evolution of superfluid turbulence in thermal counterflow, *Phys. Rev. B* **27**, 2788 (1983).

Mastracci, B., and W. Guo, Exploration of thermal counterflow in He II using particle tracking velocimetry, *Phys. Rev. Fluids* **3**, 063304 (2018).

Mastracci, B., S. Bao, W. Guo, and W. F. Vinen, Particle tracking velocimetry applied to thermal counterflow in superfluid ^{4}He: Motion of the normal fluid at small heat fluxes, *Phys. Rev. Fluids* **4**, 083305 (2019).

Mathieu, P., B. Placais, and Y. Simon, Spatial-distribution of vortices and anisotropy of mutual friction in rotating He II, *Phys. Rev. B* **29**, 2489 (1984).

Maurer, J., and P. Tabeling, Local investigation of superfluid turbulence, *Europhys. Lett.* **43**, 29 (1998).

Maxey, M. R., and J. J. Riley, Equation of motion for a small rigid sphere in a nonuniform flow, *Phys. Fluids* **26**, 883 (1983).

McCarty, R. D., *Thermophysical Properties of Helium-4 from 2 to 1500 K with Pressures to 1000 Atmospheres*, Technical Note 631, National Bureau of Standards (1972).

McKinsey, D. N., W. H. Lippincott, J. A. Nikkel, and W. G. Rellegert, Trace detection of metastable helium molecules in superfluid helium by laser-induced fluorescence, *Phys. Rev. Lett.* **95**, 111101 (2005).

Meichle, D. P., and D. P. Lathrop, Nanoparticle dispersion in superfluid helium, *Rev. Sci. Instrum.* **85**, 073705 (2014).

Melotte, D. J., and C. F. Barenghi, Transition to normal fluid turbulence in helium II, *Phys. Rev. Lett.* **80**, 4181 (1998).

Mendelssohn, K., and W. A. Steele, Quantized growth of turbulence in He II, *Proc. Phys. Soc.* **73**, 144 (1959).

Meneveau, C., and K. R. Sreenivasan, Simple multifractal cascade model for fully developed turbulence, *Phys. Rev. Lett.* 59, 1424 (1987).

Mesgarnezhad, M., R. G. Cooper, A. W. Baggaley, and C. F. Barenghi, Helicity and topology of a small region of quantum vorticity, *Fluid Dyn. Res* **50**, 011403 (2018).

Midlik, Š., *Generation and Detection of Quantum Turbulence in He II by Second Sound.* Diploma thesis, Charles University (2019).

Midlik, Š., D. Schmoranzer, and L. Skrbek, Transition to quantum turbulence in oscillatory thermal counterflow of ^{4}He, *Phys. Rev. B* **103**, 134516 (2021).

Milliken, F. P., K. W. Shwarz, and C. W. Smith, Free decay of superfluid turbulence, *Phys. Rev. Lett.* **48**, 1204 (1982).

Mocz, P., M. Vogelsberger, V. H. Robles, J. Zavala, M. Boylan-Kolchin, A. Fialkov, and L. Hernquist, Galaxy formation with BECDM I. Turbulence and relaxation of idealized haloes, *Mon. Notices Royal Astron. Soc.* **471**, 4559 (2017).

Moffatt, H. K., Degree of knottedness of tangled vortex line, *J. Fluid Mech.* **35**, 117 (1969).

Moffatt, H. K., and R. L. Ricca, Helicity and the Călugăreanu invariant, *Proc. Roy. Soc.* A **439**, 411 (1992).

Mongiovì, M. S., D. Jou, and M. Sciacca, Non-equilibrium thermodynamics, heat transport and thermal waves in laminar and turbulent superfluid helium, *Phys. Rep.* **726**, 1 (2018).

Mordant, N., A. M. Crawford, and E. Bodenschatz, Three-dimensional structure of the Lagrangian acceleration in turbulent flows, *Phys. Rev. Lett.* **93**, 214501 (2004).

Morishita, M., T. Kuroda, A. Sawada, and T. Satoh, Mean free path effects in superfluid ^{4}He, *J. Low Temp. Phys.* **76**, 387 (1989).

Müller, N. P., J. I. Polanco, and G. Krstulovic, Intermittency of velocity circulation in quantum turbulence, *Phys. Rev. X* **11**, 011053 (2021).

Murakami, M., M. Hanada, and T. Yamazaki, Flow visualization study on large-scale vortex ring in He II, *Jpn. J. Appl. Phys.* **26**, Suppl. 26–3, 107 (1987).

Murphy, P. J., J. Castiglione, and J. T. Tough, Superfluid turbulence in a nonuniform circular channel, *J. Low Temp. Phys.* **92**, 307 (1993).

Nago, Y., T. Ogawa, K. Obara, H. Yano, O. Ishikawa, and T. Hata, Time-of-flight experiments of vortex rings propagating from turbulent region of superfluid ^{4}He at high temperature, *J. Low Temp. Phys.* **162**, 322 (2011).

Navon, N., A. L. Gaunt, R. P. Smith, and Z. Hadzibabic, Emergence of a turbulent cascade in a quantum gas, *Nature* **539**, 72 (2016).

Nazarenko, S., and R. West, Analytical solution for nonlinear Schrödinger vortex reconnection, *J. Low Temp. Phys.* **132**, 1 (2003).

Nemirovskii, S. K., Diffusion of inhomogeneous vortex tangle and decay of superfluid turbulence, *Phys. Rev. B* **81**, 064512 (2010).

Nemirovskii, S. K., Quantum turbulence: Theoretical and numerical problems, *Phys. Reports* **524**, 85 (2012).

Nemirovskii, S. K., Comment on dynamics of the density of quantized vortex lines in superfluid turbulence, *Phys. Rev. B* **94**, 146501 (2016).

Nemirovskii, S. K., and W. Fiszdon, Chaotic quantized vortices and hydrodynamic processes in superfluid helium, *Rev. Mod. Phys.* **67**, 37 (1995).

Neely, T. W., A. S. Bradley, E. C. Samson, S. J. Rooney, E. M. Wright, K. J. H. Law, R. Carretero-Gonzales, P. G. Kevrekidis, M. J. Davis, and B. P. Anderson, Characteristics of two-dimensional quantum turbulence in a compressible superfluid, *Phys. Rev. Lett.* **111**, 235301 (2013).

Nichol, H. A., L. Skrbek, P. C. Hendry, and P. V. E. McClintock, Flow of He II due to an oscillating grid in the low-temperature limit, *Phys. Rev. Lett.* **92**, 244501 (2004).

Niemela, J. J., and K. R. Sreenivasan, The use of cryogenic helium for classical turbulence: Promises and hurdles, *J. Low Temp. Phys.* **143**, 163 (2006).

Niemela, J. J., and K. R. Sreenivasan, Russell Donnelly and his leaks, *Annu. Rev. Cond. Matt. Phys.* **13**, 33 (2022).

Niemela, J. J., K. R. Sreenivasan, and R. J. Donnelly, Grid generated turbulence in helium II, *J. Low Temp. Phys.* **138**, 537 (2005).

Niemetz, M., H. Kercher, and W. Schoepe, Intermittent switching between potential flow and turbulence in superfluid helium at mK tempertures, *J. Low Temp. Phys.* **126**, 287 (2002).

Niemetz, M., H. Kercher, and W. Schoepe, Stability of laminar and turbulent flow of superfluid ^{4}He at mK temperatures around an oscillating microsphere, *J. Low Temp. Phys.* **135**, 447 (2004).

Nore, C., M. Abid, and M. Brachet, Kolmogorov turbulence in low-temperature superflows, *Phys. Rev. Lett.* **78**, 3296 (1997).

Onsager, L., in discussion on paper by C. J. Gorter, *Nuovo Cimento* **6**, suppl. 2, 249 (1949); "Introductory talk", *Proc. Int. Conf. Theor. Phys., Kyoto and Tokyo*, p. 877 (September 1953); for further details see Chapter II of Donnelly (1991).

Ortiz, G., and D. M. Ceperley, Core structure of a vortex in superfluid ^{4}He, *Phys. Rev. Lett.* **75**, 4642 (1995).

Osborne, D. V., The rotation of liquid helium-II, *Proc. Phys. Soc. A* **63**, 909 (1950).

Osborne, D. R., J. C. Vassilicos, K. Sung, and J. D. Haigh, Fundamentals of pair diffusion in kinematic simulations of turbulence, *Phys. Rev. E* **74**, 036309 (2006).

Ostermeier, R. M., and W. I. Glaberson, Instability of vortex lines in the presence of axial normal fluid flow, *J. Low Temp. Phys.* **21**, 191 (1975).

Outrata, O., M. Pavelka, J. Hron, M. La Mantia, J. I. Polanco, and G. Krstulovic, On the determination of vortex ring vorticity using Lagrangian particles, *J. Fluid Mech.* **924**, A44 (2021).

Paoletti, M. S., M. E. Fisher, K. R. Sreenivasan, and D. P. Lathrop, Velocity statistics distinguish quantum turbulence from classical turbulence, *Phys. Rev. Lett.* **101**, 154501 (2008a).

Paoletti, M. S., R. B. Fiorito, K. R. Sreenivasan, and D. P. Lathrop, Visualization of superfluid helium flow, *J. Phys. Soc. Japan* **77**, 111007 (2008b).

Paoletti, M. S., M. E. Fisher, and D. P. Lathrop, Reconnection dynamics for quantized vortices, *Physica D* **239**, 1367 (2010).

Parker, N. G., *Numerical Studies of Vortices and Dark Solitons in Atomic Bose–Einstein Condensates*, Ph.D. thesis, University of Durham (2004).

Patrick, S., A. Geelmuyden, S. Erne, C. F. Barenghi, and S. Weinfurtner, Origin and evolution of the multiply-quantised vortex instability, *Phys. Rev. Res.* **4**, 043104 (2022).

Peshkov, V. P., and V. J. Tkachenko, Kinetics of the destruction of superfluidity in helium, *Sov. Phys. JETP* **14**, 1019 (1962).

Poole, D. R., C. F. Barenghi, Y. A. Sergeev, and W. F. Vinen, The motion of tracer particles in He II, *Phys. Rev. B* **71**, 064514 (2005).

Proment, D., M. Onorato, and C. F. Barenghi, Vortex knots in a Bose–Einstein condensate, *Phys. Rev. E* **85**, 036306 (2012).

Proukakis N. P., and B. Jackson, Finite-temperature models of Bose–Einstein condensation, *J. Phys. B. At. Mol. Opt.* **41**, 203002 (2008).

Pumir, A., H. Xu, E. Bodenschatz, and R. Grauer, Single-particle motion and vortex stretching in three-dimensional turbulent flows, *Phys. Rev. Lett.* **116**, 124502 (2016).

Qureshi, N. M., M. Bourgoin, C. Baudet, A. Cartellier, and Y. Gagne, Turbulent transport of material particles: An experimental study of finite size effects, *Phys. Rev. Lett.* **99**, 184502 (2007).

Qureshi, N. M., U. Arrieta, C. Baudet, A. Cartellier, Y. Gagne, and M. Bourgoin, Acceleration statistics of inertial particles in turbulent flow, *Eur. Phys. J. B* **66**, 531 (2008).

Raffel, M., C. E. Villert, S. T. Werely, and J. Kompenhans, *Particle Image Velocimetry – A Practical Guide,* 2nd ed., Springer (2007)

Rast, M. P., and J.-F. Pinton, Point-vortex model for Lagrangian intermittency in turbulence, *Phys. Rev. E* **79**, 046314 (2009).

Rayfield, G. W., and F. Reif, Quantized vortex rings in superfluid helium, *Phys. Rev.* **136**, A1194 (1964).

Reynolds, O., An experimental investigation of the circumstances which determine whether the motion of water shall be direct or sinuous and the law of resistance in parallel channels, *Phil. Trans. Roy. Soc. London* **174**, 935 (1883).

Ricca, R. L., D. C. Samuels, and C. F. Barenghi, Evolution of vortex knots, *J. Fluid Mech.* **391**, 29 (1999).

Rickinson, E., N. G. Parker, A. W. Baggaley, and C. F. Barenghi, Diffusion of quantum vortices, *Phys. Rev. A* **98**, 023608 (2018).

Rickinson, E., N. G. Parker, A. W. Baggaley, and C. F. Barenghi, Inviscid diffusion of vorticity in low temperature superfluid helium, *Phys. Rev. B* **99**, 224501 (2019).

Rickinson, E., C. F. Barenghi, Y. A. Sergeev, A. W. Baggaley, Superfluid turbulence driven by cylindrically symmetric thermal counterflow, *Phys. Rev. B* **101**, 134519 (2020).

Riley, N., Steady streaming, *Annu. Rev. Fluid Mech.* **33**, 43 (2001).

Roche, P.-E., P. Diribarne, T. Didelot, O. Francais, L. Rousseau, and H. Willaime, Vortex density spectrum of quantum turbulence, *Europhys. Lett.* **77**, 66002 (2007).

Roche, P.-E., C. F. Barenghi, and E. Leveque, Quantum turbulence at finite temperature: The two-fluids cascade, *Europhys. Lett.* **87**, 54006 (2009).

Rorai, C., J. Skipper, R. M. Kerr, and K. R. Sreenivasan, Approach and separation of quantum vortices with balanced cores, *J. Fluid Mech.* **808**, 641 (2016).

Rousset, B., P. Bonnay, P. Diribarne, A. Girard, J. M. Poncet, E. Herbert, J. Salort, C. Baudet, B. Castaing, L. Chevillard, F. Daviaud, B. Dubrulle, Y. Gagne, M. Gibert, B. Hébral, Th. Lehner, P.-E. Roche, B. Saint-Michel, and M. Bon Mardion, Superfluid high Reynolds von Kármán experiment, *Rev. Sci. Instrum.* **85**, 103908 (2014).

Rusaouen, E., B. Chabaud, J. Salort, and P.-E. Roche, Intermittency of quantum turbulence with superfluid fractions from 0% to 96%, *Phys. Fluids* **29**, 105108 (2017a).

Rusaouen, E., B. Rousset, and P.-E. Roche, Detection of vortex coherent structures in superfluid turbulence, *Europhys. Lett.* **118** 14005 (2017b).

Ruutu, V. M. H., V. B. Eltsov, A. J. Gill, T. W. B. Kibble, M. Krusius, Yu. G. Makhlin, B. Placais, G. E. Volovik, and W. Xu, Vortex formation in neutron-irradiated superfluid ^{3}He as an analogue of cosmological defect formation, *Nature* **382**, 334 (1996).

Rysti, J., M. S. Manninen, and J. Tuoriniemi, Quartz tuning fork and acoustic phenomena: Application to superfluid helium, *J. Low Temp. Phys.* **177**, 133 (2014).

Sadd, M., G. V. Chester, and L. Reatto, Structure of a vortex in superfluid ^{4}He, *Phys. Rev. Lett.* **79**, 2490 (1997).

Saffman, P. G., The large-scale structure of homogeneous turbulence, *J. Fluid Mech.* **27**, 581 (1967a).

Saffman, P. G., Note on decay of homogeneous turbulence, *Phys. Fluids* **10**, 1349 (1967b).

Saint-Michel, B., E. Herbert, J. Salort, C. Baudet, M. Bon Mardion, P. Bonnay, M. Bourgoin, B. Castaing, L. Chevillard, F. Daviaud, P. Diribarne, B. Dubrulle, Y. Gagne, M. Gibert, A. Girard, B. Hébral, Th. Lehner, B. Rousset, and SHREK Collaboration, Probing quantum and classical turbulence analogy in von Kármán liquid helium, nitrogen, and water experiments, *Phys. Fluids* **26**, 125109 (2014).

Salman, H., Breathers on quantized superfluid vortices, *Phys. Rev. Lett.* **111**, 165301 (2013).

Salman, H., Helicity conservation and twisted Seifert surfaces for superfluid vortices, *Proc. Roy. Soc. A* **473**, 20160853 (2018).

Salort, J., C. Baudet, B. Castaing, B. Chabaud, F. Daviaud, T. Didelot, P. Diribarne, B. Dubrulle, Y. Gagne, F. Gauthier, A. Girard, B. Hebral, B. Rousset, P. Thibault, and P. E. Roche, Turbulent velocity spectra in superfluid flows, *Phys. Fluids* **22**, 125102 (2010).

Salort, J., B. Chabaud, E. Leveque, and P-E. Roche, Investigation of intermittency in superfluid turbulence, *J. Phys. Conf. Ser.* **318**, 042014 (2011a).

Salort, J., P.-E. Roche, and E. Leveque, Mesoscale equaipartion of kinetic energy in quantum turbulence, *Europhys. Lett.* **94**, 24001 (2011b).

Salort, J., B. Chabaud, E. Leveque, and P-E. Roche, Energy cascade and the four-fifths law in superfluid turbulence, *Europhys. Lett.* **97**, 34006 (2012a).

Salort, J., A. Monfardini, and P-E. Roche, Cantilever anemometer based on a superconducting micro-resonator: Application to superfluid turbulence. *Rev. Sci. Instrum.* **83**, 125002 (2012b).

Salort, J., E. Rusaouën, L. Robert, R. du Puis, A. Loesch, O. Pirotte, P. E. Roche, B. Castaing, and F. Chillà, A local sensor for joint temperature and velocity measurements in turbulent flows, *Rev. Sci. Instrum.* **89**, 015005 (2018).

Salort, J., F. Chillà, E. Rusaouën, P. E. Roche, M. Gibert, I. Moukharski, A. Braslau, F. Daviaud, B. Gallet, E.-W. Saw, B. Dubrulle, P. Diribarne, B. Rousset, M. Bon Mardion, J.-P. Moro, A. Girard, C. Baudet, V. L'vov, A. Golov, and S. Nazarenko, Experimental signature of quantum turbulence in velocity spectra? *New J. Phys.* **23**, 063005 (2021).

Samuels, D. C., Response of superfluid vortex filaments to concentrated normal-fluid vorticity, *Phys. Rev. B* **47**, 1107 (1993).

Samuels, D.C. and C.F. Barenghi, Vortex heating in superfluid helium at low temperatures, *Phys. Rev. Lett.* **81**, 4381 (1998).

Sasa, N., T. Kano, M. Machida, V. S. L'vov, O. Rudenko, and M. Tsubota. Energy spectra of quantum turbulence: Large-scale simulation and modeling, *Phys. Rev. B* **84**, 054525 (2011).

Sasaki, K., N. Suzuki, and H. Saito, Benard–von Karman vortex street in a Bose–Einstein condensate, *Phys. Rev. Lett.* **104**, 150404 (2010).

Sato, A., M. Maeda, and Y. Kamioka, Normalized representation for steady state heat transport in a channel containing He II covering pressure range up to 1.5 MPa. In *Proc. of 20th Int. Cryogenic Conf.* (ICEC20) **97**, Chapter 20, 849–852 (2005).

Schmoranzer, D., M. Rotter, J. Sebek, and L. Skrbek, Experimental setup for probing a von Karman type flow of normal and superfluid helium. In *Experimental Fluid Mechanics 2009, Proc. Int. Conf.* (Technical University of Liberec, Liberec, Czech Republic), 304, 309 (2009).

Schmoranzer, D., M. La Mantia, G. Sheshin, I. Gritsenko, A. Zadorozhko, M. Rotter, and L. Skrbek, Acoustic emission by quartz tuning forks and other oscillating structures in cryogenic ^{4}He fluids, *J. Low Temp. Phys.* **163**, 317 (2011).

Schmoranzer, D., M. J. Jackson, V. Tsepelin, M. Poole, A. J. Woods, M. Clovecko, and L. Skrbek, Multiple critical velocities in oscillatory flow of superfluid ^{4}He due to quartz tuning forks, *Phys. Rev. B* **94**, 214503 (2016).

Schmoranzer, D., M. J. Jackson, S. Midlik, M. Skyba, J. Bahyl, T. Skokankova, V. Tsepelin, and L. Skrbek, Dynamical similarity and instabilities in high-Stokes-number oscillatory flows of superfluid helium, *Phys. Rev. B* **99**, 054511 (2019).

Schumacher, J., J. D. Scheel, D. Krasnov, D. A. Donzis, V. Yakhot, and K. R. Sreenivasan, Small-scale universality in fluid turbulence, *Proc. Natl. Acad. Sci. USA* **111**, 10961 (2014).

Schwarz, K. W., Generation of superfluid turbulence deduced from simple dynamical rules, *Phys. Rev. Lett.* **49**, 283 (1982).

Schwarz, K. W., Three-dimensional vortex dynamics in superfluid ^{4}He: Line–line and line–boundary interactions, *Phys. Rev. B* **31**, 5782 (1985).

Schwarz, K. W., Three-dimensional vortex dynamics in superfluid ^{4}He. Homogeneous superfluid turbulence, *Phys. Rev. B* **38**, 2398 (1988).

Schwarz, K. W., Phase slip and turbulence in superfluid ^{4}He: A vortex mill that works, *Phys. Rev. Lett.* **64**, 1130 (1990).

Schwarz, K. W., and J. R. Rozen, Transient behavior of superfluid turbulence in a large channel, *Phys. Rev. B* **44**, 7563 (1991).

Schwarz, K. W., and C. W. Smith, Pulsed-ion study of ultrasonically generated turbulence in superfluid ^{4}He, *Phys. Lett. A* **82** 251 (1981).

Sciacca, M., D. Jou, and M. S. Mongiovì, Effective heat conductivity of helium II: From Landau to Gorter–Mellink regimes, *Z. Angew. Math. Phys.* **66**, 1835 (2014).

Serafini, S., M. Barbiero, M. Debortoli, S. Donadello, F. Larcher, F. Dalfovo, G. Lamporesi, and G. Ferrari, Dynamics and interaction of vortex lines in an elongated Bose–Einstein condensate, *Phys. Rev. Lett.* **115**, 170402 (2015).

Serafini, S., L. Galantucci, E. Iseni, T. Bienaimé, R. N. Bisset, C. F. Barenghi, F. Dalfovo, G. Lamporesi, and G. Ferrari, Vortex reconnections and rebounds in trapped atomic Bose–Einstein condensates, *Phys. Rev. X* **7**, 021031 (2017).

Sergeev, Y. A., and C. F. Barenghi, Normal fluid eddies in the thermal counterflow past a cylinder, *J. Low Temp. Phys.* **156**, 268 (2009a).

Sergeev, Y. A., and C. F. Barenghi, Particle–vortex interactions and flow visualization in ^{4}He, *J. Low Temp. Phys.* **157**, 429 (2009b).

Sergeev, Y. A., and C. F. Barenghi, Turbulent radial thermal counterflow in the framework of the HVBK model, *Europhys. Lett.* **128**, 26001 (2019).

Sergeev, Y. A., C. F. Barenghi, and D. Kivotides, Motion of micron-size particles in turbulent helium II, *Phys. Rev. B* **74**, 184506 (2006).

She, Z.-S., and E. Leveque, Universal scaling laws in fully developed turbulence, *Phys. Rev. Lett.* **72**, 336 (1994).

Sherwin-Robson, L. K., A. W. Baggaley, and C. F. Barenghi, Local and nonlocal dynamics in superfluid turbulence, *Phys. Rev. B* **91**, 104517 (2015).

Shukla, V., R. Pandit, and M. Brachet, Particles and fields in superfluids: Insight from the two-dimensional Gross–Pitaevskii equation, *Phys. Rev. A* **97**, 013627 (2018).

Sikora, G., B. Krzysztof, and W. Agnieszka, Mean-squared displacement statistical test for fractional Brownian motion, *Phys. Rev. E* **95**, 032110 (2017).

Silaev, M. A., Universal mechanism of dissipation in Fermi superfluids at ultralow temperatures, *Phys. Rev. Lett.* **108**, 045303 (2012).

De Silva, I. P. D., and H. J. S. Fernando, Oscillating grids as a source of nearly isotropic turbulence, *Phys. Fluids* **6**, 2455 (1994).

Sinhuber, M., E. Bodenschatz, and G. P. Bewley, Decay of turbulence at high Reynolds numbers, *Phys. Rev. Lett.* **114**, 034501 (2015).

Skrbek, L., and K. R. Sreenivasan, Developed quantum turbulence and its decay, *Phys. Fluids* **24**, 011301 (2012).

Skrbek, L., and K. R. Sreenivasan, How similar is quantum turbulence to classical turbulence? In *Ten Chapters in Turbulence*, P. A. Davidson, Y. Kaneda, and K. R. Sreenivasan (eds.). Cambridge University Press (2013).

Skrbek, L., and S. R. Stalp, On the decay of homogeneous isotropic turbulence, *Phys. Fluids* **12**, 1997 (2000).

Skrbek, L., and W. F. Vinen, The use of vibrating structures in the study of quantum turbulence, In *Progress in Low Temp. Phys.*, vol. XVI, M. Tsubota and W. P. Halperin (eds), Elsevier (2009).

Skrbek, L., J. J. Niemela, and R. J. Donnelly, Turbulent flows at cryogenic temperatures: A new frontier, *J. Phys. Condens. Matter* **11**, 7761 (1999).

Skrbek, L., J. J. Niemela, and R. J. Donnelly, Four regimes of decaying grid turbulence in a finite channel, *Phys. Rev. Lett.* **85**, 2973 (2000).

Skrbek, L., J. J. Niemela, and K. R. Sreenivasan, Energy spectrum of grid-generated He II turbulence, *Phys. Rev. E* **64**, 067301 (2001).

Skrbek, L., A. V. Gordeev, and F. Soukup, Decay of counterflow He II turbulence in a finite channel: Possibility of missing links between classical and quantum turbulence, *Phys. Rev. E* **67**, 047302 (2003).

Skrbek, L., D. Schmoranzer, Š. Midlik, and K. R. Sreenivasan, Phenomenology of quantum turbulence in superfluid helium, *Proc. Nat. Acad. Sci. USA* **118**, e2018406118 (2021).

Smith, M. R., R. J. Donnelly, N. Goldenfeld, and W. F. Vinen, Decay of vorticity in homogeneous turbulence, *Phys. Rev. Lett.* **71**, 2583 (1993).

Smith, M. R., D. K. Hilton, and S. W. Van Sciver, Observed drag crisis on a sphere in flowing He I and He II, *Phys. Fluids* **11**, 1 (1999).

Snyder, H. A., and Z. Putney, Angular dependence of mutual friction in rotating helium 2, *Phys. Rev.* **150**, 110 (1966).

Sonin, E. B., Vortex oscillations and hydrodynamics of rotating superfluids, *Rev. Mod. Phys.* **59**, 87 (1987).

Sreenivasan, K. R., On the scaling of the turbulence energy dissipation rate, *Phys. Fluids* **27**, 1048 (1984).

Sreenivasan, K. R., On the universality of the Kolmogorov constant, *Phys. Fluids* **7**, 27788 (1995).

Sreenivasan, K. R., An update on the energy dissipation rate in isotropic turbulence. *Phys. Fluids* **10**, 528 (1998).

Sreenivasan, K. R., Turbulent mixing: A perspective, *Proc. Nat. Acad. Sci.* **116**, 18175 (2018).

Sreenivasan, K. R., and R. A. Antonia, Skewness of temperature derivatives in turbulent shear flows, *Phys. Fluids* **20**, 1986 (1977).

Sreenivasan, K. R., and R. A. Antonia, The phenomenology of small-scale turbulence, *Annu. Rev. Fluid Mech.* **29**, 435 (1997).

Sreenivasan, K. R., and R. Narasimha, Rapid distortion of axisymmetric turbulence, *J. Fluid Mech.* **84**, 497 (1978).

Sreenivasan, K. R., and V. Yakhot, Dynamics of three-dimensional turbulence from Navier–Stokes equations, *Phys. Rev. Fluids* **6**, 104604 (2021).

Sreenivasan, K. R., S. Tavoularis, R. Henry, and S. Corrsin, Temperature fluctuations and scales in grid-generated turbulence, *J. Fluid Mech.* **100**, 597 (1980).

Sreenivasan, K. R., A. Juneja, and A. K. Suri, Scaling properties of circulation in moderate-Reynolds-number turbulent wakes, *Phys. Rev. Lett.* **75**, 433 (1995).

Stagg, G. W., N. P. Parker, and C. F. Barenghi, Ultraquantum turbulence in a quenched homogeneous gas, *Phys. Rev. A* **94**, 053632 (2016).

Stalp, S. R., *Decay of Grid Turbulence in Superfluid Helium,* Ph.D. Thesis, University of Oregon (1998).

Stalp, S. R., L. Skrbek, and R. J. Donnelly, Decay of grid turbulence in a finite channel, *Phys. Rev. Lett.* **82**, 4831 (1999).

Stalp, S. R., J. J. Niemela, W. F. Vinen, and R. J. Donnelly, Dissipation of grid turbulence in helium II, *Phys. Fluids* **14**, 1377 (2002).

Švančara, P., and M. La Mantia, Flows of He-4 due to oscillating grids, *J. Fluid Mech.* **832**, 578 (2017).

Švančara, P., and M. La Mantia, Flight-crash events in superfluid turbulence, *J. Fluid Mech.* **876**, R2 (2019).

Švančara, P., P. Hrubcová, M. Rotter, and M. La Mantia, Visualization study of thermal counterflow of superfluid helium in the proximity of the heat source by using solid deuterium hydride particles, *Phys. Rev. Fluids* **3**, 114701 (2018).

Švančara, P., M. Pavelka, and M. La Mantia, An experimental study of turbulent vortex rings in superfluid ^{4}He, *J. Fluid Mech.* **889**, A24 (2020).

Švančara, P., D. Duda, P. Hrubcova, M. Rotter, L. Skrbek, M. La Mantia, E. Durozoy, P. Diribarne, B. Rousset, M. Bourgoin, and M. Gibert, Ubiquity of particle–vortex interactions in turbulent counterflow of superfluid helium, *J. Fluid Mech.* **911**, A8 (2021).

Svistunov, B. V., Superfluid turbulence in the low-temperature limit, *Phys. Rev. B* **52**, 3647 (1995).

Swanson, C. J., K. Johnson, and R. J. Donnelly, An accurate differential pressure gauge for use in liquid and gaseous helium, *Cryogenics* **38**, 673 (1998).

Swanson, C. J., B. Julian, G. G. Ihas, and R. J. Donnelly, Pipe flow measurements over a wide range of Reynolds numbers using liquid helium and various gases, *J. Fluid Mech.* **461**, 51 (2002).

Taylor, G. I., Statistical theory of turbulence, Parts I, II, III, and IV, *Proc. Roy. Soc.* **151**, 421, 444, 455, 465 (1935).

Tebbs, R., A. J. Youd, and C. F. Barenghi, The approach to vortex reconnection, *J. Low Temp. Phys.* **162**, 314 (2011).

Thomson (Lord Kelvin), W., On vortex atoms, *Phil. Mag.* **34**, 15 (1867).

Thomson (Lord Kelvin), W., Vibrations of a columnar vortex, *Proc. Roy. Soc. Edinb.* **10**, 443 (1880).

Tilley, D. R., and J. Tilley, *Superfluidity and Superconductivity*, Adam Hilger (1986).

Tizsa, L., Transport phenomena in helium II, *Nature* **141**, 913 (1938).

Toschi, F., and E. Bodenschatz, Lagrangian properties of particles in turbulence, *Annual Rev. Fluid Mech.* **41**, 375 (2009).

Tough, J. T., Superfluid turbulence, In *Progress in Low Temperature Physics* Vol. VIII. North Holland (1982).

Tough, J. T., J. F. Kafkalidis, G. Klinich, P. J. Murphy, and J. Castiglione, Stationary superfluid turbulent fronts, *Physica B* **194–196**, 719 (1994).

Touil, H., J. P. Bertoglio, and L. Shao, The decay of turbulence in a bounded domain, *J. Turb.* **3**, 049 (2002).

Tsatsos, M. C., P. E. S. Tavares, A. Cidrim, A. R. Fritsch, M. A. Caracanhas, F. E. A. dos Santos, C. F. Barenghi, and V. S. Bagnato, Quantum turbulence in trapped atomic Bose–Einstein condensates, *Phys. Reports* **622**, 1 (2016).

Tsepelin, V., A. W. Baggaley, Y. A. Sergeev, C. F. Barenghi, S. N. Fisher, G. R. Pickett, M. J. Jackson, and N. Suramlishvili, Visualization of quantum turbulence in superfluid ^{3}He-B: Combined numerical and experimental study of Andreev reflection. *Phys. Rev. B* **96**, 054510 (2017).

Tsubota, M., and H. Adachi, Simulation of counterflow turbulence by vortex filaments, *J. Low Temp. Phys.* **162**, 367 (2011).

Tsubota, M., T. Araki, and S. K. Nemirovskii, Dynamics of vortex tangle without mutual friction in superfluid ^{4}He, *Phys. Rev. B* **62**, 11751 (2000).

Tsubota, M., T. Araki, and C. F. Barenghi, Rotating superfluid turbulence, *Phys. Rev. Lett.* **90**, 205301 (2003a).

Tsubota, M., T. Araki, and W. F. Vinen, Diffusion of an inhomogeneous vortex tangle, *Physica B* **329**, 224 (2003b).

Tsubota, M., C. F. Barenghi, and T. Araki, Instability of vortex array and transition to turbulence in rotating He II, *Phys. Rev. B* **69**, 134515 (2004).

Tsubota, M., K. Fujimoto, and S. Yui, Numerical studies of quantum turbulence, *J. Low Temp. Phys.* **188**, 119 (2017).

Van Dyke, M., *An Album of Fluid Motion*, Parabolic Press, Palo Alto (1982).

Van Sciver, S. W., *Helium Cryogenics* (International Cryogenics Monograph Series), Plenum Press (1986).

Varga, E., Peculiarities of spherically symmetric counterflow, *J. Low Temp. Phys.* **196**, 28 (2019).

Varga, E., and J. P. Davis, Electromechanical feedback control of nanoscale superflow, *New J. Phys.* **23**, 113041 (2021).

Varga, E., and L. Skrbek, Dynamics of the density of quantized vortex lines in counterflow turbulence: Experimental investigation, *Phys. Rev. B* **97**, 064507 (2018).

Varga, E., and L. Skrbek, Thermal counterflow of superfluid ^{4}He: Temperature gradient in the bulk and in the vicinity of heater, *Phys. Rev. B* **100**, 054518 (2019).

Varga, E., S. Babuin, and L. Skrbek, Second-sound studies of coflow and counterflow of superfluid 4He in channels, *Phys. Fluids* **27**, 065101 (2015).

Varga, E., S. Babuin, V. S. L'vov, A. Pomyalov, and L. Skrbek, Transition to quantum turbulence and streamwise inhomogeneity of vortex tangle in thermal counterflow, *J. Low Temp. Phys.* **187**, 531 (2017).

Varga, E., W. Guo, J. Gao, and L. Skrbek, Intermittency enhancement in quantum turbulence in superfluid ^{4}He, *Phys. Rev. Fluids* **3**, 094601 (2018).

Varga, E., M. J. Jackson, D. Schmoranzer, and L. Skrbek, The use of second sound in investigations of quantum turbulence in He II, *J. Low Temp. Phys.* **197**, 130 (2019).

Varga, E., V. Vadakkumbatt, A. J. Shook, P. H. Kim, and J. P. Davis, Observation of bistable turbulence in quasi-two-dimensional superflow, *Phys. Rev. Lett.* **125**, 025301 (2020).

Vermeer, W., W. M. Van Alphen, J. F. Olijhoek, K. W. Taconis, and R. De Bruyn Ouboter, Dissipative normal fluid production by gravitational superfluid flow, *Phys. Lett.* **18**, 265 (1965).

Villois, A., D. Proment, and G. Krstulovic, Evolution of a superfluid vortex filament tangle driven by the Gross–Pitaevskii equation, *Phys. Rev. E* **93**, 061103(R) (2016).

Vinen, W. F., Mutual friction in a heat current in liquid helium II, *Proc. Roy. Soc. A* **240**, 114–127; **240**, 128–143; **242**, 493–515; **243**, 400–413 (1957).

Vinen, W. F., Detection of single quanta of circulation in liquid helium II, *Proc. Roy. Soc. A* **260**, 218 (1961).

Vinen, W. F., Classical character of turbulence in a quantum liquid, *Phys. Rev. B* **61**, 1410 (2000).

Vinen, W. F., Decay of superfluid turbulence at a very low temperature: The radiation of sound from a Kelvin wave on a quantized vortex, *Phys. Rev. B* **64**, 134520 (2001).

Vinen, W. F., Theory of quantum grid turbulence in superfluid ^{3}He-B, *Phys. Rev. B* **71**, 024513 (2005).

Vinen, W. F., and J. J. Niemela, Quantum turbulence, *J. Low Temp. Phys.* **128**, 167 (2002).

Vinen, W. F., and L. Skrbek, Quantum turbulence generated by oscillating structures, *Proc. Natl. Acad. Sci. USA* **111**, 4699 (2014).

Vinen, W. F., M. Tsubota, and A. Mitani, Kelvin-wave cascade on a vortex in superfluid ^{4}He at a very low temperature, *Phys. Rev. Lett.* **91**, 135301 (2003).

Vitiello, S. A., L. Reatto, G. V. Chester, and M. H. Kalos, Vortex line in superfluid He^4: A variational Monte Carlo calculation, *Phys. Rev. B* **54**, 1205 (1996).

Vollhardt, D., and P. Wolfle, *The Superfluid Phases of Helium 3*, Dover (2013).

Volovik, G. E., Classical and quantum regimes of superfluid turbulence. *JETP Lett.* **78**, 533 (2003).

Volovik, G. E., On developed superfluid turbulence, *J. Low Temp. Phys.* **136**, 309 (2004).

Volovik, G. E., *The Universe in a Helium Droplet*, Oxford University Press (2007).

Volovik, G. E., Quantum turbulence and Planckian dissipation, *JETP Lett.* **115**, 461 (2022).

Wacks, D. H., and C. F. Barenghi, Shell model of superfluid turbulence, *Phys. Rev. B* **84**, 184505 (2011).

Walmsley, P. M., and A. I. Golov, Quantum and quasiclassical types of superfluid turbulence, *Phys. Rev. Lett.* **100**, 245301 (2008).

Walmsley, P. M., and A. I. Golov, Coexistence of quantum and classical flows in quantum turbulence in the $T = 0$ limit, *Phys. Rev. Lett.* **118**, 134501 (2017).

Walmsley, P. M., A. I. Golov, H. E. Hall, A. A. Levchenko, and W. F. Vinen, Dissipation of quantum turbulence in the zero temperature limit, *Phys. Rev. Lett.* **99**, 265302 (2007).

Walmsley, P. M., A. I. Golov, H. E. Hall, W. F. Vinen, and A. A. Levchenko, Decay of turbulence generated by spin-down to rest in superfluid ^{4}He, *J. Low Temp. Phys.* **153**, 127 (2008).

Walmsley, P. M., D. Zmeev, F. Pakpour, and A. I. Golov, Dynamics of quantum turbulence of different spectra, *Proc. Nat. Acad. Sci. USA* **111** Suppl. 1, 4691 (2014).

Wen, X., S. R. Bao, L. McDonald, J. Pierce, G. L. Greene, Lowell Crow, Xin Tong, A. Mezzacappa, R. Glasby, W. Guo, and M. R. Fitzsimmons, Imaging fluorescence of He excimers created by neutron capture in liquid Helium II, *Phys. Rev. Lett.* **124**, 134502 (2020).

White, A. C., C. F. Barenghi, N. P. Proukakis, A. J. Youd, and D. H. Wacks, Nonclassical velocity statistics in a turbulent atomic Bose–Einstein condensate, *Phys. Rev. Lett.* **104**, 075301 (2010).

White, C. M., A. N. Karpetis, and K. R. Sreenivasan, High-Reynolds-number turbulence in small apparatus, *J. Fluid. Mech.* **452**, 189 (2002).

Wilkin, S. L., C. F. Barenghi, and A. Shukurov, Magnetic structures produced by the small-scale dynamo, *Phys. Rev. Lett.* **99**, 134501 (2007).

Williamson, C. H. K., Vortex dynamics in the cylinder wake, *Ann. Rev. Fluid Mech.* **28**, 477 (1996).

Winiecki, T., and C. S. Adams, Motion of an object through a quantum fluid, *Europhys. Lett.* **52**, 257 (2000).

Wyatt, A. F. G., Quantum evaporation and the study of exitations in liquid ^{4}He, *Physica B* **169**, 130 (1991).

Xie, Z., Y. Huang, F. Novotny, S. Midlik, D. Schmoranzer, and L. Skrbek, Spherical thermal counterflow of He II, *J. Low Temp. Phys.* **208**, 426 (2022).

Xu, H., A. Pumir, G. Falkovich, E. Bodenschatz, M. Shats, H. Xia, N. Francois, and G. Bofetta, Flight-crash events in turbulence, *Proc. Natl. Acad. Sci. USA* **111**, 7558 (2014).

Xu, T., and S. W. Van Sciver, Particle image velocimetry measurements of the velocity profile in He II forced flow, *Phys. Fluids* **19**, 071703 (2007).

Yakhot, V., Decay of three-dimensional turbulence at high Reynolds numbers, *J. Fluid Mech.* **505**, 87 (2004).

Yakhot, V., and K. R. Sreenivasan, Towards a dynamical theory of multifractals in turbulence, *Phys. A* **343**, 147 (2004).

Yamada, M., and K. Ohkitani, Lyapunov spectrum of a chaotic model of 3-dimensional turbulence, *J. Phys. Soc. Japan.* **56**, 4210 (1987).

Yang, J., and G. G. Ihas, Decay of grid turbulence in superfluid helium-4: Mesh dependence, *J. Phys. Conf. Series* **969**, 012004 (2018).

Yano, H., A. Handa, H. Nakagawa, K. Obara, O. Ishikawa, T. Hata, and M. Nakagawa, Observation of laminar and turbulent flow in superfluid ^{4}He using a vibrating wire, *J. Low Temp. Phys.* **138**, 561 (2005).

Yano, H., A. Handa, H. Nakagawa, K. Obara, O. Ishikawa, and T. Hata, Study on the turbulent flow of superfluid ^{4}He generated by a vibrating wire, *AIP Conf. Proc.* **850**, 195 (2006).

Yano, H., N. Hashimoto, A. Handa, M. Nakagawa, K. Obara, O. Ishikawa, and T. Hata, Motions of quantized vortices attached to a boundary in alternating currents of superfluid ^{4}He, *Phys. Rev. B* **75**, 012502 (2007).

Yao, J., and F. Hussain, Separation scaling for viscous vortex reconnections, *J. Fluid Mech.* **900**, R4 (2020).

Yarmchuk, E. G., and W. I. Glaberson, Counterflow in rotating superfluid helium, *J. Low Temp. Phys.* **36**, 381 (1979).

Yarmchuk, E. J., M. J. V. Gordon, and R. E. Packard, Observation of stationary vortex array in rotating superfluid helium, *Phys. Rev. Lett.* **43**, 214 (1979).

Yui, S., and M. Tsubota, Counterflow quantum turbulence of He-II in a square channel: Numerical analysis with nonuniform flows of the normal fluid, *Phys. Rev. B* **91**, 184504 (2015).

Yurkina O., and S. K. Nemirovskii, On the energy spectrum of the 3D velocity field, generated by an ensemble of vortex loops, *Low Temp. Phys.* **47**, 652 (2021).

Zhang, T., and S. W. Van Sciver, The motion of micron-sized particles in He II counterflow as observed by the PIV technique, *J. Low Temp. Phys.* **138**, 865 (2005a).

Zhang, T., and S. W. Van Sciver, Large scale turbulent flow around a cylinder in counterflow superfluid ^{4}He (He II), *Nature Phys.* **1**, 36 (2005b).

Zhang, T., D. Celik, and S. W. Van Sciver, Tracer particles for application to PIV studies of liquid helium, *J. Low Temp. Phys.* **134**, 985 (2004).

Zhang, T., D. Celik, and S. W. Van Sciver, The motion of micron-sized particles in He II counterflow as observed by the PIV technique, *J. Low Temp. Phys.* **134**, 865 (2005).

Zmeev, D. E., A method for driving an oscillator at a quasi-uniform velocity, *J. Low Temp. Phys.* **175**, 480 (2014).

Zmeev, D. E., F. Pakpour, P. M. Walmsley, A. I. Golov, W. Guo, D. N. McKinsey, G. G. Ihas, P. V. E. McClintock, S. N. Fisher, and W. F. Vinen, Excimers He-2* as tracers of quantum turbulence in ^{4}He in the $T = 0$ limit, *Phys. Rev. Lett.* **110**, 175303 (2013).

Zmeev, D. E., P. M. Walmsley, A. I. Golov, P. V. E. McClintock, S. N. Fisher, and W. F. Vinen, Dissipation of quasiclassical turbulence in superfluid ^{4}He, *Phys. Rev. Lett.* **115**, 155303 (2015).

Zuccher, S., and R. L. Ricca, Twist effects in quantum vortices and phase defects, *Fluid Dyn. Res.* **50**, 011414 (2018).

Zuccher, S., M. Caliari, A. W. Baggaley, and C. F. Barenghi, Quantum vortex reconnections, *Phys. Fluids* **24**, 125108 (2012).

Zurek, W. H., Cosmological experiments in superfluid-helium, *Nature* **317**, 505 (1985).

Index